C0-AVT-070

Automotive Antenna Design and Applications

Automotive Antenna Design and Applications

Victor Rabinovich
Nikolai Alexandrov
Basim Alkhateeb

CRC Press is an imprint of the
Taylor & Francis Group, an **informa** business

CRC Press
Taylor & Francis Group
6000 Broken Sound Parkway NW, Suite 300
Boca Raton, FL 33487-2742

Printed in the United States of America on acid-free paper
10 9 8 7 6 5 4 3 2 1

International Standard Book Number: 978-1-4398-0407-0 (Hardback)

Library of Congress Cataloging-in-Publication Data

Rabinovich, Victor.
Automotive antenna design and applications / authors, Victor Rabinovich, Nikolai Alexandrov, Basim Alkhateeb.
p. cm.
Includes bibliographical references and index.
"A CRC title."
ISBN 978-1-4398-0407-0 (hardcover : alk. paper)
1. Antennas (Electronics) 2. Automobiles--Electronic equipment. 3. Automobiles--Radio equipment. I. Alexandrov, Nikolai. II. Alkhateeb, Basim. III. Title.

TL272.5.R33 2010
629.2'77--dc22 2009047095

Visit the Taylor & Francis Web site at
http://www.taylorandfrancis.com

and the CRC Press Web site at
http://www.crcpress.com

Contents

Preface

General Organization of Book

During recent years, extensive development of wireless communication technologies has resulted in innovative services and devices are now utilized in modern vehicles: analog and digital audio broadcasting radio; satellite radio and GPS reception; cellular phone communications that operate in a few frequency ranges; short range wireless communication techniques such as remote keyless entry and remote start engines; automatic toll collection; and long and short range radar devices employed for vehicle safety and other purposes. All these services require antennas as key elements that provide good quality communications. As a result, modern vehicles commonly include multiple antennas for different applications.

The objective of this book is to provide detailed discussions of the traditional and advanced new antennas utilized in the automotive industry. The contents of the book are based on antenna designs published in numerous technical papers, patents, and patent applications. Many commercially available automotive antennas are presented as well.

Advanced new designs and techniques allow antenna dimensions to be reduced without significant degradation of quality parameters. Reduced size helical antennas, multiband compact, printed on dielectric, and patch designs in a single package have become typical for automotive applications and are covered in this book. Because receiving antennas often are designed as active devices, this book demonstrates several examples of the active antennas for automotive applications. It covers the design process from simulation through prototype fabrication and production adjustments for use in a real car environment. The final step is very important because the body of a vehicle can dramatically change antenna parameters. Numerous examples of antenna performance measurements for automotive use are demonstrated.

This book is organized into nine chapters. Chapter 1 presents an overview of traditional and more advanced antennas used in the automotive industry. It covers printed-on-window-glass AM/FM broadcasting antennas, short helical roof mounted systems, multiband cellular phone antennas, antennas intended for receiving TV in cars, satellite radio and GPS antenna systems, antennas for short range communication systems, and collision radar antenna systems intended to enhance car safety.

Chapter 2 describes conventional basic antenna parameters, parameters related to small antenna design, the future of measurements required for car mounted antennas, and notes about decibels that reflect conversion ratios between absolute measurement values and logarithmic units.

Because most receiving car antennas are active devices, their development is of significant interest. Chapter 3 discusses the main equations related to active and diversity antenna systems. Signal and noise analyses of the active antennas are presented as are important diversity techniques for improving reception quality. Expressions for the main diversity parameters such as gain, correlation coefficient, and radiation patterns are presented for two or more antennas utilized in the diversity system.

Chapter 4 demonstrates antennas for automotive analog and digital audio broadcasting reception. It describes traditional whip antennas, antennas printed-on-window-glass, short helical roof-mounted designs, and a few examples of antennas intended for digital audio broadcasting applications. Antenna simulation results using FEKO and NEC software programs are demonstrated. Antenna amplifiers for AM/FM frequency bands with special protection circuits that provide good quality reception in cars in the vicinity of high power transmission stations are also discussed.

Chapter 5 describes cellular phone antennas for automotive applications: single, dual, triple, and quadruple band systems designed for roof mounting, printed-on-window-glass, or mounted on bumpers. Simulations are confirmed by measurement results. Interesting solutions have been achieved by combining systems: FM/PCS, AM/FM/AMPS/PCS, PCS/remote keyless entry, and cellular/Wi-Fi systems. Detailed explanation is devoted to diversity cellular phone antennas and their parameters.

Chapter 6 describes antennas for terrestrial and satellite TV application. It discusses antenna requirements and satellite antenna systems based on phase antenna arrays. Data concerning printed-on-glass antenna configurations taken from various papers and patents are included.

Chapter 7 discusses active satellite radio antennas: requirements, simulations, measurements, and practical aspects of automotive use. The chapter also discusses amplifiers for these antennas.

Chapter 8 demonstrates GPS antenna design including simulations and combined GPS/SDARS topologies. It also investigates combined cellular phone and GPS design for the popular OnStar automotive application.

Finally, Chapter 9 is devoted to antennas for short range wireless communications. Compact 315 MHz and 433.92 MHz antenna design is demonstrated. A few antenna topologies with extended communication ranges (patents pending) are also described.

We hope this book will serve as a very useful design reference for readers who are interested in automotive antenna design.

Acknowledgments

It is a pleasure to express my appreciation to the people who helped in some way in the completion of this project. As the lead author I would first like to express my gratitude to Nikolai Alexandrov and Basim Alkhateeb, my coauthors, for the use of their materials in several chapters of the book.

I would like to thank my wife for her remarkable work in designing printed antennas for short range wireless communications. Without the help of my son, Dmitri Rabinovich, who prepared all illustrations for this book and who has written several software programs to control the outside turntable intended to measure antennas mounted on and in automobiles, this text could not have been written. Dmitri also participated in measuring antennas for short range automotive communications.

I am grateful to Elizabeth Albert, a former vice president of Tenatronics Ltd., for her understanding and support during my early career in Canada after immigration from Russia.

Thanks also to Steve Hyde, Chrysler's lead engineer, for his understanding and support during work on the RKE antenna project. Also, I would like to thank two great antenna engineers and scientists, Drs. A. Lemanski and V. Kashin, who introduced me to the secrets of antenna design and supported me during 20 years of working together in Russia. Finally, it is a pleasure to acknowledge Nora Konopka, Amy Blalock, and Prudy Taylor Board of Taylor & Francis Group, who helped bring this text to you.

Victor Rabinovich
Toronto, Canada

Lead Author

Victor Rabinovich was born in Moscow, Russia and earned an MS in electronic engineering from the Moscow Institute of Physics and Technology. In 1976, he was granted a PhD in the fields of electromagnetics and antennas. For more than 20 years he was employed in the research and development department of a leading Moscow scientific industrial corporation. He participated in a variety of projects, including the development of phased antenna arrays with electronically controlled beams and smart adaptive antenna arrays. Since 1995, he has worked in North America, designing antennas for various automotive applications. He has published more than 40 journal articles and holds more than 30 patents. His current research and design interests are automotive antennas including small antennas for multiband operations.

List of Abbreviations

ACC: Adaptive Cruise Control
AMP: Audio Amplifier
AM: Amplitude Modulation
AMPS: Advanced Mobile Phone System uses 824–849 MHz to send information from the mobile station to the base station (uplink) and 869–894 MHz for forward channels (base to mobile)
AR: Axial Ratio
ATC: Automatic Temperature Control
CCN: Cluster Control Node
CDMA: Code Division Multiple Access
CPW: Coplanar Waveguide
CST: Computer Simulation Technology
CW: Continuous Wave
DAB: Digital Audio Broadcasting
DBS: Direct Broadcast Satellite
DC: Direct Current Produced By Battery
DCS: Digital Cellular Service
DSRC: Dedicated Short Range Communication
DVB: Digital Video Broadcasting
EMI: Electromagnetic Interference
ETC: Electronic Toll Collection
FEKO: The name is derived from a German acronym which can be translated as "field computation for objects of arbitrary shape"
FM: Frequency Modulation
FR-4: Dielectric Circuit Board Name
GPS: Global Position System
GSM: Global System for Mobile Communication
GSM 850: Global System for Mobile Communication System uses 824–849 MHz to send information from the mobile station to the base station (uplink) and 869–894 MHz for the other direction (downlink)
GSM 900: Global System for Mobile Communication uses 890–915 MHz to send information from the mobile station to the base station (uplink) and 935–960 MHz for the other direction (downlink)
GSM 1800: Global System for Mobile Communication uses 1710–1785 MHz to send information from the mobile station to the base station (uplink) and 1805–1880 MHz for the other direction (downlink)
GSM 1900: Global System for Mobile Communication uses 1850–1910 MHz to send information from the mobile station to the base station (uplink) and 1930–1990 MHz for the other direction (downlink)
HP: Horizontal Polarization

HVAC: Heat Ventilation Air Condition
IF: Intermediate Frequency
IM: Intermodulation Product
LHCP: Left Hand Circular Polarization
LNA: Low Noise Amplifier
LOS: Line of Sight
LP: Linear Polarization
LRR: Long Range Radar
MoM: Method of Moments
MRC: Maximum-Ratio Combining
NASA: National Aeronautics and Space Administration
NLOS: Non Line of Sight
OEM: Original Equipment Manufacturer
PCB: Printed Circuit Board
PCS: Personal Communication Service
RF: Radio Frequency
RHCP: Right Hand Circular Polarization
PIFA: Planar Inverted F-type Antenna
RK: Remote Controllable
RKE: Remote Keyless Entry
RKES: Remote Keyless Entry Service
RSE: Remote Start Engine
SARSAT: Search and Rescue Satellite Aided Tracking
SDARS: Satellite Digital Audio Radio Service
SNR: Signal to Noise Ratio
SRR: Short Range Radar
TMC: Traffic Message Channel
TPMS: Tire Pressure Monitoring System
TV: Television
UHF: Ultra High Frequency
UMTS: Universal Mobile Communication System
USB: Universal Serial Bus
VHF: Very High Frequency
VP: Vertical Polarization
VSTR: Vehicle Shielded Test Room
WCM: Wireless Control Module
Wi-Fi: The Name of Wireless Network that Provides High-Speed Internet

1

Automotive Antennas Overview: Patents, Papers, and Products

1.1 Introduction

In recent years, the automotive wireless communication market has expanded greatly. It is well known that the antenna is the key element in determining communication system performance and size. Modern cars may contain multiple antennas that capture AM/FM broadcasts, satellite digital audio radio service (SDARS) signals, global positioning system (GPS) data, cellular phone communication, digital audio broadcasting (DAB), remote keyless entry (RKE) and remote start engine (RSE) systems, television reception, electronic toll collection (ETC), tire pressure sensors, and automotive radar.

Figure 1.1 shows the frequency spectrum used by electronic devices for automotive applications. The spectrum covers frequencies from 0.5 MHz (AM radio) up to 77 GHz (Radar Collision Systems).

Figure 1.2 demonstrates locations that typically are utilized for mounting antennas on or in modern vehicles. Position number 3 or position number 9 is used typically for mounting an AM/FM whip fender antenna. An AM/FM short helical antenna can be mounted on a car roof in position 2 or position 8. AM/FM printed-on-glass antennas are mounted on side glass (position 7), rear glass near the heater, or sometimes on front glass at position 4. GPS, SDARS, and cellular phone antennas are usually mounted on the roof of a car at position 2 or on the trunk at position 1. A satellite TV antenna is mounted on a car roof; for a terrestrial application, an antenna can be printed-on-glass and mounted in a single package with a printed-on-glass AM/FM antenna. Antennas for short range wireless communication devices usually are mounted out of sight inside cars, for example at position 10 (under car front panel) or in the interior of a door at position 5. Radar collision antennas are mounted from the front (position 6) or rear.

This overview chapter describes antennas that operate at different frequency bands up to 77 GHz. Recent achievements and future trends in antenna designs cited in numerous patents and papers are discussed. Many commercially available automotive antenna designs are presented.

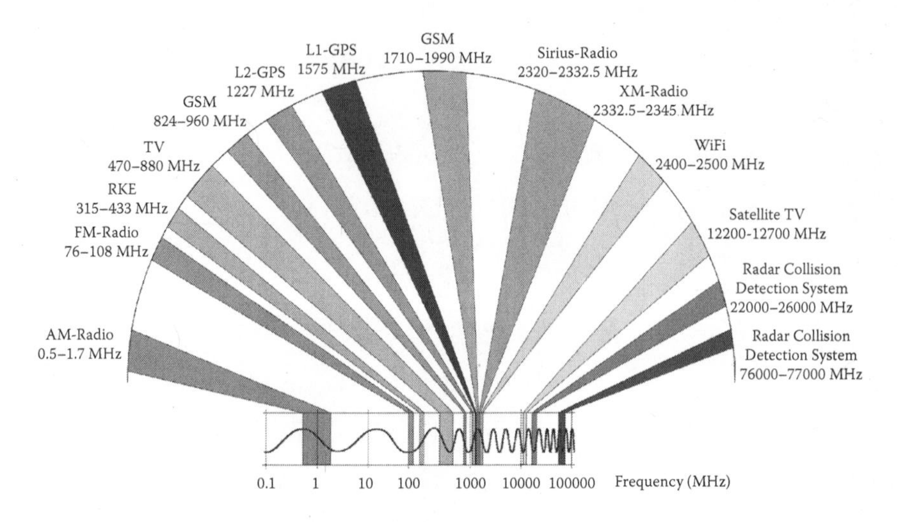

FIGURE 1.1
Frequency spectrum used in RF equipment for automotive industry.

FIGURE 1.2
Locations of car antennas for different applications.

1.2 AM/FM Antenna Systems

1.2.1 AM/FM Broadcasting Frequency Range

Automotive amplitude modulated (AM) radio carrier frequencies are located in the 530 to 1700 KHz band. Frequency modulated (FM) radio covers a range from 87.5 to 108 MHz in North America and 76 to 90 MHz in Japan. The polarization of the transmitted signal in the AM frequency band is vertical, while polarization in the FM frequency band depends on the region. In Europe, the polarization of an FM transmitted signal is horizontal, whereas power stations in the United States are capable of transmitting signals with vertical, horizontal, or circular polarization. The automotive industry is phasing out the traditional whip antennas (quarter wavelength monopoles mounted on the front or rear fender) used for more than 50 years and replacing them with popular printed-on-glass antennas and short flexible mast antennas mounted on car roofs.

1.2.2 Printed-on-Glass Antennas

Printed-on-glass antennas present some advantages over traditional whip antennas. A major flaw of a whip antenna is that it can break or bend easily upon contact with an interfering object. Printed-on-glass antennas are not noisy under windy conditions, hinder vandalism, and resist damage.

A typical antenna printed on a glass car window is constructed of a few linear conductive strips arranged on the front (windshield antenna), side or rear glass of a car. AM and FM antennas can be designed as two separate units, one representing each frequency band and containing its own input; AM and FM output are combined. A single glass window antenna designed for AM and FM reception typically has only one input and one output. Rear antennas can be mounted on rear glass, separate from the defroster or combined with the defroster lines. If integrated with a defroster, a printed antenna includes a special filter to prevent noise interference caused by the defroster power line.

Many patents and theoretical papers are devoted to printed-on-glass antennas. More than half of the patents (References 1–5 serve as recent examples) describe rear window glass antennas; other publications [6–8] discuss side glass antennas. Only a few [9,10] relate to front glass applications.

High quality antenna devices require comprehensive simulation and prototype measurements. The Munich Technical University research group in Germany, led by Professor Heinz Lindenmeier, was one of the first to design and introduce printed-on-glass window antennas [11]. Professor Eric K. Walton and his partners from The Ohio State University Electro Science Laboratory published several interesting articles [12,13] dedicated to the mathematical simulation of printed-on-glass car antennas.

Walton and Abou-Jaoude studied the designs of conformal automobile antennas using numerical techniques based on the method of moments (MoM)—a method that requires that the entire modeled structure be broken down into wire segments (each segment must be small compared to the wavelength) and/or metal plates. The MoM technique determines the current on every wire segment and surface resulting from the source and all other currents. After the currents are identified, the electromagnetic field at any space point can be determined by a summation of all the electromagnetic fields from the wire segments and surface patches. The research paper [13] focuses on the simulation of the body gap's effect on conformal automotive antenna parameters.

Richard Langley and his research group from the University of Kent (UK) [14,15] numerically examined mast and printed-on-glass antennas using the commercial FEKO and CST Microwave Studio programs. They concluded that both packages produced similar results. The 3D FEKO code is based on the MoM technique combined with diffraction theory and physical optics. CST Microwave Studio on the other hand, is based on the finite integration method. Other papers [16,17] present antenna simulation results using the popular NEC-2 numerical code (MoM method).

Theoretical simulation and experimental results show that passive printed-on-glass antennas exhibit lower FM gain (averaged over 360 degrees around a vehicle) compared with whip antennas. A windshield antenna gain may be 5 to 10 dB lower than the gain of a whip antenna (depending on polarization of received wave). For an antenna located on a side or rear glass and separated from the defogger grid, the difference can exceed these values. An antenna mounted on the front windshield of a car receives noise generated by the engine and dashboard electronic components. Therefore, side or rear glass window mountings are preferable for printed-on-glass antennas.

Generally, a printed-on-glass antenna is used with a low noise amplifier that increases the weak signal received by the radio. The amplifier gain value is chosen to provide the equivalent seek/stop performance of a car radio when using printed or whip antennas. A properly designed amplifier does not degrade car radio sensitivity. A typical vehicle radio has a sensitivity of about –5 dBμV [11], based on the assumption that a receiver's bandwidth is equal to 120 kHz (–112 dBm in a 50 ohm system). Antenna amplifiers usually have two separate branches: one for the AM signal and the other for the FM signal. They have a gain value of 6 to 10 dB and a noise figure value between 3 and 5 dB over the FM frequency range. The input impedance of an amplifier should match the antenna impedance in FM band to minimize the output noise value [18]. One of the key parameters of an amplifier is intermodulation distortion. An amplifier with an automatic gain control or overload protection circuit can provide the best (lowest) intermodulation distortion [18,19].

In comparison with a whip antenna, a printed-on-glass antenna has a more directional FM radiation pattern. The difference between the maximum and minimum received FM signal levels of a whip antenna system over 360 degrees of vehicle rotation does not exceed 10 to 15 dB. On the other hand, the variation for a printed-on-glass antenna can be more than 20 to 30 dB. To overcome such printed antenna performance distortions, designers proposed a well known technique called space diversity whereby two or more antennas are utilized to improve the signal reception [20,21]. A space diversity antenna element occupying a small window area has low radiation impedance that is not matched with car radio impedance. However, one research paper [22] shows that low radiation resistance (not less than 10 ohms) combined with high amplifier input impedance may produce high quality broadcasting reception. A number of interesting results devoted to analog and digital diversity techniques were recently published [23–26]. In a correctly designed diversity system, the combined "overlap" radiation pattern from two or more diversity antenna radiation patterns is more omnidirectional in comparison with the directionality of individual antennas. Another concept is implementation of an FM antenna diversity system with a single output feed [27]; switching of the load of a parasitic antenna mounted near the main antenna provides the diversity pattern. The same idea can be used for designing diversity antenna systems with electronically (pin diode switch) reconfigurable antenna shape and single output.

1.2.3 Short Mast Helical Roof Antennas

In recent years, short mast AM/FM automobile antennas [28–30] have become more popular. Typically, a short mast car roof antenna is a helical monopole, with a height of 15 to 40 cm, wound on a flexible insulating dielectric rod and embedded into an insulating resin rod. Kraus [31] was one of the first investigators of helical monopole antennas. Such antennas can operate in axial mode, with radiation in the direction of the helix axis, or in normal mode, where the helix diameter is small compared to the wavelength and the radiation is concentrated perpendicular to the helix axis.

A normal mode roof helical antenna has a radiation pattern similar to a short straight vertical monopole for a car roof. The coil parameters of a helical antenna [28] with a length of 15 cm are: a coil conductive line having a diameter of 0.5 mm is wound so that its outer diameter is approximately 6 mm; the turn number of this wound coil is approximately 100; and the antenna effective length is approximately 1 m. Figure 1.3 shows an AM/FM active helical antenna (without coating) designed by Receptec LLP (Division of Laird Technologies) currently in mass production for automotive industry applications. The performance of a helical antenna with an amplifier is identical to the performance of active printed-on-glass antennas. A significant advantage of a helical roof-mounted antenna over

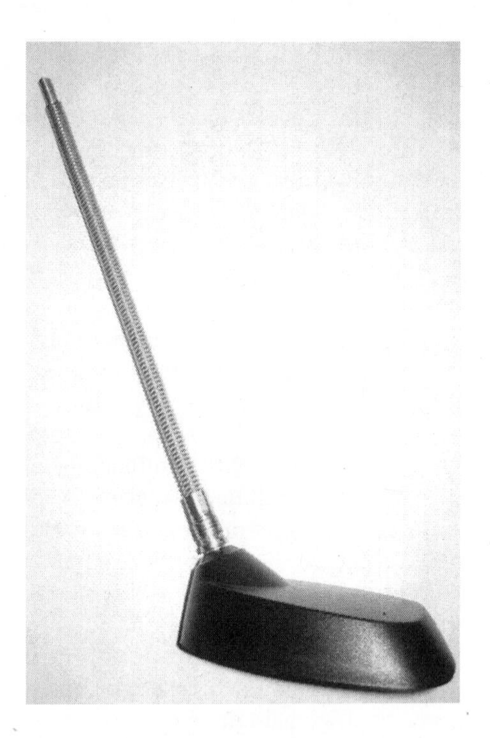

FIGURE 1.3
AM/FM helical antenna geometry.

whip or printed-on-glass types is their almost omnidirectional radiation pattern—the shape of a car body does not affect antenna parameters. Such antennas can be used on any kind of vehicle (excluding a convertible roof top) without significant modifications. The base of the antenna may be attached to the antenna helix with a swiveling mechanism [32]. An antenna with a swiveling mechanism can be folded easily, thereby preventing it from damage by an external force and it can be returned easily to a predetermined attachment angle.

1.3 Cellular Phone Antennas

Early cellular mast antennas were designed to operate in the 800 to 900 MHz frequency bands. Simple cellular vehicle antennas were built as half, quarter, or three quarter wavelength monopoles with air-wound phasing coils [33]. These antennas have omnidirectional, vertically polarized radiation patterns in the horizontal plane and various gain values based on length.

Current cellular phones operate in two, three, or even four frequency bands [33]. They use one of two major network technologies: code division multiple access (CDMA) or the global system for mobile communications (GSM) operating at the same carrier frequency bands. For example, North American dual-band GSM 850 and GSM 1900 phones operate at 824 to 894 MHz and 1850 to 1990 MHz bands, respectively. The European dual-band cell phone standards are GSM 900 (890 to 960 MHz) and GSM 1800 (1710 to 1880 MHz). Tri-band phones operate in the GSM 900, 1800, and 1900 MHz range in Europe or the GSM 850, 900, and 1900 format used in North America. Quad-band phones have the ability to operate in four different frequency bands (GSM 850, 900, 1800, and 1900) and are now common.

Generally, compact cellular antennas meet automotive aesthetics requirements. For example, dual-band cell phone monopoles designed by known Yokowo Technology [34] are only 7 cm in length, with minimum gain value of 2.5 dBi. Many planar structure designs including cellular printed circuit board antennas are described in K. L. Wong's excellent reference book [35]. The commercially available Stealth Blade antenna introduced by Radiall-Larsen Antenna Technologies (Figure 1.4) has a length of 12.5 cm, width of 2 cm, depth of 0.5 cm, and 2 dBi gain. The Stealth Blade is intended for dual-frequency band operation in the 806 to 894 MHz and 1710 to 1900 MHz ranges. The unique printed circuit board meandered antenna design makes it ideal for applications where a traditional mast antenna is not practical.

FIGURE 1.4
Stealth Blade cellular phone antenna design.

1.4 Car TV Antennas

1.4.1 Terrestrial Systems

Terrestrial TV reception covers the VHF band from 54 to 216 MHz (excluding 88 to 174 MHz) and the UHF band from 470 to 806 MHz. Numerous types of antennas are utilized with TV equipment. Dipoles, whip, mast, and printed-on-glass window antennas are the most popular types for VHF and UHF TV frequency band applications. To obtain the best quality terrestrial TV reception, dipole and whip antennas must be oriented toward the transmitting station and tuned by increasing or shortening the antenna length. Of course, this tuning process is ineffective in a moving car. The diversity technique similar to AM/FM schemes uses a few printed-on-glass antennas [36–38] on quarter side, rear window, or front glass. A diversity system combines (based on feedback from a TV receiver) the signals from receiving antennas to provide the best quality reception.

1.4.2 Satellite TV Antennas

Direct broadcasting satellite TV antennas operate in microwave Ku-band frequency range (12.2 to 12.7 GHz). Recently, KVH Industries, Inc. [39] introduced a revolutionary system called Track Vision A7. Car passengers can watch DirectTV satellite programs while their vehicle is stationary or in motion. The smart antenna used with this system is positioned on the car roof as shown in Figure 1.5a. The main component of the antenna system is

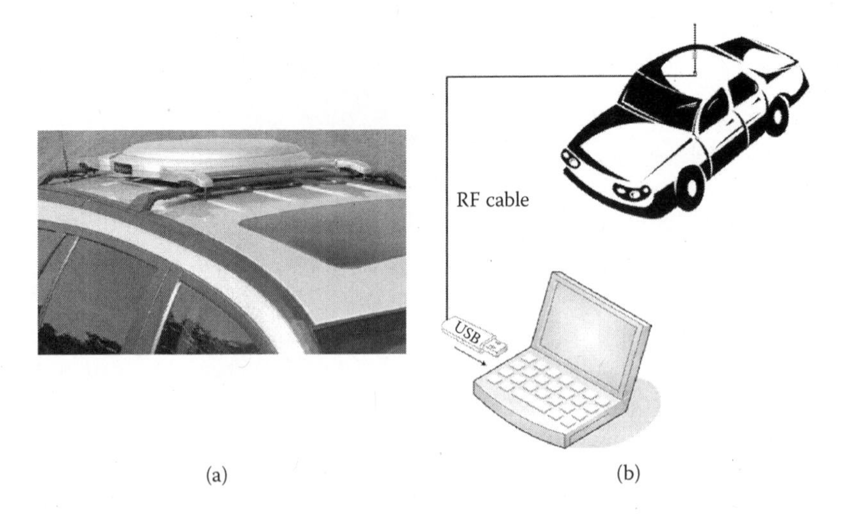

FIGURE 1.5
Automotive TV design. (a) roof satellite TV antenna. (b) Internet TV connection topology.

a phased array. Such antennas have been used in military applications for more than 30 years. Classic phased arrays consist of hundreds or even thousands of tiny antenna elements. Relative phases of the different elements vary to enable the array to radiate maximum energy in a predetermined angle direction.

Today, low cost microstrip antenna arrays [40] are used in commercial and military radar installations, satellite systems, and mobile phones. KVH Industries, Inc engineers designed a microstrip antenna array to receive Ku-band signals from DirectTV satellites. A Track Vision A7 antenna consists of 280 separate flat receiving elements. Signals received by each element are combined into a single high-strength signal to produce the TV picture and sound. Such array design significantly reduces antenna system depth in comparison with a traditional dish devices. Most TV satellite dish antennas have depths of 40 to 50 cm in comparison with a microstrip array with a depth of about 13 cm. Antenna elements are designed to pick up signals with both right- and left-hand circular polarization (RHCP and LHCP, respectively).

It is necessary to emphasize that an A7 antenna array does not have an electronically controlled beam. In a moving car, an A7 array may be rotated mechanically over 360 degrees in the horizontal plane while also tilting ±15 degrees in the vertical plane. Such electro-mechanical design allows orientation of the main antenna beam toward a DirectTV satellite. The maximum radiation pattern oriented toward the satellite is controlled by a GPS system integrated into the tracking mechanism of the antenna system. This antenna system operates in the 12.2 to 12.7 GHz frequency range; its dimensions are $82 \times 79 \times 13.4$ cm^3 and its weight is 20.4 kg.

1.4.3 Internet TV in Vehicles

Thousands of people and companies now can connect their laptops to the Internet using wireless cell phone networks. Figure 1.5b is a block diagram of the Internet connection for a car application. The laptop in the car is connected via a specially designed USB wireless adapter coupled with an internal or external antenna mounted on the car roof as shown in the figure. The USB key provides Internet service to the laptop anywhere a cell phone network is available. In areas with strong received and transmitted cellular band signals, the internal USB adapter antenna guarantees stable Internet connection. In areas with a poor signal strength, the gain of the internal antenna is inadequate for stable connection and therefore the external antenna mounted on the car roof (may be standard dual-band cellular phone monopole) is preferable. Such topology allows passengers to watch Internet TV and listen to Internet radio continuously while the vehicle is in motion.

It is necessary to emphasize that the electronic components (USB adapter and external antenna) are far less costly than the TV antenna array described in Section 1.4.2. The USB Internet key is the smallest modem on the market today. However monthly payments to acquire such service depend on the monthly message-carrying capacity (bandwidth) usage and TV viewing may be fairly expensive.

1.5 Satellite Radio Antennas

A few years ago, the new satellite digital audio radio service (SDARS) became available in the United States and Canada. Sirius Satellite Inc. and XM Satellite Radio Inc. provide 24 hours of satellite audio broadcasting over the North America Continent.* A microwave satellite left hand circular polarized transmitting signal (LHCP) covers a 2320 to 2332.5 MHz band for Sirius and a 2332.5 to 2345 MHz frequency covers the XM system. SDARS also uses terrestrial ground–base transmitting towers positioned in urban areas to transmit vertically polarized (VP) signals providing terrestrial broadcasting at the same frequency bands. Therefore, SDARS antennas receive both LHCP signals from satellites and VP signals from ground-based towers. Early antenna systems described in Reference 41 had two antennas, two amplifiers, and two outputs: one for satellite reception and another one for terrestrial reception.

Today, the automotive antenna industry produces very compact SDARS printed-on-dielectric-patch antennas with single output. Figure 1.6 shows a current Sirius OEM mass production patch antenna designed by Receptec LLC. This design is based on a single active antenna containing one radio frequency (RF) output. The dielectric patch dimensions are $33 \times 33 \times 3.7$ mm^3 and the antenna receives LHCP and VP signals as described above. A Tyco Company design uses a $25 \times 25 \times 4$ mm^3 antenna. Experiments show that patch antennas yield excellent performance for satellite reception but VP terrestrial gain value is relatively poor (less than –5 dBi).

Engineers now suggest new designs [42–45] that increase VP antenna gain at low elevation angles, thus improving terrestrial reception. One such idea proposed by Delphi Technology was discussed in a patent application [45]. This system includes a regular patch antenna and parasitically enhanced perimeter (a metal "fence") extending from the circuit board and encompasses a patch antenna. Adjusting the distance between the "fence" and the metal patch makes it possible to increase VP antenna gain. According to the graphs in the patent application [45], the gain of VP signal could reach –1.5 to –0.5 dBi, depending on the elevation angle. An amplifier circuit is placed under the ground of the antenna. The total depth of such an antenna does not exceed the regular patch (without-a-fence) antenna depth.

An SDARS antenna is mounted generally on a car roof—the best location for receiving signals from a satellite. For a car with a convertible top, the SDARS antenna is mounted inside the car or in an externally located

* They recently merged.

FIGURE 1.6
Sirius satellite patch geometry.

vehicular mirror housing [46]. The weakened reception abilities of antennas positioned in this manner may be strengthened by diversity techniques [47,48] proposed for satellite applications.

1.6 GPS Antenna Systems

A classic antenna for a GPS is a microstrip patch type mounted on the vehicle roof or interior. According to the requirements, an antenna designed for receiving right hand circular polarization (RHCP) signals operates at a 1575 MHz ± 2 MHz frequency band. The gain of the passive antenna is determined by ground plane size. For example, an antenna mounted on a 2.5 cm diameter ground plane has a maximum gain of about 0 dBic; a patch antenna with a 10 cm ground plane mounting has a maximum gain of about 5 dBic. These values correspond to the performance of a metal patch with dimensions of $21 \times 21 mm^2$ and a ceramic dielectric substrate (dielectric constant $\varepsilon_r = 20$) with dimensions of $25 \times 25 \times 4 mm^3$.

GPS technology can be used in tandem with cellular technology (OnStar system in North America or eCall in Europe) to define vehicle location. This automatic emergency call system notifies a call center of accidents via a mobile network. When a driver presses the emergency button of the OnStar system, current vehicle data and GPS location are immediately gathered and transmitted to a special call center that operates 24 hours a day. This makes it possible to initiate an emergency response, very rapidly. A typical system that includes both GPS and cellular phone antennas mounted in one package is shown in Figure 1.7. The low noise GPS antenna amplifier is placed on the bottom side of the patch antenna ground plane.

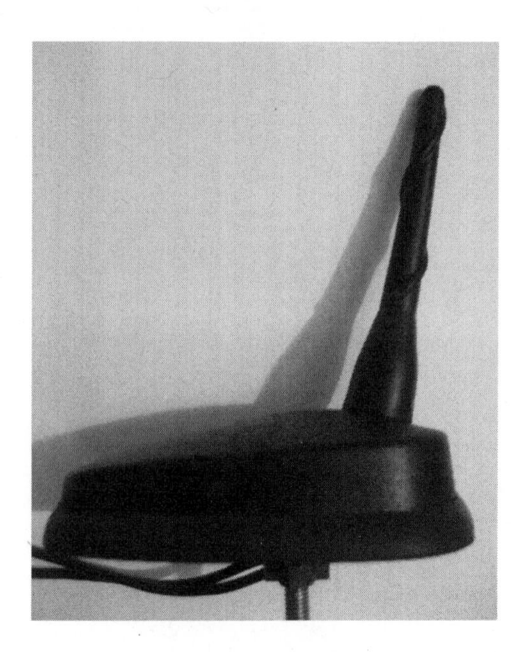

FIGURE 1.7
Combined GPS/cell phone design.

A GPS antenna can be combined with a SDARS antenna to function as a single dual-frequency unit [50,51]. Receptec LLP proposed in a patent [50] an integrated SDARS/GPS patch antenna that has the ability to receive both LHCP and RHCP signals (a more detailed description of the patent appears in Chapter 8). The antenna includes two coplanar concentric patches. The inner patch is substantially square and the outer patch has square inner and outer edges. The two patches are not in physical contact. A single feed connected to the inner patch receives the LHCP signal, and the two patches together receive the RHCP signal. Patent materials show that the relative sizes, shapes, and orientations of the patches can be tuned through a trial-and-error process. The single feed is connected only to the inner patch—off center as is conventional for antennas gathering circularly polarized signals.

1.7 Antennas for Short Range Communications

1.7.1 Introduction

Wireless devices such as remote start engine (RSE), remote keyless entry (RKE), tire pressure monitoring (TPMS), and electronic toll collection (ETC) systems now are considered "classic" for short range vehicle wireless communication [52]. RSE, RKE, and TPMS technologies use the 315 MHz frequency band in the U.S., Canada, and Japan, and the 433.9 MHz band in Europe. A few frequency ranges are used in ETC systems including 866 to 869 MHz in Europe, 902 to 928 MHz in North and South America, and 950 to 956 MHz in Japan and some other Asian countries.

1.7.2 Remote Start Engine and Remote Keyless Entry Antennas

A typical RSE/RKE system consists of an antenna and a control module. Currently, two types of antennas are produced: (1) an internal antenna mounted in a single case with a control module and (2) an external antenna separate from the control module and hidden in the interior of a vehicle. The integration of RF and digital electronic components with the internal receiving antenna in a single case reduces the number of wires and connectors, thus reducing system cost, but presents a significant disadvantage. Parasitic radiation emissions from circuit board electronic components near the receiving antenna can reduce the communication range up to several meters.

Figure 1.8 depicts two mass production internal antennas (loop and bended monopole). They feature reduced size (in comparison with wavelength), low gains (–15 to –20 dBi), and narrow bandwidths. Measured communication ranges (as a rule below 40 m) depend on the electronic component interference noise level and key fob effective radiation power. The key fob power is the product of transmitting power and antenna gain and typically varies from –15 to –20 dBm everywhere but Japan which prohibits radiation

FIGURE 1.8
Internal wireless control module with antennas for short range communication.

FIGURE 1.9
External printed-on-dielectric board design for short range communication.

exceeding –43 dBm of unlicensed power in this frequency range. Therefore, several meters represent the typical communication range for Japan.

RSE/RKE systems with external antennas based on circuit board [53,54], simple dipole [55], whip, printed-on-glass [56], and cable pigtail [57] designs have increased communication range greatly in comparison with systems using internal antennas.

Figure 1.9 illustrates two mass produced printed-on-circuit-board antennas with asymmetrical and symmetrical designs. These antennas have compact (less than wavelength) dimensions: 50×70 mm^2. The symmetrical meandered dipole has a higher gain (–6 to –7 dBi) in comparison with the asymmetrical type (–8 to –9 dBi). Asymmetrical antennas function together with RF cables in a design similar to a pigtail. The total gain of such an antenna is determined by the cable route in the car and is almost equivalent to the half wave dipole [58]. Generally, an antenna has a one-stage amplifier circuit. The gain and noise level of the amplifier depends on control module design. Typical values of the gain vary from 8 to 15 dB, and noise figure is about 2 dB. Such external circuit board antennas allow receivers to achieve a communication range of approximately 100 m or more (depending on electronic component noise level in a car). Of course, antennas located inside cars do not have omnidirectional radiation patterns as noted in some research papers [53]. A few methods have been proposed to overcome poor quality reception at the dip angle direction. The first is to use a well known diversity technique [59,60] without a feedback from the control module. The second idea is

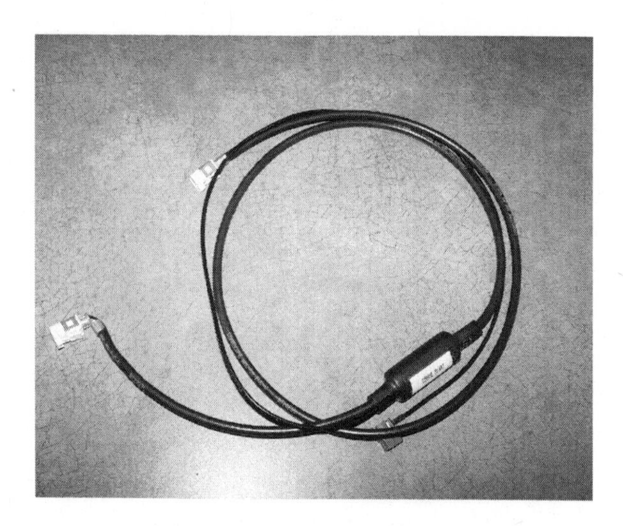

FIGURE 1.10
Splitter system for use with regular whip antenna.

to use an existing car antenna (for example, an AM/FM whip or short roof helix) for RSE/RKE applications [61]. In this case, a special splitter inserted into the RF cable between the whip and car radio allows separation of AM/FM and RSE/RKE signals.

Whip antennas used for RSE/RKE applications have more omnidirectional radiation patterns as compared with antennas mounted inside cars. A splitter is designed as a three-port circuit. The input port is connected with a whip antenna; one output port is connected to the car radio and the second output port is connected to an RKE control module. A splitter design (Figure 1.10) installed in a JK Chrysler car RKE system was tested and its communications range almost exceeded 200 m (splitter circuit is described in Chapter 9).

1.7.3 Tire Pressure Sensors

Tire pressure monitoring systems (TPMSS) are used for measuring and monitoring the pressures and temperatures inside vehicle tires. An air pressure sensor transmitting antenna positioned in each wheel periodically transmits individual air pressure levels to a receiver via a radio link. The transmitter data includes information about tire air pressure and a code identifying the particular tire. The receiver control module collects data from the sensor transmitters via a receiving antenna. If tire pressure drops below a certain level in one or more tires, the monitoring system illuminates a warning signal in the instrumental panel. The transmitting antenna operates in two frequency bands: 315 MHz (North America and Japan) and 433.92 MHz (Europe).

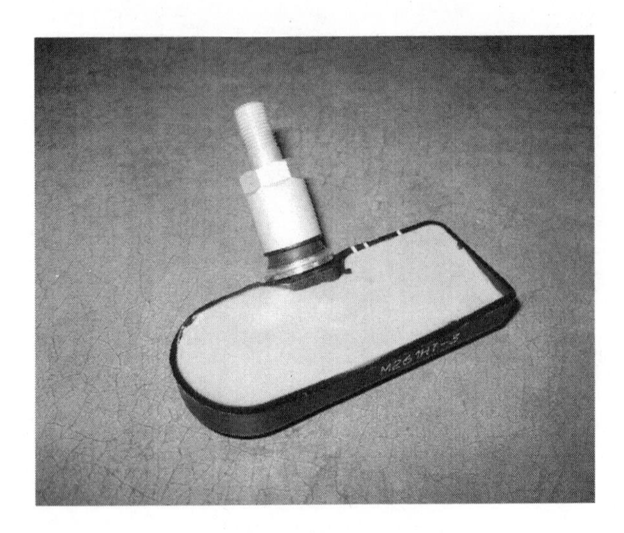

FIGURE 1.11
Valve stem with incorporated tire pressure sensor antenna.

Generally, RSE/RKE and TPMS circuits are mounted in the same housing and use the same receiving antenna. One popular tire pressure transmitting antenna is incorporated into a valve stem used to fill a tire with air. A mass production valve stem with an incorporated antenna [62] is shown in Figure 1.11. Patents [63,64] describe a printed circuit board antenna design mounted on or near a metal wheel rim for TPMS application. Well designed transmitting antennas provide sufficient power for reception regardless of tire angle orientation during rotation. Typical sensitivity of the receiving module is about –103 to –108 dBm.

1.7.4 Electronic Toll Collection Systems

Electronic toll collection systems (ETCs) are two-way radio communication systems that identify vehicles and/or their owners. An FM signal transmitted by a roadside antenna is received by an active transponder (tag) mounted on the vehicle windshield. The transponder responds by generating a microwave signal containing data stored in the tag. The transmitted signal is received by a roadside system that decodes the data and relays it to a computer for vehicle identification. Current RF tags utilize a few frequency ranges including 866 to 869 MHz in Europe, 902 to 928 MHz in North and South America, and 950 to 956 MHz in Japan and some other Asian countries.

The maximum detectable communication range between the roadside system and tag is usually 5 to 10 m. One of the active mass production transponder antennas in Figure 1.12 is a low profile printed circuit board bent monopole that operates in the 902 to 928 MHz frequency band; it was designed for the Ontario Highway 407 ETC system. The antenna radiation

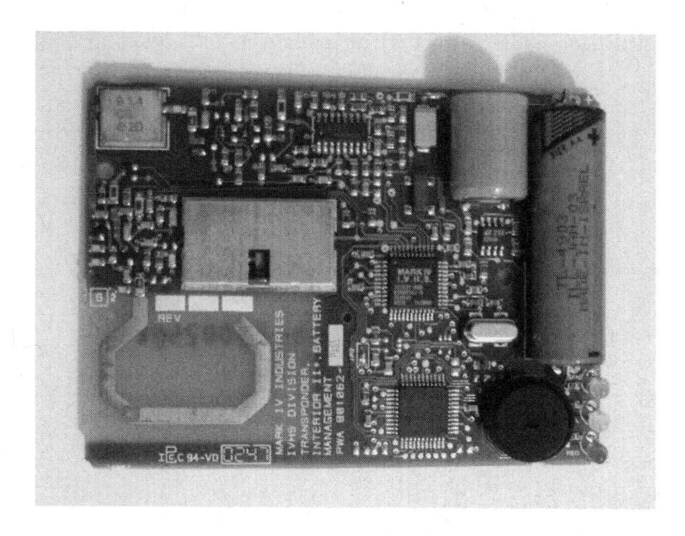

FIGURE 1.12
Electronic toll collection antenna geometry.

pattern is almost omnidirectional with a gain of –8 to –12 dBd and polarization is linear. Commercially produced U.S. transponders that operate at 902 to 928 MHz use five different dedicated short range communication (DSRC) protocols for electronic toll collection. The new uniform standard uses a 5.9 GHz DSRC band in North America and 5.8 GHz in Europe. The new frequency range provides 75 MHz of frequency bandwith (915 MHz frequency has only 12 MHz of bandwidth) and permits much higher data transmission rates (6 to 27 MBps) than the lower frequency 915 MHz band (0.5 MBps). Potential DSRC applications include:

- Electronic toll collection
- Vehicle safety inspection
- Electronic parking payments
- Approaching emergency vehicle warning
- Highway–rail intersection warning
- Commercial vehicle clearance and safety inspections
- In-vehicle signing
- Transit or emergency vehicle signal priority
- Intersection collision avoidance

Presently, the 5.9 GHz systems market focuses on research and testing. The commercial introduction of 5.9 GHz systems is expected in 2010 when the Institute of Electrical and Electronics Engineers (IEEE) ratifies Standard

802.11p. A simple 5.8 GHz circular polarized patch antenna with a gain of 5 dBic can be utilized for DSRC vehicle applications. Another possible solution is a printed-on-circuit-board FR-4 monopole about 0.28 of wavelength [65].

A reader base station antenna for DSRC applications, mounted at lane side or overhead in each traffic lane should have the proper main lobe width and low side lobe level to provide a stable communication link between a vehicle transponder and the reader system. A microstrip antenna array [66] is a good candidate for such applications. This antenna, with dimensions of 70×75 mm^2, has four circular polarized elements; the distance between elements is 0.7 of wavelength, with a 13 dBic gain in total. The design is implemented on the base of a printed circuit board (PCB) with a dielectric constant equal to 2.55 and thickness of 1.5 mm. The size of each element is about 16×16 mm^2 and the main lobe angle width is about 70 degrees (–3 dB level).

1.8 Thin Film Antennas

Thin film antennas are utilized in communication systems at frequencies from the AM band to a few GHz. These easily removable antennas can be mounted on the front, rear, or side glass of a vehicle. Two different designs are available on the aftermarket. The first is a thin strip line of conductive material deposited on one face of a sheet of nonconductive transparent plastic film such as polyester. Usually a copper or silver strip about 0.2 to 0.7 mm wide and 5 to 50 μm thick is used to fabricate the conductive strip. The DC electrical resistance of a strip line antenna is less than 0.2 ohm/cm. An example of such a design (Figure 1.13) is a GPS film antenna to receive RHCP

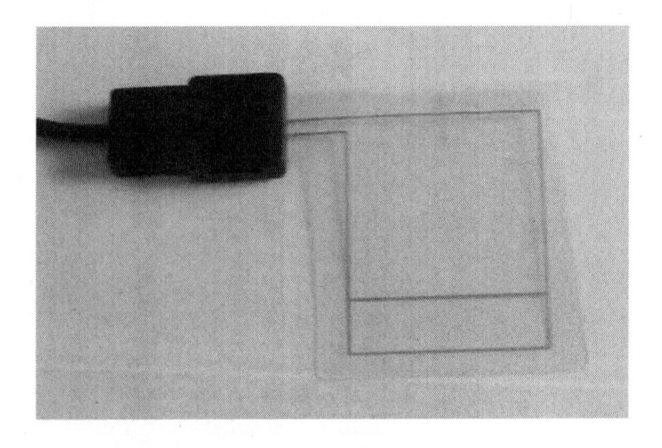

FIGURE 1.13
Thin film antenna.

provided by an asymmetrical feed point arrangement in a loop antenna and an additional parasitic element. The maximum gain of this antenna is equal to 0 dBic.

The second design option is an antenna containing a transparent conductive material. For example, NASA uses patch antennas for 2.3 GHz or 19.5 GHz applications [67]. This type of antenna is made of a thin, clear polyester sheet with an AgHT-8 optically transparent conductive coating deposited on one face. The patch and feed strip are arranged by cutting the coated film in the required pattern. The transparency of the material used for window-mounted thin film antennas is an important consideration. According to automotive original equipment manufacturer (OEM) requirements, a vehicle window is not sufficiently transparent if its transmittance is below 70% for visible light. Commercially available antennas built with transparent conducting films with acceptable transparency (above 70%) typically have DC resistance values several orders of magnitude greater than values of copper strip line antennas whose efficiency is reduced due to ohmic loss in higher resistance films. As a result, antenna gain can be reduced by 3 to 7 dB, depending upon the type.

1.9 Digital Audio Broadcasting Design

Digital audio broadcasting (DAB) service is being introduced throughout the world. A DAB receiver along with an antenna installed in a car provides high quality audio reception whether the car is in motion or stopped. Different frequency bands are designated for DAB service. For example, Canadian DAB currently operates in the L band (1452 to 1492 MHz) while European DAB operates in Band III (174 to 240 MHz). Terrestrial towers that provide the transmitting power for DAB reception operate with VP signals. Good candidates for receiving are vertical monopole antennas or printed-on-glass antennas for vehicle applications. Paper [68] investigates two versions, both for vehicle use. The first is a vertical active monopole mounted on a car roof and the second is a flat active antenna printed on a dielectric circuit board. (Detailed investigations of these antennas are described in Chapter 4.)

The active gain value of these antennas (passive antenna gain plus amplifier gain) exceeds 10 dBi and the noise figure of an amplifier is about 1.5 dB. The roof-mounted antenna is almost omnidirectional, while the directionality of a printed-on-circuit-board (PCB) antenna mounted in a car has a few angle dips that reduce the communication range. However, it is known that space diversity utilizing two or more antennas mounted in a vehicle increases the quality of reception.

1.10 Automotive Radar Antenna Systems

Two commercially available automotive systems are long range radar (LRR) [69] and short range radar (SRR) [70]. LRR operates at 77 GHz and has a resolution of about 3 m at a distance of 150 m. The 77 GHz radar technology is used in adaptive cruise control (ACC) applications. One or a few LRR main lobes control the driving path in front of the car to determine the constant minimum safety distance relative to the vehicle in front. The SRR system operates at 24 GHz with a range below 30 m. The SRR system is used to provide side, rear, front, intersection, blind spot, and other types of safety enhancements. The automotive manufacturers that utilize radar devices include Daimler-Benz, BMW, Jaguar, Nissan, Toyota, Honda, Volvo, and Ford. Typical performance specifications of an ACC system with an antenna are [69–71]:

- Transmit frequency 76 to 77 GHz
- Mean transmitted power level <50 dBm (peak level 55 dBm)
- Target range 2 to 150 m
- Range resolution ±1 m
- Velocity resolution ±1 km/h
- Azimuth angular coverage ±8 degrees (3 degree minimum resolution)
- Elevation angular coverage 3 to 4 degrees (single beam)
- Antenna gain 26 to 34 dBi

Frequency bandwidth of the SRR is 4 to 5 GHz. Antenna specifications for such ultrawide bandwidth devices include [70,71]:

- Transmit frequency 22 to 26 GHz
- Mean transmitted power level <0 dBm (peak limit 20 dBm)
- Target detection range 0.05 to 25 m
- Azimuth angular coverage 55 degrees (3 dB typical beamwidth)
- Typical elevation 3dB beamwidth 15 degrees
- Range resolution 0.05 to 0.2 m (depends on application)
- Range accuracy ± 5 cm

Millitech Corporation developed and built a 77 GHz antenna system based on a folded optic design. It is 145 mm in diameter and 80 mm long [72]. This antenna has three outputs that collimate the energy in three angle directions covering 8 degrees in azimuth. This LRR can be located at the car front bumper area. Detailed antenna design is described in patents referenced by this work [72–74].

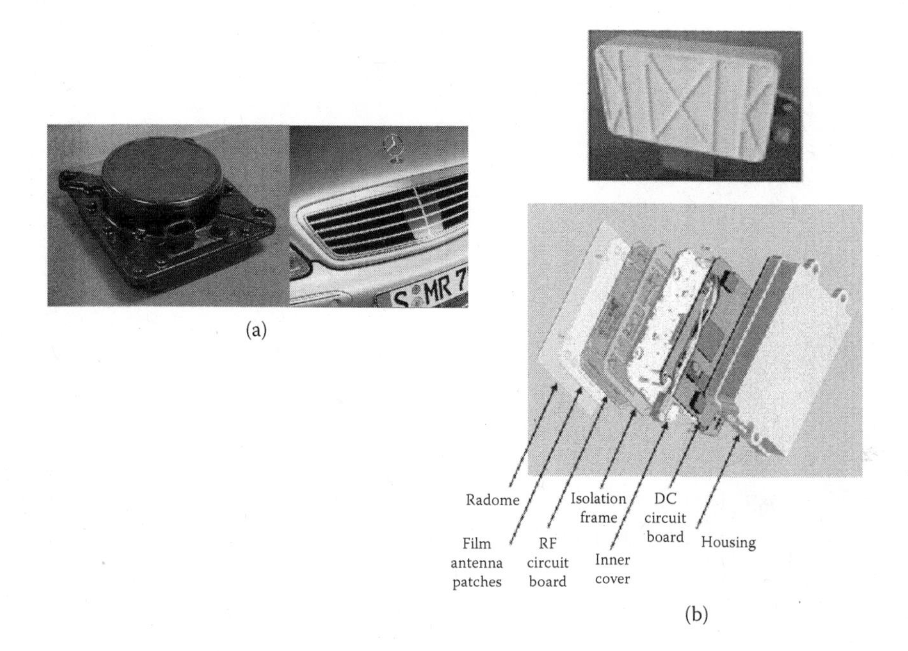

FIGURE 1.14
Automotive radar antennas, (a) Mercedes long range radar. (From Wenger, J., *IEEE Compound Semiconductor Integrated Circuit Symposium*. 2005. Copyright 2005 IEEE. With permission.) (b) Short range antenna system. (From Gresham, I., et al. *IEEE Radio Frequency Integrated Circuits Symposium*, 8–10 June 2003. Copyright 2004 IEEE. With permission.)

An example of a front bumper-mounted ACC system with LRR radar is the Mercedes Distronic radar sensor [73] shown in Figure 1.14a. A typical SRR is presented in Figure 1.14b. The antenna system includes two separate antennas: one to transmit and one to receive. Each antenna presents a microstrip slot-coupled antenna array. The antenna [70,71] provides 60 degree azimuth beamwidth, 12 to 15 degree elevation beamwidth, and 11 to 14 dBi gain, depending on specific implementation. Side lobe level of the elevation plane is less than –25dB. One researcher [70] commented that the side lobes are very sensitive to the assembly of an SRR antenna module.

1.11 Antenna Packaging Issues

Modern cars may be equipped with antennas operating simultaneously in different frequency bands. For example, popular design includes AM/FM with SDARS; cell phone with GPS; or AM/FM with cell phone and GPS. Some antennas, such as those for cellular phones, must operate in a few

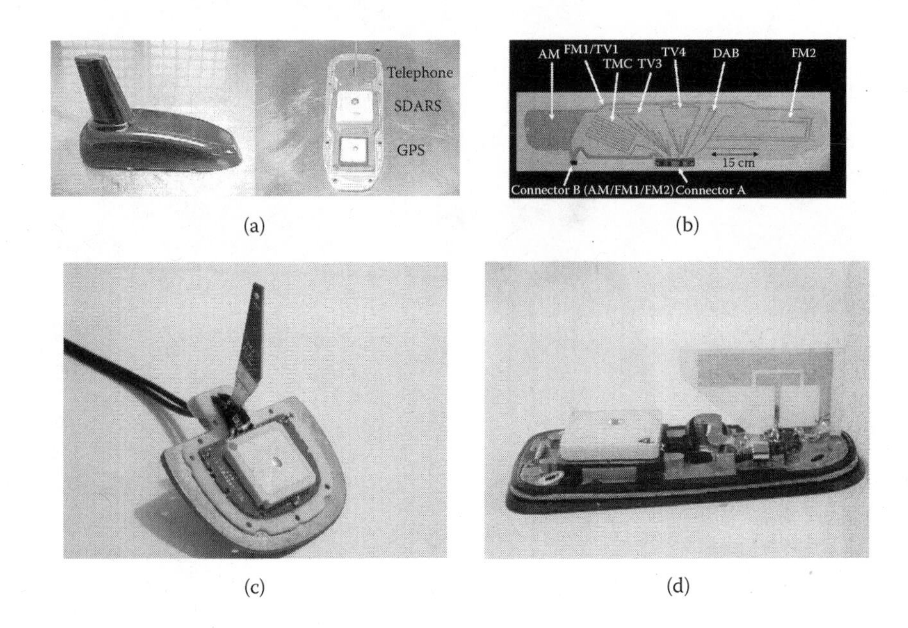

(a) (b) (c) (d)

FIGURE 1.15
Multiband antenna packages: (a) Combined cellular phone satellite radio, and GPS antenna system (From E. Perri, *IEEE International Symposium on Antennas and Propagation Digest*, July 2006. Copyright 2006 IEEE. With permission.) (b) Printed on transparent thin film. (From L. Low et al., *IEEE Transactions on Antennas and Propagation*. 2006. With permission.) (c) Example of combined cellular phone and GPS antenna system. (d) PIFA cellular phone combined with GPS and AM/FM antenna system (AM/FM is not shown). (From R. Leelaratne and R. Langley, *IEEE Transactions on Vehicular Technology*. 2005. With permission.)

frequency ranges. Therefore, miniaturization is becoming an important issue related to providing well-built, aesthetically pleasing antennas for the automotive industry. Figure 1.15 presents a few packages combining antennas operating in multiple frequency ranges.

Figure 1.15a shows cell phone, SDARS, and GPS antennas [75]. Figure 1.15b shows a printed-on-film package with multiple elements including AM, two FM, TV, DAB, and toll collection antennas [76]. Figure 1.15c shows a commercially available antenna system combining GPS and cell phone. A PIFA cell phone antenna on a roof-mounted base with a GPS patch and AM/FM mast is presented in Figure 1.15d (mast not shown) [77].

1.12 Summary

This review clearly shows that the development of low profile, multifrequency, hidden systems is a promising trend in the design of automotive antennas. Antennas intended for short range automotive communications covering frequencies up to 6 GHz including vehicle-to-vehicle communication, intelligent transportation service, and electronic toll collection systems are becoming more popular. A very interesting solution combining SDARS and GPS antennas allows the sizes of multifrequency antenna packages to be reduced.

Improved SDARS patch antennas can be used to receive circular and linear polarized waves. Splitter antenna systems allow one antenna to receive different frequency band signals. Antenna mounting space is restricted and therefore the development of multiband antenna systems including a few antennas in a single package (e,g., AM/FM, cell phone, GPS, SDARS, and RKE) is a very promising direction that is challenging automotive antenna design engineers. Forward-looking antenna designs based on arrays have attractive applications for automotive radar systems and satellite TV. Automotive radar devices operating at 24 GHz (short range) and 77 GHz (long range) with antenna arrays are used for adaptive cruise control (LRR) and crash safety (SRR) systems.

References

1. T. Tokuda and Y. Baba, Antenna Device for Motor Vehicle, U.S. Patent 7,227,503, June 5, 2007.
2. N.F. Bally et al., AM/FM Dual Grid Antenna, U.S. Patent 7,038,630, May 2, 2006.
3. H. Lindenmeier, Active Broad-band Reception Antenna with Reception Level Regulation, U.S. Patent 6,888,508, May 3, 2005.
4. E.K. Walton et al., Layout for Automotive Window Antenna, U.S. Patent 6,693,597, February 17, 2004.
5. A.D. Fuchs, R. Lindackers, and R.A. Duersch, Vehicle Window Antenna System, U.S. Patent 6,239,758, May 29, 2001.
6. Y. Nagayama and M. Maegawa, Glass Antenna for Side Windshield of Automotive Vehicle, U.S. Patent 6,437,749, August 20, 2002.
7. J. Ooe and T. Imoto, Vehicular Antenna Device, U.S. Patent Application 20030156070, August. 21, 2003.
8. H. Fujii and M. Uemura, Antenna for Vehicle, U.S. Patent Application 20060176227, August 10, 2006.
9. P.T. Dishart et al., Antenna On-Glass, U.S. Patent 6,384,790, May 7, 2002.
10. P.T. Dishart and D. A. Saar, Glass Antenna System with an Impedance Matching Network, U.S. Patent 5,999,134, December 7, 1999.
11. H.K. Lindenmeier et al. *Mobile Antenna Systems Handbook*, Boston, Artech, 1994, p 293.
12. R. Abou-Jaoude and E.K. Walton, Numerical modeling of on-glass conformal automobile antennas, *IEEE Transactions on Antennas and Propagation*, 46, 845–852, 1998.
13. Y. Kim and E.K. Walton, Effect of body gaps on conformal automotive antennas, *Electronics Letters*, 40, 1161–1162, 2004.
14. L. Low and R. Langley, Modeling automotive antennas, *IEEE International Symposium on Antennas and Propagation Digest*, 3, 3171–3174, June 2004.

15. R. Langley and J. Batchelor, Hidden antennas for vehicles, *Electronics and Communication Engineering Journal*, 14, 253–262, 2002.
16. J. Batchelor, R. Langley, and H. Endo, On-glass mobile antenna modeling, *IEEE Proceedings: Microwaves, Antennas and Propagation*, 148, 233–238, 2001.
17. J. Shim et al., Selection of optimum feeding point of heater-grid antennas using NEC-2 Moment Method Code, *Microwave and Optical Technology Letters*, 22, 310–314, 1999.
18. G. Gonzalez, *Microwave Transistor Amplifiers: Analysis and Design*. Upper Saddle River, NJ: Prentice Hall, 1996.
19. V. Rabinovich, I. Rabinovich, and T.R. Reardon, Active Window Glass Antenna System with Automatic Overload Protection Circuit, U.S. Patent 6,553,214, April 22, 2003.
20. H. Lindenmeier et al., Antenna for Radio Reception with Diversity Function in a Vehicle, U.S. Patent 7,564,416, July 21, 2009.
21. H. Lindenmeier, J. Hopf, and L. Reiter, Scanning Diversity Antenna System for Motor Vehicles, U.S. Patent 6,611,677, August 26, 2003.
22. L. Reiter et al., Compact antenna with novel high impedance amplifier diversity module for common integration into narrow dielectric parts of a car's skin. University of the Bundeswehr, Munich, Retrieved October 20, 2007 from http://rp.iszf.irk.ru/hawk/URSI2005/pdf/D08.7(0919).pdf
23. S. Lindenmeier et al., Integrated microwave antenna systems in mobile applications. University of the Bundeswehr, Munich, Retrieved October 20, 2007, from http://www.ursi.org/Proceedings/ProcGA05/pdf/D08.6(0897).pdf
24. H. Lindenmeier, Scanning Antenna Diversity System for FM Radio for Vehicles, U.S. Patent 7,127,218, October 24, 2006.
25. R. Shatara and J.J. Marrah, Automotive FM diversity systems 2: analog systems, presented at Society of Automotive Engineers World Congress, Detroit, April 2007, *SAE Technical Papers*, Document 2007-01-1732 2007.
26. R.S. Shatara, M.A. Boytim, and J.J. Marrah, Automotive FM diversity systems 3: digital systems, presented at Society of Automotive Engineers World Congress, Detroit, April 2007, *SAE Technical Papers*, Document 2007-01-1734 2007.
27. L. Low and R.J. Langley, Single feed antenna with radiation pattern diversity, *Electronics Letters*, 40, 975–976, 2004.
28. W. Yanagisawa, et al., Antenna for Mounting on Vehicle, Antenna Element and Manufacturing Method Thereof, U.S. Patent 6,259,411, July 10, 2001.
29. O.M. Garay and Q. Balzano, Wide-Band Helical Antenna, U.S. Patent 4,772,895, September 20, 1988.
30. E.B. Perri, Continuously loaded mast antennas for vehicular applications, *IEEE International Symposium on Antennas and Propagation Digest*, 23576–2360, July 2006.
31. J.D. Kraus, Helical beam antenna, *Electronics*, 20, 109–111, 1947.
32. H. Maeda et al., Vehicle Roof Mount Antenna, U.S. Patent 6,791,501, September 14, 2004.
33. L.L. Nagy, Automobile antennas. In *Antenna Engineering Handbook*, 4th Ed. J. Volakis, Ed. New York: McGraw-Hill, 2007, Chap. 39.
34. http://www.yokowo.co.jp/english/product/pdf/vccs_02e.pdf.
35. K.L. Wong, *Planar Antennas for Wireless Communications*. Hoboken, NJ: Wiley Interscience, 2003.
36. J. Ohe and H. Kondo, Automobile Antenna System for Diversity Reception, U.S. Patent 4,845,505, July 4, 1989.

37. H. Toriyama et al., Development of printed-on-glass TV antenna system for car, *IEEE 37th Vehicular Technology Conference*, pp. 334–342, June 1987.
38. Y. Noh, Wideband Glass Antenna for Vehicle, U.S. Patent 7,242,358, July 10, 2007.
39 http://www.kvh.com
40. D. Pozar and D. Scaubert, *Microstrip Antennas: The Analysis and Design of Microstrip Antennas and Arrays*, IEEE Press, New York, 1995.
41. A. Petros, I. Zafar, and S. Licul, Reviewing SDARS antenna requirements, *Microwaves & RF*, 51–62, September 2003.
42. A. Duzdar et al., Radiation efficiency measurements of a microstrip antenna designed for reception of XM satellite radio signals, presented at Society of Automotive Engineers World Congress, Detroit, April 2006, *SAE Technical Papers*, Document 2006-01-1354.
43. A. Riza, Modular Patch Antenna Providing Antenna Gain Direction Selection Capability, U.S. Patent 7,167,128, January 23, 2007.
44. K. Yegin et al., Directional Patch Antenna, U.S. Patent 7,132,988, November 7, 2006.
45. K. Yegin et al., Patch Antenna with Parasitically Enhanced Perimeter, U.S. Patent Application 20050280592, December 22, 2005.
46. K. Yegin et al., Vehicular Mirror Housing Antenna Assembly, U.S. Patent 7,248,225, July 24, 2007.
47. H. Lindenmeier et al., A new design principle for a low profile SDARS antenna including the option for antenna diversity and multiband application, presented at the Society of Automotive Engineers World Congress, Detroit, March 2002, *SAE Technical Papers* Document 2002-01-0122.
48. H. Lindenmeier et al., SDARS antenna diversity schemes for a better radio link in vehicles, presented at the Society of Automotive Engineers World Congress, Detroit, April 2005, *SAE Technical Papers* Document 2005-01-0568.
49. P. Argy, Headphone Antenna Assembly, U.S. Patent Application 20060050893, March 9, 2006.
50. D. Ayman and F. Andreas, Single-Feed Multi-Frequency Multi-Polarization Antenna, U.S. Patent 7,164,385, January 16, 2007.
51. K. Yegin et al., Integrated GPS and SDARS Antenna, U.S. Patent 7,253,770, August 7, 2007.
52. A. Bensky, *Short-Range Wireless Communications.* LLH Technologies, Eagle Rock, VA, 2000.
53. B. Al-Khateeb, V. Rabinovich, and B. Oakley, An active receiving antenna for short range wireless automotive applications, *Microwave and Optical Technology Letters*, 43, 293–297, 2004.
54. V. Rabinovich et al., Compact planar antennas for short-range wireless automotive communication, *IEEE Transactions on Vehicular Technology*, 55, 1425–1435, 2006.
55. H. Blaese, Inside Window Antenna, U.S. Patent 5,027,128, June 25,1991.
56. L. Nagy et al., Automotive Radio Frequency Antenna System, U.S. Patent 6,266,023, July 24, 2001.
57. M. Wiedmann, M. Pfletschinger, and D. Wendt, Antenna for a Central Locking System of an Automotive Vehicle, U.S. Patent 6,937,197, August 30, 2005.

58. V. Rabinovich et al., Small printed meandered symmetrical and asymmetrical antenna performances, including the RF cable effect, in 315 MHz frequency band, *Microwave and Optical Technology Letters*, 48, 1828–1833, 2006.
59. V. Rabinovich et al. Compact diversity antenna system for remote control automotive applications, *IEEE International Symposium on Antennas and Propagation Digest*, 2B, 379–381, July 2005.
60. V. Rabinovich et al., Antenna System for Remote Control Automotive Application, U.S. Patent Application 20060170610, August 3, 2006.
61. T. Desai, Vehicle Antenna System for Multiple Vehicle Electronic Components, U.S. Patent 6,339,403, January 15, 2002.
62. K. Yin et al., Package for a Tire Pressure Sensor Assembly, U.S. Patent Application 20060272758, December 7, 2006.
63. L. Lin et al., RF Tire Pressure Signal Sensor Antenna and Method of Packaging, U.S. Patent 6,958,684, October 25, 2005.
64. A. Nanz et al., Antenna for Tire Pressure Monitoring Wheel Electronic Device, U.S. Patent 6,933,898, August 23, 2005.
65. W. Duerr et al., A low noise active receiving antenna using a SIGE HBT, *IEEE Microwave and Guided Wave Letters*, 7, 63–65, 1997.
66. W. Liu, RFID Antenna design of highway ETC in ITS, *Seventh International Symposium on Antennas, Propagation, and Electromagnetic Theory*, October 2006, pp. 1–4.
67. R. Simons and R. Lee, Feasibility study of optically transparent patch antenna, *IEEE International Symposium on Antennas and Propagation Digest*, 4, 2100–2103, July 1997.
68. V. Rabinovich et al., L-band active receiving antenna for automotive applications, *Microwave and Optical Technology Letters*, 39, 319–323, 2003.
69. R. Jaoude, ACC Radar sensor technology, test requirements, and test solutions, *IEEE Transactions on Intelligent Transportation Systems*, 4, 115–122, 2003.
70. I. Gresham et al., Ultra-wideband radar sensors for short-range vehicular applications, *IEEE Transactions on Microwave Theory and Techniques*, 52, Part 1, 2105–2122, 2004.
71. Electromagnetic Compatibility and Radio Spectrum Matters: Radio Equipment to be Used in the 24 GHz Band. System Reference Document for Automotive Collision Warning Short Range Radar, Technical Report ETSI TR 101 982 V1.2.1, 2002-2007.
72. R. Hguenin et al., Compact Microwave and Millimeter Wave Radar, U.S. Patent 5,680,139, October 21, 1997.
73. J. Wenger, Automotive radar status and perspectives, *IEEE Compound Semiconductor Integrated Circuit Symposium*, 2005, pp. 21–24.
74. I. Gresham et al., Ultrawide band 24GHz automotive radar front-end, *IEEE Radio Frequency Integrated Circuits Symposium*, June 2003, pp. 505–508.
75. E. Perri, Dual band cellular antenna in a multifunction platform for vehicular applications, *IEEE International Symposium on Antennas and Propagation Digest*, 9, 2361–2364. July 2006.
76. L. Low et al., Hidden automotive antenna performance and simulation, *IEEE Transactions on Antennas and Propagation*, 54, 3707–3712, 2006.
77. R. Leelaratne and R. Langley, Multiband PIFA Vehicle Telematics antennas, *IEEE Transactions on Vehicular Technology*, 54, 477–485, 2005.

2

Basic Antenna Parameters and Definitions

2.1 Introduction

As noted in Chapter 1, the automotive industry uses three types of antennas. The first type radiates (or receives) energy from 360 degrees around a car (radiation pattern must be omnidirectional). These antennas typically are used for terrestrial AM/FM and TV reception, cellular phones, RKE or RSE systems, and ETC devices. The second group used for SDARS and GPS systems receive energy only from the upper half of the sphere space. A third antenna class including satellite TV or radar collision systems must receive energy in a narrow angle space. This chapter will introduce the main antenna parameters that will be explained further in subsequent chapters. In addition, this chapter will demonstrate measurement techniques specific to antennas mounted on and in vehicles.

2.2 Far Zone and Radiation Pattern

The far field antenna region or far zone distance R_0 is defined as the distance from the antenna where angular field distribution is independent of the distance from the antenna. Typically, a far zone is determined by the minimum distance from the antenna equal to $2L^2/\lambda$ where λ = wavelength and L = maximum linear antenna size. For example, for an FM whip antenna combined with car body far zone distance equals more than 17 m; for a half wave FM dipole, this distance is about 2 m.

Far zone antenna radiation pattern or far zone directionality is the parameter that determines angle dependence of the power radiated (or received) by an antenna versus angle coordinates θ and φ shown in Figure 2.1. The IEEE standard for a spherical coordinate system determines that angle θ is measured from the z-axis (zenith). The azimuth angle φ is measured from the projection of the radius vector $\bar{r}$ to the xy (horizontal) plane with $\varphi = 0$ at the x-axis increasing counterclockwise. Radiation pattern is a measure of the capability of an antenna to capture signals more favorably in one angle

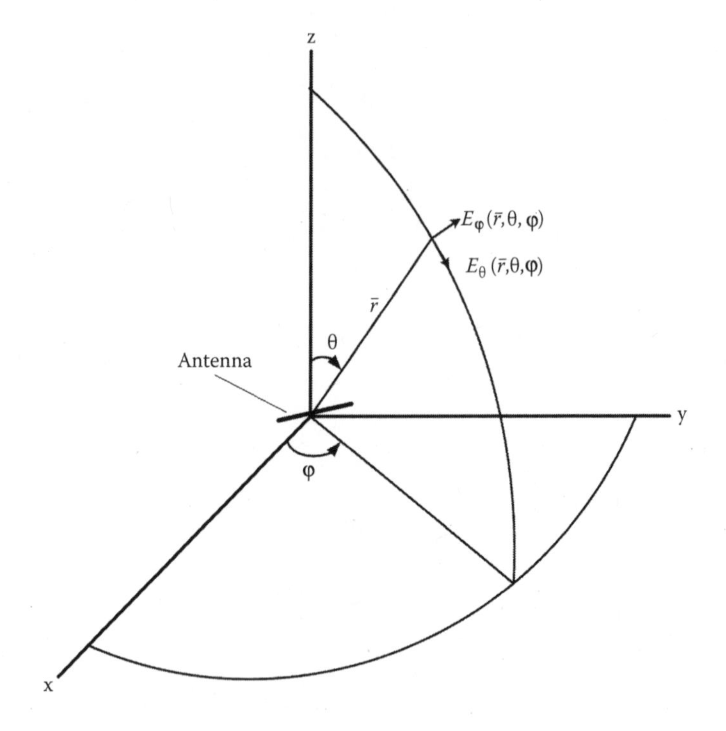

FIGURE 2.1
Coordinate system for antenna analysis.

direction than another. Radiation pattern $F(\theta, \varphi)$ can be presented as a sum of two orthogonal components (Reference [1], pp. 38, 44).

$$F(\theta,\varphi) \cong \frac{B \cdot r^2}{2 \cdot W_0}\left[|E_\theta(\bar{r},\theta,\varphi)|^2 + |E_\varphi(\bar{r},\theta,\varphi)|^2\right] = F_\theta(\theta,\varphi) + F_\varphi(\theta,\varphi) \tag{2.1}$$

where B is a constant, $E_\theta(\bar{r},\theta,\varphi)$ and $E_\varphi(\bar{r},\theta,\varphi)$ as shown in Figure 2.1 are antenna far zone electric field components, W_0 is the intrinsic impedance of the medium, and r equals the distance from the antenna to the observation point.

Equation (2.1) can be applied to both transmitting and receiving antennas. The horizontal and vertical planes through which antenna radiation cuts are the main performance factors for investigating antennas for automotive applications. Generally, radiation pattern cut in a horizontal plane $\theta = 90°$ includes two components $F_\theta(\theta = 90°, \varphi)$ and $F_\varphi(\theta = 90°, \varphi)$; a radiation pattern in a vertical plane is a sum of $F_\theta(\theta, \varphi = const)$ and $F_\varphi(\theta, \varphi = const)$.

A simple monopole antenna mounted on an infinite metal screen has only an omnidirectional radiation pattern $F_\theta(\theta, \varphi)$ that does not depend on the φ angle value ($F_\varphi(\theta, \varphi) = 0$). A whip antenna mounted on a fender has components $F_\theta(\theta, \varphi)$ and $F_\varphi(\theta, \varphi)$ in horizontal and vertical planes due to the effects of car body shadow. Both components in horizontal plane (θ = 90 degrees) are not omnidirectional. The important parameter γ that equals the maximum-to-minimum ratio of the radiation pattern value in horizontal plane when measuring 360 degrees around a car determines the quality of the whip antenna design. Whip antennas usually meet the requirement when γ is less than 12 to 15 dB for the $F_\theta(\theta = 90°, \varphi)$ component. Antennas printed on window glass suffer from more dip variations of directionality over a 360 degree span (sometimes more than 25 dB). However, these dips are generally very narrow and do not exceed a few degrees. As a rule, radiation pattern levels are relative values with respect to certain reference values and expressed in decibels.

2.3 Polarization and Radiation Pattern Measurements

Spatial orientation of the electric field radiated by a transmitting antenna determines polarization. A linear polarized (LP) antenna radiates an electric field that does not change the angle orientation along the direction of propagation. For example, a monopole on the infinite ground plane transmits only a vertically polarized (VP) wave $E_\theta(\theta, \varphi)$ (radiation field component ($E_\varphi(\theta, \varphi) = 0$). According to the reciprocity for radiation patterns (Reference [1], pp. 130–132), the same monopole operating in receiving regime responds only to the vertically polarized transmitting wave. This means that the receiving monopole on the metal screen is also a vertically polarized antenna.

Therefore an antenna under test can be measured in transmitting or receiving regime. This fundamental principle can be applied to any antenna shape and polarization type for the linear media of wave propagation. A simple horizontally oriented dipole transmits only $E_\varphi(\theta, \varphi)$ ($E_\theta(\theta, \varphi) = 0$) and constitutes a horizontally polarized (HP) antenna. A similarly oriented dipole operating in receiving mode also responds to an HP transmitting antenna. Elliptically polarized electric fields are comprised of two linear components that are orthogonal to one another. Each component has a different magnitude and phase. At any fixed point along the direction of propagation, the total electric field traces an ellipse as a function of time. A circular polarized (CP) wave is comprised of two linearly polarized electric field components E_x and E_y; both are orthogonal, have equal amplitudes, and are 90 degrees out of phase.

Right hand circular polarization (RHCP) is produced if the electrical vector rotates clockwise. If rotation is counterclockwise, polarization is left circular (LHCP). Ideally, an antenna designed to receive RHCP waves does not respond to LHCP wave; conversely an antenna for LHCP waves does not respond to RHCP waves. However, in reality, any antenna designed for

RHCP or LHCP waves receives cross components. High quality CP antenna design utilizes a ratio of main and cross component values exceeding 10 dB.

Axial ratio (AR) is an important parameter that characterizes CP antenna design. AR value is the ratio between E_x and E_y components (Reference [2], p. 44).

$$AR = \sqrt{\frac{1 + |E_x/E_y|^2 + T}{1 + |E_x/E_y|^2 - T}} \tag{2.2}$$

where

$$T = \sqrt{1 + \left|\frac{E_x}{E_y}\right|^4 + 2\left|\frac{E_x}{E_y}\right|^2 \cdot \cos(2\alpha)} \tag{2.3}$$

and α = phase of E_x/E_y. The AR can be presented as a combination of two circular polarized waves: RHCP and LHCP [3]:

$$AR = \frac{|E_{RHCP} + E_{LHCP}|}{|E_{RHCP} - E_{LHCP}|} \tag{2.4}$$

To achieve circular polarization, the magnitude of axial ratio must be unity and the phase must be ±90 degrees.

The typical polarization types for automotive antennas are linear (for example, broadcasting radio, cell phone radio), right hand circle (GPS) and left hand circle (satellite radio). To transfer maximum energy between transmitting and receiving antennas, both have the same polarization sense and AR.

Typically, the site for measuring horizontal plane antenna radiation patterns is an outdoor turntable that can rotate an antenna 360 degrees around a car. A computer controls a spectrum analyzer and proper turntable rotation while transferring measured data to a hard drive and printer. Generally, the data points are taken on a 0 degree elevation amplitude pattern every 1 or 2 degrees as the turntable rotates through a full 360 degree azimuth.

The vertical plane directionality components $F_\theta(\theta, \varphi = const)$ and $F_\varphi(\theta, \varphi = const)$ of the mounted antenna can be measured using the special arch shown in Figure 2.2. A transmitting antenna can move along the arch with a radius about 5 to 10 m while changing the elevation angle of the radio wave incident to the vehicle.

To measure directionality of a receiving antenna for specific polarization (horizontal, vertical, or circular), the transmitting antenna must be used with the same requested polarized wave orientation. For example, measuring whip antenna directionality for vertical polarization in horizontal plane ($F_\theta(\theta = 90°, \varphi)$, the transmitting antenna must be turned to supply a vertical polarized radiated wave; a vertically oriented mast transmitting antenna can be used

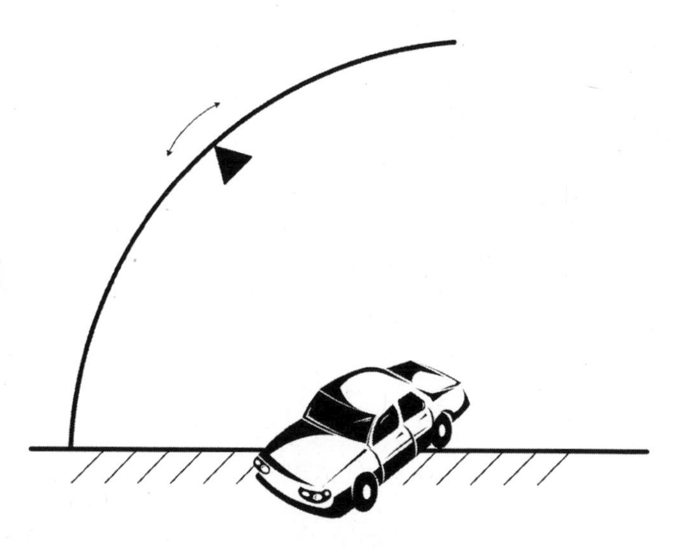

FIGURE 2.2
Special arch for measuring antenna radiation pattern in elevation plane.

to serve this purpose. When measuring directionality for horizontal polarization ($F_\varphi(\theta = 90°, \varphi)$), the transmitting antenna must be turned 90 degrees with respect to the vertical polarization. Antennas intended to receive RHCP or LHCP waves must be measured by using transmitting antennas with the required polarization orientation.

2.4 Directivity, Gain, Average Gain, and Antenna Beamwidth

Antenna directivity is defined as the ratio of the radiation intensity in a given angle direction (as a rule in the angle of maximum radiation) to the intensity of the same antenna as it radiates in all directions. Based on Reference [1], p. 44, mathematical directivity can be expressed as:

$$D(\theta_0, \varphi_0) = \frac{4\pi \cdot F(\theta_0, \varphi_0)}{\int\int F(\theta, \varphi) \sin\theta \, d\theta \, d\varphi} \tag{2.5}$$

Integration limits by φ are from 0 to 360 degrees and by θ from 0 to 180 degrees.

The maximum value of Equation (2.5) can be simplified (Reference [1], p. 593) as:

$$D_{\max} \approx \frac{4 \cdot \pi \cdot S}{\lambda^2} \tag{2.6}$$

where S = physical area of aperture.

For narrow beam antenna systems, Equation (2.5) can be simplified [1, p. 46]:

$$D_{max} \approx \frac{32400}{\Omega_1 \cdot \Omega_2} \tag{2.7}$$

where Ω_1 = half power beamwidth in one plane (degrees) and Ω_2 = half power beamwidth in the orthogonal plane (degrees).

Directivity is measured in dBi or on a dBd scale. A value of 0 dBi corresponds to the isotropic radiator that emits uniformly in all angle directions and 0 dBd is the maximum radiation level of a half wave dipole in free space. The ratio of dBi and dBd is 1 dBd = 2.17 dBi.

Often, the automotive industry uses simple monopole antennas mounted on car roofs (cellular phone) or fender-mounted whips for AM/FM applications. The maximum directivity value of a monopole mounted on an infinite perfectly conductive metal screen is 5.17 dBi—3 dB larger in comparison to a half wave dipole. This result is explained by the fact that a monopole antenna radiates only to the upper half space of the ground plane. However, a metal car roof is not infinite and since the monopole antenna is not mounted at the center of the roof, its directivity is less than 5.17 dBi and depends on location on the car. Directivity $D(\theta_0, \varphi_0)$ (D below) and gain $G(\theta_0, \varphi_0)$ (G below) values are related by the ratio:

$$G = \eta \cdot D \tag{2.8}$$

where η = antenna efficiency. According to Reference [1], pp. 60 and 61, total antenna efficiency can be presented as:

$$\eta = \eta_1 \cdot \eta_2 \cdot \eta_3 \tag{2.9}$$

where $\eta_1 = (1 - |\Gamma|^2)$ reflection efficiency, Γ = reflection coefficient to be determined in Section 2.5, η_2 = conduction efficiency, and η_3 = dielectric efficiency. The conduction–dielectric efficiency $\eta_{23} = \eta_2 \cdot \eta_3$ (Reference [1], pp. 78 and 79) determines the ratio of radiation resistance R_{rad} and loss resistance R_{loss}:

$$\eta_{23} = \frac{R_{rad}}{R_{rad} + R_{loss}} \tag{2.10}$$

where R_{loss} = loss resistance of the antenna. Loss resistance due to the skin effect for a wire antenna with wire diameter equal to d and total length L can be estimated by:

$$R_{loss} = \frac{L}{\pi \cdot d \cdot \delta_S \cdot \sigma} \tag{2.11}$$

where σ = conductivity of the wire (for copper σ = 5.8 Siemens/m). δ_S = skin depth is determined by:

$$\delta_S = \frac{1}{\sqrt{\pi \cdot f \cdot \mu \cdot \sigma}} \tag{2.12}$$

where f = frequency (Hz) and $\mu = 4 \cdot \pi \cdot 10^{-7}$ (H/m) is the permeability for nonmagnetic materials. For example, a copper conductor has a skin depth at 1 GHz of about 2.1 μm or 0.11 mil. For a whip antenna with 0.75 m height and 3.5 mm diameter, loss resistance for 100 MHz equals 0.18 ohm. For a strip line antenna (printed on dielectric substrate) with a total length L, width W, and depth of T, loss resistance is:

$$R_{loss} = \frac{L}{2 \cdot (W + T) \cdot \delta_S \cdot \sigma} \tag{2.13}$$

A hypothetical lossless antenna perfectly matched with load has gain G equal to directivity D. A traditional and convenient method of measuring gain is a relative technique that includes measurements of an antenna under test with a reference-calibrated antenna. A calibrated antenna has a predetermined gain value of G_{ref}. First, the power level P_1 received by the antenna under test is measured for the given angle. Second, the power level P_2 received by the reference antenna is measured for the same angle position. The transmitting antenna does not change position during measurement. The gain of the antenna under test can be calculated by:

$$G_{ant} = G_{ref} \cdot \frac{P_1}{P_2} \tag{2.14}$$

The gain of the antenna under test expressed in dBi or dBd is given by:

$$G_{ant}(dBi) = G_{ref}(dBi) + P_1(dBm) - P_2(dBm) \tag{2.15}$$

$$G_{ant}(dBd) = G_{ref}(dBd) + P_1(dBm) - P_2(dBm) \tag{2.16}$$

where the $P_1(dBm)$ and $P_2(dBm)$ powers are measured in dBm. Often the gain of an antenna designed for automotive applications is compared with a receiving quarter wave monopole that has an almost omnidirectional radiation pattern when mounted on the middle of a car roof.

Average gain (over 360 degrees around a car) is calculated using the arithmetic mean of 360 degrees uniformly distributed as linear gain values (not dB scale values). Normally, the antenna beamwidth (half power beamwidth) is measured as an angle space between two angle directions where radiation intensity is one half the maximum value of the beam. For engineering estimation, the following beamwidth (BW) expression can be used for narrow beam antennas:

$$BW \approx \frac{\lambda}{L} \tag{2.17}$$

2.5 Impedance, Voltage Standing Wave Ratio, Bandwidth, and Quality Factor

Antenna impedance is determined by three in-series connected elements: radiation resistance, loss resistance, and reactance:

$$Z_a = R_{rad} + R_{los} + j \cdot X_a \tag{2.18}$$

Radiation resistance R_{rad} is responsible for the energy portion radiated or received by an antenna. For example, power radiated by an antenna with uniform current distribution I equals $P_{rad} = I^2 \cdot R_{rad}$. Loss resistance R_{loss} as noted previously reduces antenna gain. Reactance, X_a determines the parasitic power portion that is not radiated into the far antenna zone. Generally, antenna impedance is a complex value. Network analyzers can measure complex impedance (real and imaginary) and it is necessary to emphasize that the network analyzer measures radiation and loss resistor together, not separately. Impedance of a half wave dipole equals 73 + j42.5 ohm (radiation resistance is 73 ohm); impedance of the quarter wave monopole is equal to half that of the half wave dipole or 36.5 + j21.25 ohm (radiation resistance is 36.5 ohm).

Voltage standing wave ratio (VSWR) is the ratio of the amplitude of the standing wave at a maximum to the amplitude at the minimum in an electrical transmission line. For example, VSWR 2:1 denotes a maximum standing wave amplitude two times greater than the minimum standing wave amplitude. The amplitude reflection coefficient is related to the VSWR according to the following expression:

$$\rho = |\Gamma| = \frac{V_r}{V_f} = \frac{(1 + VSWR)}{(1 - VSWR)} \tag{2.19}$$

where V_f and V_r are amplitudes of forward and reflected waves in a transmission line. $|\Gamma|$ indicates absolute Γ value. When $\Gamma = 0$, no reflection is present. $\Gamma = -1$ means maximum reflections where the line is shorted; $\Gamma = 1$ means maximum reflection with open transmission line. VSWR is measured the same way as impedance using a network analyzer.

According to Balanis [1], p. 63, the bandwidth of an antenna "*BANDW* is defined as the range of frequencies within the performance of the antenna which, with respect to some characteristic, conforms to a specified standard." For example, in an impedance bandwidth antenna, VSWR does not exceed a predetermined VSWR ratio, for example, VSWR ≤2:1:

$$BANDW = \frac{(Fh - Fl)}{Fc} \cdot 100\% \tag{2.20}$$

where *Fl* and *Fh*, respectively, are the lower and upper frequency limits where the magnitude of VSWR ≤ 2:1; F_C is the center frequency. For example, patch or electrically small antenna typically have narrow bandwidth, while whip and printed-on-glass antenna bandwidths are much wider.

The quality factor Q for an antenna tuned to have zero reactance at frequency ω_0 ($X(\omega_0)=0$) is determined by the ratio of the average reactive power stored around the antenna to the total power accepted by an antenna including radiating power plus power loss [5]:

$$Q = Q(\omega_0) = \frac{\omega_0 \cdot |W(\omega_0)|}{P_A(\omega_0)} \tag{2.21}$$

The quality factor is a useful parameter for estimating bandwidth:

$$BANDW = \frac{VSWR - 1}{Q \cdot \sqrt{VSWR}} \cdot 100\% \tag{2.22}$$

The half power bandwidth (VSWR approximately 5:1) is given by:

$$BANDW_{1/2} \approx \frac{2}{Q} \cdot 100\% \tag{2.23}$$

The quality factor and effective volume are related by

$$a = \frac{\lambda}{2 \cdot \pi} \cdot \left(\frac{9}{2 \cdot Q} \right)^{1/3} \tag{2.24}$$

where a is the radius of a sphere defining the effective volume of the antenna. The lower bound on the quality factor for a linear polarized antenna is given by [5]:

$$Q_{lb} = \eta_{23} \cdot \left(\frac{1}{k \cdot a} + \frac{1}{(k \cdot a)^3} \right) \tag{2.25}$$

The lower bound on the quality factor for a circular polarized antenna is expressed as:

$$Q_{lb} = \frac{\eta_{23}}{2} \cdot \left(\frac{2}{k \cdot a} + \frac{1}{(k \cdot a)^3} \right) \tag{2.26}$$

If the antenna has a matching circuit and if the matching circuit is not lossless (quality factor Q_m) combined antenna and matching circuit quality factor Q_S equals:

$$\frac{1}{Q_S} = \frac{1}{Q} + \frac{1}{Q_m} \tag{2.27}$$

The total efficiency [6] can be expressed as:

$$\eta_S = \eta_{23} \cdot \eta_m = \frac{\eta}{1 + \frac{Q}{Q_m}} \tag{2.28}$$

2.6 Impedance Matching between Antenna and Car Receiver

When a passive antenna with an RF cable is power matched with a car receiver, the power received by the antenna is transferred to the receiver load without reflections between antenna and receiver. Under power matching conditions, the real antenna portion of the impedance (along with RF cable impedance) must equal the input portion of receiver impedance. Furthermore, an imaginary portion of antenna impedance is out of phase 180 degrees with the input imaginary portion of the car receiver impedance. Typical OEM car receivers have only real input impedance values of 50, 75, 93, or 125 ohm. In AM frequency range, a car radio input has a capacitance of about 15 pF and radiation resistance far below 1 ohm. Therefore, power matching for AM frequency range is difficult and only voltage gain is available.

Noise matching provides minimum residual noise at the output of an amplifier and does not coincide with the maximum power matching of an antenna and an amplifier.

2.7 Electrically Small Antennas

Generally, an electrically small antenna is defined [4] as having maximum dimension $2 \cdot a$ determined by:

$$k \cdot a < 1 \tag{2.29}$$

where $k = 2 \cdot \pi / \lambda$. The theoretical estimation of maximum small antenna gain [7] is:

$$G_{small} \approx 2 \cdot (k \cdot a) + (k \cdot a)^2 \tag{2.30}$$

Paper [8] presents the antenna Q factor for the following electrically small structures:

A small dipole or monopole above infinite ground plane:

$$Q \approx \frac{6 \cdot \left[\ln\left(\frac{a}{a_0}\right) - 1 \right]}{(ka)^3} \tag{2.31}$$

A small loop:

$$Q \approx \frac{6 \cdot \ln\left(\frac{a}{a_0}\right)}{\pi \cdot (ka)^3} \tag{2.32}$$

A small inverted antenna with vertical length portion equal to horizontal length:

$$Q \approx \frac{3 \cdot \left[\ln\left(\frac{a}{a_0}\right) - 0.653 \right]}{(ka)^3} \tag{2.33}$$

In deriving the above expressions, it was assumed that wire radius a_0 is much smaller than the total antenna length; inverted antenna is mounted on the infinite ground plane.

Some important conclusions related to electrically small antennas are presented in Reference [9]. First, an impedance matching antenna does not significantly improve the quality factor; second, the radiation properties of electrically small antennas are essentially independent of geometry. The radiation resistance of an electrically small self-resonance antenna is primarily established by antenna height relative to resonant wavelength.

2.8 Radio Frequency Cables and Connectors

Radio frequency (RF) cables connect antennas to receivers. Cable length varies, depending on the antenna location on or in a car. An RF cable may be very long (5 to 6 m for an antenna mounted on the rear of a car for a receiver mounted on the front panel). The automotive industry generally uses coaxial RF cables. Such a cable consists of an inner conductor surrounded by a dielectric insulator with dielectric constant ε_r surrounded by an outer cylindrical conductor (metallic shield) and plastic jacket (Figure 2.3). The inner conductor may be copper- or silver-plated iron wire and the insulator surrounding the inner conductor may be solid polyethylene or Teflon (PTFE).

The outer cylindrical conductor is a braided copper wire that forms a shield. Theoretically, the electromagnetic energy (transverse electric magnetic mode or TEM wave) is concentrated only in the space between the inner

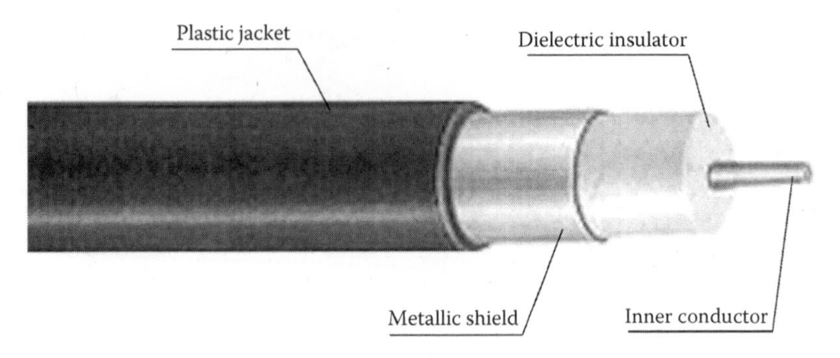

FIGURE 2.3
Radio frequency cable structure.

and output conductors. The electrical parameters of a typical cable are shunt capacitance per unit length; series inductance per unit length; series resistance per length; characteristic impedance in ohms; attenuation losses per unit length; velocity of propagation; and cutoff frequency that determines practical use in a certain frequency range.

Typical automotive industry RF cables have values of impedance of 50, 75, 93, and 125 ohms. Table 2.1 shows parameters of a typical RF cable made for automotive use. The diameters noted in the second column represent outer diameters of plastic jackets. Loss values (dB/m) are noted at 100, 500, 1000, 2000, and 3000 MHz frequencies.

RF connectors are used to connect the RF cable with an antenna and car receiver. Generally, for OEM applications, one end of the RF cable is soldered directly to the antenna and the other end is connected to the receiver with an RF connector. The most popular RF connectors utilized by North America automotive OEMs are Fakra connectors (Figure 2.4). These connectors are built on the bases of tiny SMBs with insertion losses below 0.3 dB in DC to 3 GHz in the frequency band. The applications include AM/FM radio, GPS and SDARS systems, and Bluetooth and Internet access devices. The

TABLE 2.1
Typical Radio Frequency Cable Parameters

Cable Type	Diameter (mm)	Impedance (ohm)	Loss 100	Loss 500	Loss 1000	Loss 2000	Loss 3000
RG 58	5	50	0.13	0.34	0.52	0.82	1.1
RG 59	6.1	75	0.1	0.22	0.32	0.46	0.6
RG 62	6.15	93	0.09	0.12	0.28	0.41	0.51
RG 174	2.55	50	0.28	0.72	1.1	1.74	2.3
RG 178	1.8	50	0.52		1.7		2.6

FIGURE 2.4
Fakra connectors.

industry produces connectors in different plastic housings. Blue, brown, yellow, white, or black housings with different locks provide connections in different frequency ranges. Black Fakra connectors do not have locks and can be used with any receivers. Popular SMA precision connectors for microwave applications up to 25 GHz generally are used for most test measurements.

2.9 Notes on Decibels

The Decibel (dB) is an important unit used to measure antenna and propagation parameters. A decibel is a logarithmic unit of measurement that expresses the magnitude of a physical quantity relative to a specified reference level. When referring to measurements of power P_1, a ratio may be expressed in decibels by evaluating ten times the base 10 logarithm of the ratio of the measured quantity to the reference level P_0.

$$\text{Power ratio (dB)} = 10 \cdot \log\left(\frac{P_1}{P_0}\right) \tag{2.34}$$

When referring to measurement of a power radiation pattern $F_{\theta,\varphi}(\theta, \varphi)$:

$$\text{Radiation pattern (directionality; dB)} = 10 \cdot \log\left(\frac{F_{\theta,\varphi}(\theta,\varphi)}{F^0_{\theta,\varphi}(\theta,\varphi)}\right) \tag{2.35}$$

where $F^0_{\theta,\varphi}(\theta,\varphi)$ typically determines maximum radiation pattern value $F_{\theta,\varphi}(\theta,\varphi)$ within the given observation angles. When measuring azimuth cut radiation patterns, $(F_{\theta,\varphi}(90°,\varphi)$ values are typically normalized to a maximum value measured in the 360 degrees by azimuth. Elevation radiation pattern $F_{\theta,\varphi}(\theta,\varphi = const)$ values are normalized to the maximum value measured in the elevation plane.

RF power levels are measured typically as milliwatts (mW) and dBm. Measuring RF power in mW is not always convenient because power value varies inversely as the square of distance, not linearly. A more convenient representation in dBm format is related to the mW as:

$$\text{Power (dBm)} = 10 \cdot \log\text{ (power in mW)} \tag{2.36}$$

Thus, 1 mW = 0 dBm, 10 mW = 10 dBm, 100 mW = 20 dBm, 1W or 1000 mW = 30 dBm and so forth. Similarly:

$$\text{Voltage (dB}\mu\text{V)} = 20 \cdot \log\text{ (voltage in }\mu\text{V)} \tag{2.37}$$

Therefore 1 μV corresponds to 0 dBμV, 10 μV to 20 dBμV, and 100 μV to 40 dBμV. The ratio of power and voltage depends on the load resistor value:

$$Power = \frac{(Voltage)^2}{Resistor\ Value}$$

Let us assume that load resistor value is 50 ohms. The ratio between power and voltage (dB) is given by:

$$P[dBm] = V[dB\mu V] - 107dB \quad \text{for 50 ohm load resistor} \tag{2.38}$$

2.10 Converting Field Strength to Power

The U.S. Code of Federal Regulations, Title 47, Part 15, includes tables showing emission level limits for some RF devices expressed as field strength in microvolts per meter at a 3 m distance from the radiator. Sometimes the emission level limits are expressed in the term of effective radiation power of a device under test. Therefore, it is important to establish a relationship between field strength and effective radiation power (ERP) of a device. ERP is a product of the transmitted power of the device and the gain of the transmitting antenna. Power density P_R from the device with the transmitting antenna at distance R is given by:

$$P_R = \frac{ERP}{4 \cdot \pi \cdot R^2} \tag{2.39}$$

For free space wave impedance = 377 ohm electric field can be represented as:

$$E = \sqrt{\frac{377 \cdot ERP}{4 \cdot \pi \cdot R^2}} \tag{2.40}$$

Therefore the ratio (in dB) between the ERP and field strength for distance $R = 3\ m$ is given by:

$$ERP[dBm] = E[dB\mu V/meter] - 95.22 \text{ dB} \tag{2.41}$$

To briefly summarize some of these notes:

- Power value A in dB scale: A (dB) = 10*log (A)
- Absolute power level in dBm scale: P (dBm) = 10*log (P in milliwatts)
- Voltage V in dBuV scale: V (dBuV) = 20*log (V in microvolts)
- V (dBuV) = P (dBm) + 107dB for 50 ohm load impedance
- V (dBuV) = P (dBm) + 108.8 for 75 ohm load impedance
- Effective radiation power of transmitting antenna (P) in dBm scale: field strength E (dBuV/m) + 95.2dB, based on assumption that field strength is measured from antenna at a distance of 3 m

References

1. C. Balanis, *Antenna Theory, Analysis, and Design.* New York: John Wiley & Sons, 1997.
2. R. Bancroft, *Microstrip and Printed Antenna Design.* Atlanta: Noble, 2004.
3. W. Kunysz, Effect of antenna performance on GPS signal accuracy, *Proceedings of National Technical Meeting of Institute of Navigation*, Long Beach, CA, pp. 575–580, January 1998.
4. H.A. Wheeler, Fundamental limits of small antennas, *Proceedings of IRI (IEEE)*, pp. 1479–1484, December 1947.
5. S.R. Best, Low Q electrically small linear and elliptical polarized spherical dipole antennas, *IEEE Transactions on Antennas and Propagation*, 53, 1047–1053, 2005.
6. G.A. Thiele, P.L. Detweiler, and R.P. Penno, On the lower bound of radiation Q for electrically small antennas, *IEEE Transactions on Antennas and Propagation*, 51, 1263–1269, 2003.
7. R.F. Harrington, Effect of antenna size on gain, bandwidth, and efficiency, *Journal of Research of the National Bureau of Standards: Radio Propagation*, 64D, 1--12, 1960.
8. W. Geyi, Method for the evaluation of small Antenna Q, *IEEE Transactions on Antennas and Propagation*, 51, 2124–2129, 2003.
9. S.R. Best, Discussion on the quality factor of impedance matched electrically small wire Antennas, *IEEE Transactions on Antennas and Propagation*, 53, 502–508, 2005.

3

Active and Diversity Receiving Antenna Systems

3.1 Introduction

Active antennas in automotive systems increase the reception sensitivity and dynamic range of a receiving system. Low noise amplifiers (LNAs) are used in most receiving automotive systems: SDARS, GPS, FM/AM printed-on-glass antennas, digital audio broadcasting (DAB) and other types of antennas.

A low noise amplifier (LNA) integrated with a passive antenna portion provides the first level of signal amplification by an electronic system. An LNA must amplify extremely weak signals with minimal added noise, providing the signal-to-noise ratio of the system at a predetermined required level. The weakest signal received without noise distortion by an LNA defines receiver system sensitivity. The upper power signal level that can be amplified without nonlinear distortion establishes the linear dynamic range of the amplifier. The dynamic range of the amplifier (difference between the strongest and weakest received signals) defines the quality of the amplifier circuit. Proper LNA design is crucial in today's communication solutions. Due to the complexity of modern communications, design considerations must be investigated very carefully.

The diversity antenna system uses two or more spaced antennas with one receiver to improve reception quality. Often in urban environments, no clear line of sight is possible between the transmitter and receiver. The signal to the receiver is distorted by multiple reflections before it is finally received. These distortions result in amplitude attenuations and phase shifts of the incoming signal. In such an environment, each of several antennas experiences a different interference. Thus, if one antenna experiences a deep fade, it is likely that another has a sufficient signal. Therefore, reception quality could be improved by combining (according to a special algorithm) the signals received by several antennas.

Of course, the diversity technique requires more hardware than a single antenna, but the benefits in some applications are significant. The diversity system can also be efficient when the radiation pattern of a single antenna is not omnidirectional (for example, due to car body effect). If two diversity antennas have dips at different angle directions, the combined diversity overlap directionality becomes more omnidirectional, increasing the received signal. The diversity technique can be employed with both passive and active receiving antennas.

3.2 Signal and Noise Analysis of Active Antenna

3.2.1 Signal-to-Noise Ratio

Now we will estimate the signal-to-noise ratio (SNR) that can be obtained with an active rather than passive antenna. Figure 3.1 shows an active automotive system that includes an antenna with noise temperature T_A and gain G_a, an amplifier, RF cable, and receiver (radio, GPS, or SDARS, etc.). According to Kraus ([1], pp. 401–412) the noise value of such a system at the output receiver terminal is given by:

$$N_{act} = k \cdot T_0 \cdot G_{amp} \cdot G_{cab} \cdot G_{rec} \cdot F_{act} \cdot RECBW \tag{3.1}$$

It is assumed in the above expression that a lossless antenna system in a car is placed in an anechoic chamber with a temperature equal to T_0. The noise figure F_{act} is given by:

$$F_{act} = F_{amp} + \frac{F_{cab} - 1}{G_{amp}} + \frac{F_{rec} - 1}{G_{amp} \cdot G_{cab}} \tag{3.2}$$

where F_{amp} = amplifier noise figure, F_{cab} = cable noise figure (equal to loss value), F_{rec} = receiver noise figure, $G_{cab} = \frac{1}{F_{cab}}$ = cable gain, G_{rec} = receiver gain, G_{amp} =

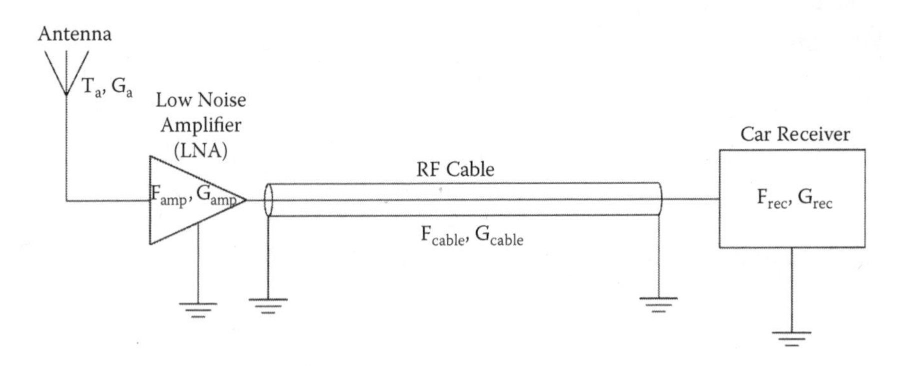

FIGURE 3.1
Block diagram of active antenna system.

gain of amplifier, $RECBW$ = receiver bandwidth, $T_0 \approx 290$ K = room temperature, and K = Boltzmann constant. The noise figure F_{amp} of the amplifier is determined as follows:

$$F_{amp} = \frac{\mathrm{SNR}_{\mathrm{in}}}{\mathrm{SNR}_{\mathrm{out}}}$$

where $\mathrm{SNR}_{\mathrm{in}}$ = the SNR at the amplifier input terminals and $\mathrm{SNR}_{\mathrm{out}}$ = the SNR at the output amplifier terminal. The same definition can be applied to any active or passive device such as a receiver or RF cable. The noise factor of an amplifier or receiver is related to its noise temperature T_{amp} via the following expressions:

$$F_{amp} = 1 + \frac{T_{amp}}{T_0}$$

$$F_{rec} = 1 + \frac{T_{rec}}{T_0}$$

The signal received by the system with an active antenna is equal to:

$$P_{act} = \frac{\Pi_r \cdot G_{act} \cdot \lambda^2 \cdot G_{cab} G_r}{4\pi} \tag{3.3}$$

where Π_r = power density from the transmitter at the receiving antenna input, $G_{act} = G_{a1} \cdot G_{amp}$ = a product of the receiving passive antenna part gain G_{a1} and antenna amplifier gain G_{amp}, and λ = wavelength. The same formulas for a system with a passive antenna may be expressed as:

$$N_{pas} = k \cdot T_0 \cdot G_{cab} \cdot G_{rec} \cdot F_{pas} \cdot RECBW \tag{3.4}$$

$$F_{pas} = F_{cab} + \frac{F_{rec} - 1}{G_{cab}} = F_{cab} \cdot F_{rec} \tag{3.5}$$

$$P_{pas} = \frac{\Pi_r \cdot G_{a2} \cdot \lambda^2 \cdot G_{cab} G_r}{4\pi} \tag{3.6}$$

where G_{a2} = the passive antenna gain.

Let us assume that the coefficient Q_1 determines the improvement in the SNR of the receiving system with an active antenna in comparison with a passive system (SNR improvement factor):

$$Q_1 = \frac{P_{act}}{P_{pas}} \cdot \frac{N_{pas}}{N_{act}} \tag{3.7}$$

The ratio $\frac{N_{pas}}{N_{act}}$ is expressed as:

$$\frac{N_{pas}}{N_{act}} = \frac{F_{cab} \cdot F_{rec}}{F_{amp} \cdot G_{amp} - 1 + F_{cab} \cdot F_{rec}} \tag{3.8}$$

The ratio $\frac{P_{act}}{P_{pas}}$ is equal to:

$$\frac{P_{act}}{P_{pas}} = G_{amp} \cdot \frac{G_{a1}}{G_{a2}} \tag{3.9}$$

Thus,

$$Q_1 = \frac{G_{a1}}{G_{a2}} \cdot \frac{G_{amp} \cdot F_{cab} \cdot F_{rec}}{F_{amp} \cdot G_{amp} - 1 + F_{cab} \cdot F_{rec}} \tag{3.10}$$

For $G_{amp} >> 1$, Equation (3.10) is simplified:

$$Q_1 \approx \frac{G_{a1}}{G_{a2}} \cdot \frac{F_{cab} \cdot F_{rec}}{F_{amp}} \tag{3.11}$$

If $G_{a1} = G_{a2}$ (we compare two identical antennas; one has an amplifier and the other does not), the ratio (3.11) depends only on the noise figure of an amplifier, a receiver, and the RF cable:

$$Q_1 \approx \frac{F_{cab} \cdot F_{rec}}{F_{amp}} \tag{3.12}$$

When expressed in decibel scale:

$$10 \cdot \log(Q_1) \approx 10 \cdot \log(F_{cab}) + 10 \cdot \log(F_{rec}) - 10 \cdot \log(F_{amp})$$

or

$$Q_1(\text{dB}) \approx F_{cab}(\text{dB}) + F_{rec}(\text{dB}) - F_{amp}(\text{dB})$$

Let us examine this ratio for a roof-mounted AM/FM antenna. Typically, such an antenna is located on the rear edge of a car roof and the radio is on the front panel. Assume that we use 5 m of an RG-58 50 ohm RF cable to connect the antenna with the radio. A cable of this length has a loss of about 0.7 dB in the FM frequency range (F_{cab}(dB) = 0.7 dB). Assume that near the antenna, we place an amplifier with noise figure $F_{amp} \approx 3$ dB in the entire FM frequency range and a gain G_{amp}(dB) of about 10 dB. It is known [2] that a typical car radio receiver has noise figure of F_{rec}(dB) $\approx$ 6 dB. The signal-to-noise improvement factor Q_1(dB) for this example is about 3.7 dB. Thus an amplifier with such parameters is a reasonable choice. However, the $Q_1\,(dB)$ ratio is only about 0.7 dB if the noise figure values of an amplifier and a receiver are the same.

Equation (3.10) was obtained by assuming that the amplifier was mounted near the antenna. Now we estimate the signal-to-noise improvement factor when an amplifier is mounted near a receiver. In this case, (3.2) can be written as:

$$F_{act} = F_{cable} + \frac{F_{amp} - 1}{G_{cab}} + \frac{F_{rec} - 1}{G_{amp} \cdot G_{cab}} \tag{3.13}$$

But $\frac{1}{G_{cab}} = F_{cab}$, and therefore:

$$F_{act} = F_{amp} \cdot F_{cab} + \frac{(F_{rec} - 1) \cdot F_{cab}}{G_{amp}} \tag{3.14}$$

The SNR improvement factor Q_2 is given by:

$$Q_2 = \frac{G_{a1}}{G_{a1}} \cdot \frac{F_{rec\cdot} \cdot F_{cab} \cdot G_{amp}}{F_{amp} \cdot F_{cab} \cdot G_{amp} + F_{cab} \cdot F_{rec} - F_{cab}} \tag{3.15}$$

If $G_{a1} = G_{a2}$ and G_{amp} >>1, Equation (3.15) is simplified:

$$Q_2 \approx \frac{F_{rec}}{F_{amp}} \tag{3.16}$$

This formula depends only on the ratio between the receiver noise and the amplifier noise. As we can see, when cable losses are small, the $F_{cab} \approx 1$ formulas (3.12) and (3.16) show almost the same results. However, when the cable losses are high, the integration of the amplifier with an antenna is preferable. The improvement factor q that determines the location of the amplifier near the antenna equals:

$$q = Q_1 / Q_2 \approx F_{cab} \tag{3.17}$$

For example, we design an LNA for satellite radio (G_{amp}(dB) ≈ 30 dB, F_{amp}(dB) ≈ 1 dB), and use 6 m of RG-174 coaxial cable (4 dB losses). In this case, the improvement factor q equals 2.5. An amplifier mounted near the antenna shows an increase in SNR equal to 4 dB compared to an amplifier mounted close to the receiver. By using an amplifier near an antenna, we compensate cable loss if the amplifier gain G_{amp} >> 1.

3.2.2 Antenna Noise Temperature

Antenna noise temperature is defined as the temperature of a hypothetical resistor at the input of an ideal noise-free receiver that would generate the same output noise power per unit bandwidth as that at the antenna output at a specified frequency. An antenna mounted on or in a car receives noise contributions from three main sources:

- External (galaxy, sun, and Earth) radiation
- External electrical devices (including car electronic component noise sources)
- Itself

The antenna noise temperature is an important factor that determines the minimal level detectable by the receiver signal. The antenna noise temperature T_A [3, pp. 99] is given by:

$$T_A = T_a + T_p \cdot \left(\frac{1}{\eta_2} - 1 \right) \tag{3.18}$$

where T_a = noise temperature contributions from external sources, T_p = physical antenna temperature (generally $T_0 \approx 290$ K), and η_2 = thermal efficiency of the antenna. The operating noise temperature of the receiving system with an active antenna can be expressed as $T_{sys} = T_A + T_{amp} + T_r/G_{amp}$ where T_r is the noise temperature of a car receiver with an RF cable.

3.3 Low Noise Amplifier Parameters

3.3.1 Introduction

The main parameters of a low noise amplifier are power gain, input and output impedances, noise figure, amplifier stability, frequency bandwidth, 1 dB compression power point level, and intermodulation distortion. An LNA design presents considerable challenge due to its simultaneous requirements for high gain, low noise, good input and output matching, and unconditional stability at the lowest possible current draw from the amplifier.

Transistor selection is the first and the most important step when designing an LNA. If the antenna has very low radiation resistance (less than a few ohms), impedance matching of antenna and amplifier input is difficult. In this case, a transistor with very high input impedance is used and provides the maximum transmission voltage value from antenna to receiver and a reasonable noise output value.

3.3.2 S Parameters

Typically, scattering (S) parameters ([4], pp. 23 and 24) are used by electronic engineers to describe main amplifier characteristics such as gain, input and output impedance, and VSWR. Every transistor or amplifier circuit can be characterized for any specific operation frequency point by four complex S-parameters:

- S_{11} or input port voltage reflection coefficient
- S_{21} or forward voltage gain
- S_{12} or reverse voltage gain
- S_{22} or output port voltage reflection coefficient

3.3.3 Stability Analysis

Stability analysis should be the first priority when designing an LNA. Unconditional stability is the goal: for any load presented to the input or output of a device, the circuit will not become unstable (will not oscillate). Instabilities are primarily caused by three phenomena: internal feedback of the transistor, external feedback around the transistor caused by an external circuit, and excess gain at frequencies outside the band of operation. S transistor parameters generally provided by the manufacturer will aid in stability analysis of the LNA circuit. The main technique in S parameter stability analysis involves calculating a term called the Rollett stability factor K (Reference [4], pp. 217–223]). To estimate the final equation for the K factor, an intermittent quantity called delta Δ should be calculated first:

$$\Delta = S_{11} \cdot S_{22} - S_{21} \cdot S_{12} \tag{3.19}$$

Subsequently,

$$K = \frac{1 + |\Delta|^2 - |S_{11}|^2 - |S_{22}|^2}{2 \cdot |S_{21}| \cdot |S_{12}|} \tag{3.20}$$

When the K factor is greater than unity and $|\Delta| < 1$, a circuit will be unconditionally stable for any source and load impedance value. When K is less than unity, the circuit is potentially unstable and oscillation may occur with a certain combination of source and/or load impedance presented to the transistor. A sweeping of K factor over the wide frequency band for a given biasing point has to be performed to ensure unconditional stability outside the band of operation. The goal is an LNA circuit that is unconditionally stable for the complete range of frequencies at which the device has a substantial gain.

3.3.4 Amplifier Gain

The S parameters of a transistor help in designing a matching system that transfers maximum power from an antenna to a receiver. Two different matching methods are applied to a transistor: one provides the maximum

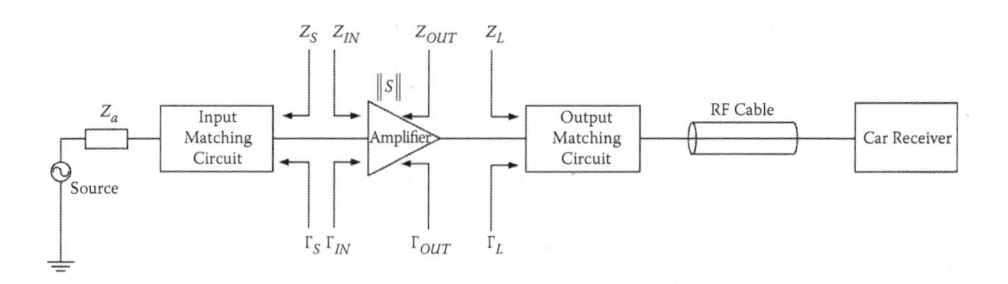

FIGURE 3.2
Generic amplifier system.

gain of the amplifier system and the other provides the minimum noise that the transistor brings to the receiver.

To aid in finding a matching solution that maximizes amplifier gain, let us examine Figure 3.2, a block diagram with the following elements: an antenna as a power source with impedance Z_a (for example, $Z_a = 50$ ohm), an input matching circuit, a transistor with S parameters, output matching circuit, RF cable, and car receiver (for example, a 50 ohm cable with a 50 ohm receiver). The reflection coefficient Γ_{IN} is an input to the transistor, Γ_{OUT} is the output reflection coefficient, Γ_s is the source reflection coefficient seen from the output side of the input matching circuit, and Γ_L is the load reflection coefficient seen from the input side of the output matching circuit. Impedance values Z_S, Z_{IN}, Z_{OUT}, and Z_L are related to the corresponding reflection coefficients in a Z_0 system with the following simple expression:

$$\Gamma_i = \frac{Z_i - Z_0}{Z_i + Z_0} \quad (i = S \text{ or IN or OUT or } L)$$

A typical system uses $Z_0 = 50$ impedance. The matrix $||S||$ in Figure 3.2 is determined to be:

$$||S|| = \begin{vmatrix} S_{11} & S_{12} \\ S_{21} & S_{22} \end{vmatrix}$$

Let us assume that the amplifier in Figure 3.2 does not have input and output matching circuits ($\Gamma_S = 0$, $\Gamma_L = 0$). The gain of the amplifier, called the transducer power gain, is equal to:

$$G_T = \frac{\textit{Power delivered to the load}}{\textit{Power available from the source}} = |S_{21}|^2 \tag{3.21}$$

The conditions ($\Gamma_s = 0$, $\Gamma_L = 0$) correspond to the 50 ohm source and load impedances. When input and output matching circuits are used as shown in Figure 3.2, the input and output reflection coefficients can be expressed ([4], p. 214) as:

$$\Gamma_{\text{IN}} = S_{11} + \frac{S_{12} \cdot S_{21} \cdot \Gamma_L}{1 - S_{22} \cdot \Gamma_L} \tag{3.22}$$

$$\Gamma_{\text{OUT}} = S_{22} + \frac{S_{12} \cdot S_{21} \cdot \Gamma_S}{1 - S_{11} \cdot \Gamma_S} \tag{3.23}$$

where the reflection coefficient Γ_S is the source reflection coefficient seen from the output side of the input matching circuit and Γ_L is the load reflection coefficient seen from the input side of the output matching circuit. The gain value (3.21) in this case is given by the following

$$G_T = \frac{1 - |\Gamma_S|^2}{|1 - \Gamma_{\text{IN}} \cdot \Gamma_S|^2} \cdot |S_{21}|^2 \frac{1 - |\Gamma_L|^2}{|1 - S_{22} \cdot \Gamma_L|^2} \tag{3.24}$$

3.3.5 Matching for Maximum Amplifier Gain

To maximize the gain of the amplifier system (transistor + input matching circuit + output matching circuit), the matching circuit design must meet certain conditions ([4], pp. 240 and 241):

$$\Gamma_S = \Gamma_{\text{in}}^*; \quad \Gamma_L = \Gamma_{\text{out}}^* \tag{3.25}$$

This conjugate matching provides maximum gain, determined by:

$$G_{\text{TMAX}} = \frac{|S_{21}|}{|S_{12}|} \cdot \left(K - \sqrt{K^2 - 1}\right) \tag{3.26}$$

This equation assumes that the amplifier is unconditionally stable ($K > 1$ and $|\Delta| < 1$). If $S_{12} \approx 0$, we have a so-called unilateral network (S_{12} is small enough and ignored), and the formula for the gain value is modified ([4], p. 228) as follows:

$$G_{TU} = \frac{1 - |\Gamma_S|^2}{|1 - S_{11} \cdot \Gamma_S|^2} \cdot |S_{21}|^2 \cdot \frac{1 - |\Gamma_L|^2}{|1 - S_{22} \cdot \Gamma_L|^2} \tag{3.27}$$

Optimized values Γ_S and Γ_L to provide maximum gain are given by:

$$\Gamma_S = S_{11}^*; \quad \Gamma_L = S_{22}^* \text{ and } G_{TU\text{MAX}} = \frac{1}{1 - |S_{11}|^2} \cdot |S_{21}|^2 \cdot \frac{1}{1 - |S_{22}|^2} \tag{3.28}$$

A number of computer programs can help find the values of reflection coefficients Γ_S and Γ_L to a certain $P_{\max}$. One powerful program is Genesys from Eagleware Corporation [5].

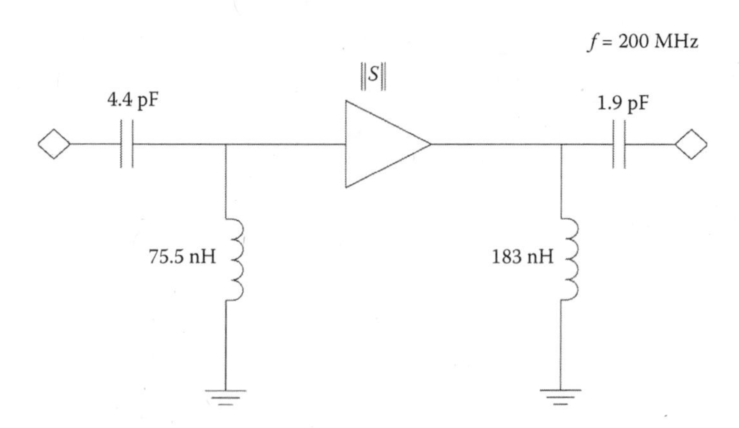

FIGURE 3.3
Maximum power matching circuit (conjugate).

As an example, let us choose a transistor with the following complex S parameters at 200 MHz frequency: S_{11} = 0.92/–29; S_{21} = 1.39/154; S_{12} = 0.002/72; and S_{22} = 0.97/–11. *S* parameters are expressed in magnitude and phase values. The results of computer simulations are the following: $G_{\max}$ = 23.5 dB, $K = 1.7$ (transistor is unconditionally stable); $\Gamma_S = 0.932/31.8$; and $\Gamma_L = 0.975/12$.

Figure 3.3 presents a matching circuit tuned for maximum power (conjugate) matching for the transistor parameter example. The basic equations (3.22) to (3.28) and simulation results show how to find a solution that provides the maximum power delivered from the source to the load using an amplifier with matching circuits, but this solution does not provide minimal noise value at the amplifier output.

3.3.6 Noise Matching Design

A typical approach in LNA design is to devise an input matching circuit that terminates the transistor with a reflection coefficient, $\Gamma_S = \Gamma_{opt}$ representing the optimum terminating impedance of the transistor for the best noise match (see Reference [4], pp. 234 and 235). In many cases, this means an optimum noise match with maximum SNR: the input return loss of the LNA will be sacrificed due to the mismatching of the input impedance to 50 ohm. A typical method used in designing an input matching network displays noise circles and gain/loss circles of the input network on the same Smith chart ([4], pp. 295–308). The noise figure of an amplifier can be estimated by the following expression:

$$F_{amp} = F_{\min} + \frac{4 \cdot R_n \cdot |\Gamma_S - \Gamma_{opt}|^2}{\left(1 - |\Gamma_S|^2\right) \cdot |1 + \Gamma_{opt}|^2} \quad (3.29)$$

F_{min}, Γ_{opt}, and R_n (equivalent noise resistance of a transistor) are generally specified in the transistor data sheet. Therefore, by choosing different Γ_S values, we can control the value of the noise at the amplifier output. When $\Gamma_S = \Gamma_{opt}$, the noise figure value F_{amp} is minimal and equal to F_{min}. If we lack such data for noise optimization, we may use a noise figure meter to find the optimal value experimentally.

3.3.7 Output Gain Matching for Noise Matched LNA

After we obtain a specific LNA noise value, then based on the procedure described in Section 3.3.6 covering noise minimization, we can use an additional procedure as the last step in designing an output matching circuit. This design procedure provides a reasonable noise figure and acceptable amplifier gain. The noise minimization procedure transforms the existing source reflection coefficient Γ_S to the new value Γ_{1S} that allows us to calculate the resulting output reflection coefficient Γ_{1out}

$$\Gamma_{1out} = S_{22} + \frac{S_{12} \cdot S_{21} \cdot \Gamma_{1s}}{1 - S_{11} \cdot \Gamma_{1s}} \tag{3.30}$$

Because the maximum gain approach is based on a conjugate-matched output port, we must create an output circuit to transform the system termination to the required conjugate-matched source. The final step is to transform the system load termination to the complex conjugate of this new $\Gamma_{1L} = \Gamma^*_{1out}$. This method is known as the available gain design outline [6]. Amplifiers designed on this principle achieve a perfectly matched output port with a mismatched input port.

The mismatch loss at the input determines the magnitude of the input reflection coefficient. For example, if we have to sacrifice a 1 dB gain at the input port for the best noise value, the 1 dB mismatch loss converts to a 0.45 input reflection coefficient magnitude. A 2 dB mismatch loss leads to an input reflection coefficient magnitude of 0.6—a poor input match.

3.3.8 Low Noise Amplifier Distortion Parameters

Active RF devices are ultimately nonlinear in operation. When driven with a large RF signal, a device will generate undesirable spurious signals. A 1 dB gain compression point parameter is a measure of the linearity of a device and is defined as the input power that causes a 1 dB drop in linear gain due to device saturation. A 1 dB gain output compression point is related to the 1 dB gain input compression point as follows (dBm scale):

$$P_{1dB}(output) = P_{1dB}(input) + (GAIN - 1) \tag{3.31}$$

When two or more harmonic signals are applied to a nonlinear amplifier, the output contains additional frequency components called intermodulation products. For example, if two sinusoidal signals of frequencies f_1 and f_2 are applied to an amplifier, the output signal contains frequency components $n \cdot f_1 \pm m \cdot f_2$ (n and m are any integers). The frequencies $2f_1$ and $2f_2$ are the second harmonics, $f_1 \pm f_2$ are the second-order intermodulation products, and $2f_1 \pm f_2$ and $2f_2 \pm f_1$ are the third-order intermodulation products. The special problem with third-order products is that they can fall within the operating frequency band. We know ([4], p. 363) that the ratio of a 1 dB output power compression point and a third-order product output power level can be estimated (dBm scale) as follows:

$$P_{IP3}(output) \approx P_{1dB}(output) + 10\,\text{dB} \tag{3.32}$$

Also, it can be shown that:

$$P_{2f_1} = 3P_{f_1} - 2P_{IP3} \tag{3.33}$$

where P_{2f_1} and P_{f_1} are power levels of harmonics with frequencies of $2f_1$ and f_1, respectively.

The spurious free dynamic range *DRANGE* of an amplifier is defined as:

$$DRANGE = \frac{2}{3}[P_{IP3}(\text{dB}m) + 174\text{dB}m - 10\log AMPBW - F(\text{dB}) - X(\text{dB}) - G_{amp}(\text{dB})].$$

where *AMPBW* is the amplifier frequency bandwidth defined as the difference between the frequency limits of the amplifier that corresponds to 3 dB signal attenuation. A typical value of *X(dB)* is 3 dB.

3.3.9 Measurement Set-Up to Estimate Third-Order Intermodulation Distortions

Two continuous wave (CW) signal generators with equal power levels are set a few megahertz apart, near the low end of the specified frequency band. The test is repeated in mid-band and near the high end. The fundamental tone levels are set so as not to saturate the amplifier tested. The test equipment has not contributed too much of its own IM (intermodulation). The power combiner and attenuators ahead of the amplifier promote isolation between RF generators that could produce IM internally. Low-pass filters reduce harmonics that could add to distortions generated by the amplifier. The output third-order intercept point (Figure 3.4) is computed from the measurements as:

$$P_{IP3}(output) = P + \frac{P - \text{MAX}(A1, A2)}{2} \tag{3.34}$$

where A1 and A2 are the power levels of harmonics $2 \cdot f_1 - f_2$ and $2 \cdot f_2 - f_1$ and P is the amplified power of the CW signal generator. A1, A2, and P are read on a spectrum analyzer in dBm. The input third-order intercept point is P_{IP3} *(input)* = P_{IP3} *(output)* – G (dB scale); G is the gain of the amplifier tested.

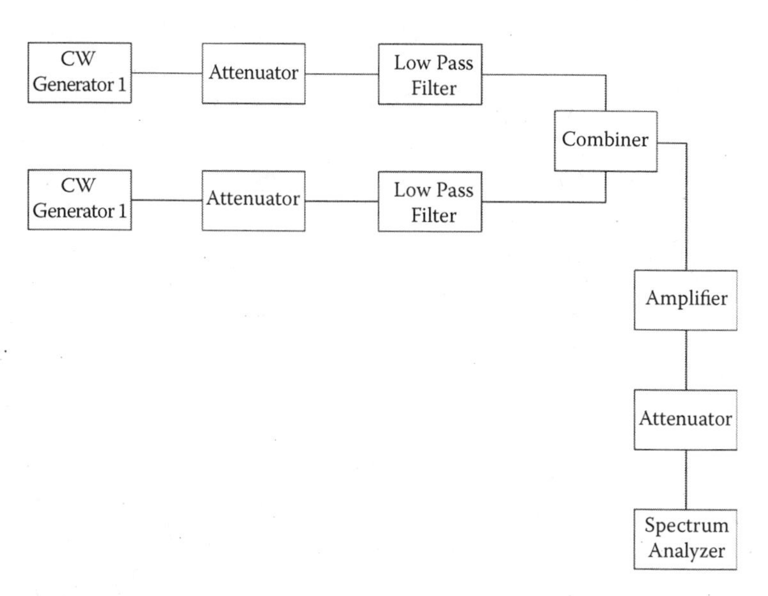

FIGURE 3.4
Arrangement for third-order intermodulation measurements.

3.4 Active Antenna Gain

The gain of an active antenna can be defined [7] as follows

$$G_{act}(\theta_0, \varphi_0) = G_t \cdot G_a(\theta_0, \varphi_0) \tag{3.35}$$

where G_t is the transducer power gain of the active circuit with the antenna as the source and the car receiver as the load. The transducer gain value is a function of the frequency and does not depend on the angle direction. $G_a(\theta_0, \varphi_0)$ is the gain of the passive antenna portion from the standard gain definition.

Active antenna gain can be measured with the standard gain transfer (gain comparison) method by a reference antenna with a known gain value. An amplifier gain measured with a network analyzer that has a 50 ohm input and output port impedances has the same value as the transducer gain portion (3.35) only when the antenna impedance and car receiver (with RF cable) impedance equal 50 ohm. Otherwise, to achieve branch amplifier gain equivalent to the transducer gain portion (3.35), it is necessary to transform (using a passive network circuit) the 50 ohm input port impedance of the network analyzer to the impedance equivalent passive antenna portion impedance and 50 ohm output port impedance to the car receiver impedance.

3.5 Low Noise Amplifiers for Electrically Small Antennas

Electrically small antennas used widely in communication systems exhibit antenna impedance below a few ohms (sometimes much less than 1 ohm). For example, the AM broadcasting frequency range corresponds to a wavelength of a few hundred meters, and the radiation impedance of a common whip antenna in this frequency range is far smaller than 1 ohm. It has an approximate input of a 10 to 15 pF capacitor. Car radio input does not match with an antenna in this frequency range. Instead of matching antenna and amplifier, a different concept is used. The input impedance of the amplifier chosen is much greater than the antenna output impedance [8,9]. Such a design provides the maximum voltage transferred from the antenna to the amplifier, a wide bandwidth for the receiving active antenna, and reasonable residual amplifier noise.

3.6 Diversity Techniques

As noted in the introduction, reception quality is degraded due to multipath fading in urban environments and car body effects. For example, the directionality of an antenna inserted in car glass is not omnidirectional. As a rule, it has a few directionality dips over 360 degrees. Toward the angle dip in directionality, the antenna–receiver SNR can be dramatically reduced. Fast fading characterized by deep fades within a fraction of the wavelength is caused by multiple reflections and a mobile structure [10]. This means that the antenna element often receives reflecting waves as well as direct waves from the broadcasting transmitting antenna.

The reflecting waves are emitted from the ground, buildings, and surrounding vehicles. When two radio waves with opposite phases are received, the received signal weakens. Sometimes, several reflecting waves reach the antenna element from several reflection paths. Mathematical estimation of these phenomena is based on the assumption that the resultant amplitude signal at the receiving end is random and has a Rayleigh distribution over relatively small distances, and the phase of the signal has a random value with uniform distribution.

If we have two spaced antennas with reasonably uncorrelated received signals, we can connect the receiver to the antenna with the stronger received signal level and thus improve reception quality. A low correlation factor corresponds to better quality reception. The correlation coefficient ρ_e can be calculated simply according to the following ratio:

$$\rho_e = |\rho|^2 = \frac{\left|\Sigma_{i=1}^{N}(X_i - X_{av})\cdot(Y_i - Y_{av})^*\right|^2}{\left(\Sigma_1^N(|X_i|^2 - |X_{av}|)^2\right)\cdot\left(\Sigma_1^N(|Y_i|^2 - |Y_{av}|^2)\right)} \tag{3.36}$$

where X_i = the measured signal level by the first antenna at discrete time moments i (i = 1, 2, 3...N), Y_i = measured signal level received by the second antenna at the discrete moment i, X_{av} = average value of the received signal calculated as $X_{av} = (\Sigma_{i=1}^{N} X_i)/N$, and $Y_{av} = (\Sigma_{i=1}^{N} Y_i)/N$ = the average value of the received signal received by the second antenna.

Additional de-correlation of signals received by the first and second antennas can be provided by the polarization factor. Polarization of the received signal under conditions of multipath long or short term fading is also a critical factor, for example, a wave with pure vertical or horizontal polarization can be converted at the receiving end to an elliptically polarized wave (due to multipath reflections [10]).

If we have two orthogonally oriented antennas with de-correlated received signals, again we can choose the strongest signal that increases reception quality. Therefore, space and mutual orientation factors are critical when designing a space diversity system:

- *Spatial (or space) diversity* employs at least two antennas, usually with identical parameters, that are physically separated.
- *Polarization diversity* combines a pair of antennas with orthogonal or near orthogonal polarizations (horizontal/vertical, ± slant 45 degree linear, RHCP, LHCP, or elliptical). The reflected signals in the multipath propagation area can undergo polarization changes. By pairing two polarizations, a signal processing circuit can maximize the SNR of the incoming signal.
- *Pattern diversity* consists of two co-located antennas (or one with a switched parasitic element) with different radiation patterns.

Antenna outputs are combined in signal processing advice provided at the output (according to different algorithms), increasing the SNR. The main signal processing algorithms are:

- The switching technique uses a switching receiver; the signal from one antenna is fed to the receiver as long as the signal remains above a predetermined signal. When the signal drops below the threshold, another antenna is switched in. This is the simplest and easiest technique.
- Selective processing presents the signal of one antenna to the receiver at a given time. The antenna chosen is based on the best SNR estimation among the signals. This method requires estimation of the SNR.
- In the combining technique, all antennas are connected with a signal processing receiver.

The combining method with weighted signals provides the maximum SNR and exhibits the best improvement in SNR, but is the most complicated technique to implement.

Diversity gain is one of the main parameters that determines the efficiency of a diversity system. Diversity gain quantifies the improvement in SNR of a received signal obtained using different receiver branches. Diversity gain for a given cumulative probability p is:

$$G_{div} = \gamma_{div}(p) - \gamma_1(p) \tag{3.37}$$

where γ_{div} is the SNR with diversity and γ_1 is the SNR of the single branch without diversity combining. Thus, diversity gain is an improvement in SNR based on the diversity technique for a given cumulative probability or reliability. Analytical expressions for various combining techniques in Rayleigh fading channels as functions of branch correlation and power balance between branches (signals received individual antennas) were derived by C. Dietrich et al. [11]. Theoretically, the envelope of the correlation coefficient is a function of two identical antennas separated by distance d, assuming the random multipath with a uniform angle arrival distribution in azimuth and an antenna with an omnidirectional radiation pattern derived by Clarke [12].

$$\rho_e \cong J_0^2\left(\frac{2 \cdot \pi \cdot d}{\lambda}\right) \tag{3.38}$$

where $J_0(x)$ = the Bessel function of the first kind with order zero. This result is valid for a uniform angle of arrival distribution in azimuth and identically polarized omnidirectional receiving antennas matched to the polarization of the incoming wave. All multipath components are assumed to lie in the horizontal plane.

A rough estimation of the improving signal-to-distortion ratio (when a few diversity antennas are used) can be based on the following expression [13]:

$$\text{SNR}(M) \approx \text{SNR} \cdot M \tag{3.39}$$

where SNR = the signal-to-distortion ratio for a system with one antenna (dB scale) and M = number of diversity antennas. When two diversity antennas are used, the signal-to-distortion ratio doubles. The formula for the diversity gain when the selection diversity technique is used is the following [14] (two-term approximation) :

$$G_{div1} \cong 10 \cdot \log_{10}\left(\frac{\sqrt{\alpha \cdot (1 - |\rho|^2)}}{-\log(1 - \alpha)}\right) \tag{3.40}$$

In the case of maximal ratio combining:

$$G_{div2} \cong 10 \cdot \log_{10}\left(\frac{\sqrt{2 \cdot \alpha \cdot (1 - |\rho|^2)}}{-\log(1 - \alpha)}\right) \tag{3.41}$$

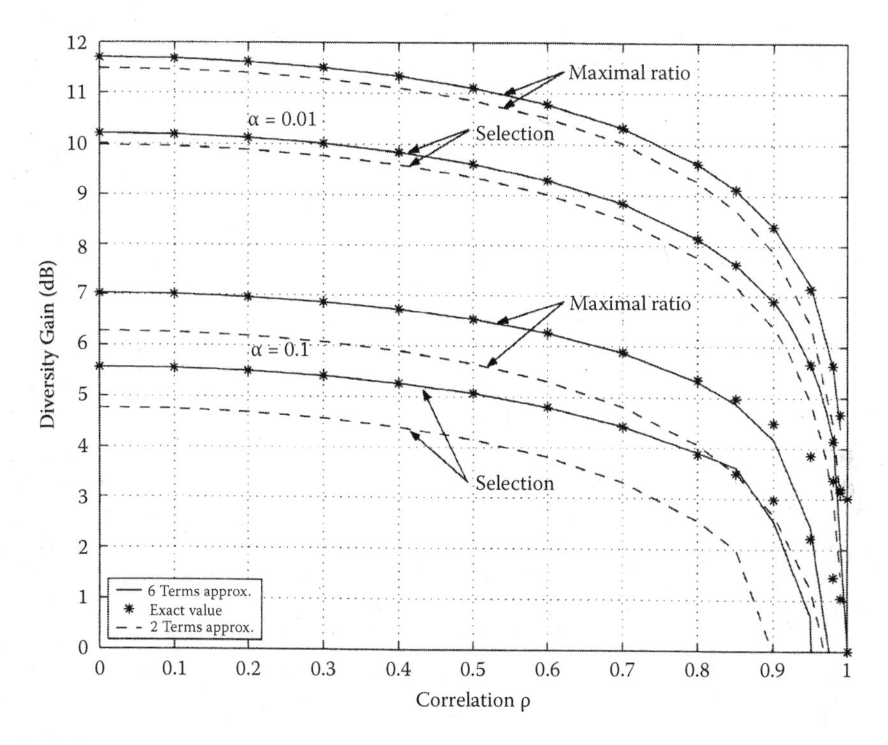

FIGURE 3.5
Diversity gain versus correlation for maximal ratio combining and selection for two antennas, including six-term approximation. (From P. Mattheijssen et al., *IEEE Transactions on Vehicular Technology*, Vol. 53, No. 4. 2004. Copyright 2004 IEEE. With permission.)

where α = the given outage. Figure 3.5 [14] shows two- and six-term approximation curves that determine diversity gain as a function of the correlation coefficient ρ for 0.1 and 0.01 outage values. Using these plots, it is easy to estimate diversity gain according to the known value of the correlation coefficient. Roughly, the correlation coefficient can be estimated using the given distance (ratio) between the diversity antennas according to the ratio (3.38).

Figure 3.6 shows a simplified block diagram of the simplest selection diversity system with two antennas, two low-noise amplifiers, an electronically controlled switch, and a control logic circuit. The output of the receiver detects the intermediate frequency signal and the control logic circuit (according to a specific algorithm) connects the first or second antenna to the car radio. The result of operating such a system is an increase in SNR at the receiver output and thus an improvement in reception quality. The simplest algorithm includes signal estimation, and when the signal level received by the operating (first) antenna falls below the predetermined level, the switch connects the second antenna to the radio. A more complicated algorithm includes estimations of four signal levels: from the first antenna, from the second antenna, from the first and second

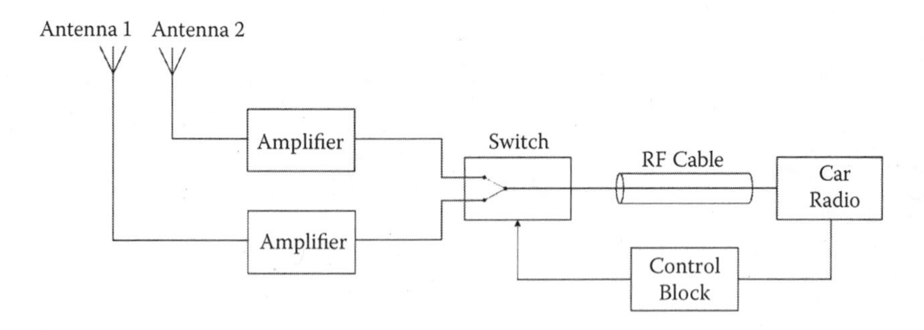

FIGURE 3.6
Block diagram of scanning diversity system.

antennas together in phase, and from the first and second antennas out of phase 180 degrees.

A few different control logic algorithms can provide these options. Detailed circuits devoted to the selection of antennas combining space diversity techniques are described in References [2], [15], and [16]. Because the space diversity technique allows the choice of an operating antenna that receives a higher signal diversity, the radiation pattern [17] is more unidirectional compared to that of a single antenna.

References

1. J.D Kraus, *Antennas for all Applications*, Boston: McGraw-Hill, 2002.
2. K. Fujimoto and J.R. James, *Mobile Antenna Systems Handbook*, Boston: Artech, 1994.
3. C. Balanis, *Antenna Theory Analysis, and Design*. New York: John Wiley & Sons, 1997.
4. G. Gonzalez, *Microwave Transistor Amplifiers: Analysis and Design*, Upper Saddle River, NJ: Prentice Hall, 1997.
5. Eagleware RF and Microwave Design software, Eagleware Corporation, www.eagleware.com
6. R. Gilmore and L. Besser, *Practical RF Circuit Design for Modern Wireless Systems*, Vol. II, *Active Circuits and Systems*, Boston: Artech 2003.
7. H. An et al., A novel measurement technique for amplifier type active antennas, *IEEE Microwave Symposium Digest*, 3, 1473–1476, 1994.
8. L. Reiter et al., Compact Antenna with Novel High Impedance Amplifier Diversity Module for Common Integration into Narrow Dielectric Parts of CA's Skin, Institute of High Frequency Technology and Mobile Communication, University of Bundeswëhr Munich.

9. A. Negut et al., Performance of a 20 cm short active AM/FM monopole antenna for automotive application, Antennas and Propagation, 2009. EUCAP. 3rd Europian Conference on 23–27 March 2009, pp. 2708–2712.
10. T. Taga, Analysis for mean effective gain of mobile antennas in land mobile radio environment, *IEEE Transactions on Vehicular Technology*, 39, 117–131, 1990.
11. C.B. Dietrich et al., Spatial, polarization, and pattern diversity for wireless handheld terminals, *IEEE Transactions Antennas and Propagation*, 49, 1271–1281, 2001.
12. R.H. Clarke, A statistical theory of mobile radio reception. *Bell Systems Technical Journal* 47, 957–1000, 1968.
13. H. Lindenmeier et al., A new design principle for a low profile SDARS antenna including the option for antenna diversity and multiband application, Publication 2002-01-0122, Society of Automotive Engineers World Congress, Detroit, March 2002.
14. P. Mattheijssen et al., Antenna pattern diversity versus space diversity for use at handhelds, *IEEE Transactions on Vehicular Technology*, 53, 1035–1042, 2004.
15. W.C.Y. Lee, *Mobile Communications Engineering Theory and Applications*, 2nd ed., New York: McGraw-Hill, 1998.
16. T.S. Rappoport, *Wireless Communications Principles and Practice*, 2nd ed., Upper Saddle River, NJ: Prentice Hall, 2002.
17. V. Rabinovich et al., Compact diversity antenna system for remote control, *IEEE Automotive Applications, Antennas and Propagation*, 2B, 379–382, 2005.

4

Audio Broadcasting Antennas

4.1 AM/FM Whip Antenna

Traditional whip antennas for receiving analog AM/FM signals have been used in the automotive industry for more than 50 years. Usually, a whip antenna about 75 cm long (a quarter of the wavelength in the FM band) is mounted on a front or rear fender. Figure 4.1 shows the radiation pattern in the horizontal plane for right slant (45 degree) linear polarization of the transmitting antenna. The measurements on slant linear polarization reflect the need for an FM broadcasting antenna to meet the requirements of receiving signals with different spatial-oriented polarizations, not simply vertical or horizontal.

Broadcasting stations use transmitting antennas with vertical polarization, horizontal polarization, both vertical and horizontal polarizations, and circular polarization. Part 73 of the regulations of the Federal Communications Commission (FCC) covers radio and television broadcasting rules. Paragraph 73.316 says: "It shall be standard to employ horizontal polarization; however, circular or elliptical polarization may be employed if desired. Clockwise or counterclockwise rotation may be used." According to this regulation, any polarization of a transmitting signal is acceptable. Additionally the polarization of the signal may be changed when propagating due to multipath reflections from buildings, cars, trees, and other obstructions. The measured radiation pattern shown in Figure 4.1 is not omnidirectional (car body effect). The maximum-to-minimum ratio:

$$\gamma = \frac{\max(F_{slant}(\theta = 90^{\circ}, \varphi))}{\min(F_{slant}(\theta = 90^{\circ}, \varphi))}, \quad (\varphi = 0 \div 360^{\circ}) \tag{4.1}$$

is about 10 dB. Gain values averaged over FM for 360 degrees around a car are less than the reference dipole gain of 3 to 5 dB, depending on the frequency point.

Typically, a horizontal reference dipole antenna used for FM gain measurements is a folded dipole tuned to the broadcasting frequency range and mounted on a wooden tripod located at the center of a vehicle test site; the

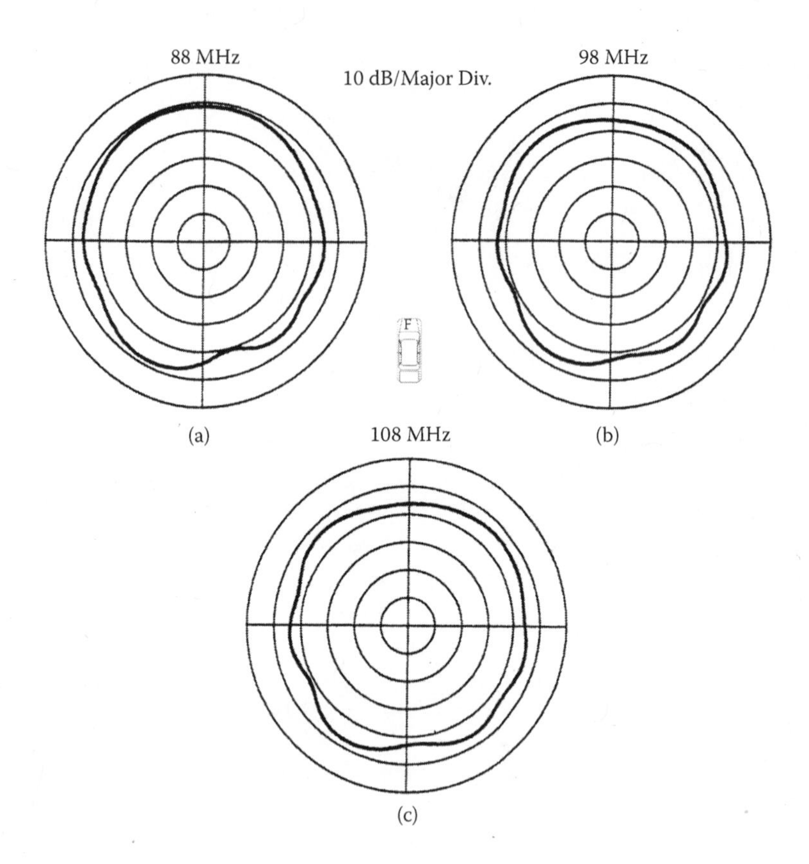

FIGURE 4.1
Measured right slant radiation pattern for whip antenna, Sonoma pick-up truck.

midpoint of the reference antenna is 1.5 m above the site surface and faces the transmitting antenna which, during reference and test measurements is inclined +45 or −45 degrees to the horizon. When measuring the radiation pattern of a receiving antenna for vertical or horizontal polarization, the transmitting antenna must be oriented vertically or horizontally.

4.2 Printed-on-Glass AM/FM Antennas

4.2.1 Introduction

Today, various printed-on-glass AM/FM antenna systems are available in the automotive market sector. Window-printed antennas can be located on front, rear, or side glass. Some antennas are designed only as passive units;

however, most have low-noise amplifiers to improve sensitivity and increase the dynamic range of the car radio system. The various window glass antennas present certain advantages and disadvantages.

4.2.1.1 Rear Glass Systems

One version consists of a single antenna for AM/FM reception mounted separately from the defogger. This compact antenna encounters low-noise interference from the defogger and typically is used with an amplifier. The passive portion gain of such an antenna is relatively low.

In two-antenna systems, one is intended for AM and the other for FM reception. Generally, the defroster serves as the AM antenna. The gain in the AM frequency range is high. The structure requires a special filter circuit to suppress AM interference noise caused by the heater system. The FM antenna is mounted above or below the heater. This system requires two glass connectors: one for the AM portion and another for the FM system, which, as a rule, is used with an amplifier.

4.2.1.2 Side Glass Systems

A single antenna or two separate antennas may be utilized for AM/FM reception. The antenna with an amplifier system requires reasonable glass area to achieve acceptable antenna gain.

4.2.1.3 Front (Windshield) Systems

A single antenna handles combined AM/FM transmissions. The antenna exhibits high gain and the system can be used without an amplifier; however, it is noisy due to engine and electronics components.

4.2.1.4 Diversity Window Antennas

This system usually consists of two antennas printed on a rear or side glass. One operates as the AM antenna and the second functions with the first to constitute a diversity FM system. The rear glass system includes one antenna mounted above the heater. The second antenna is located below the heater (for example, in the Toyota Camry). The side glass system uses two rear-side windows, with one antenna mounted from the driver side and a second from the passenger side (for example, in a Volvo). The diversity antenna system improves reception quality under multipath fading conditions. Such a system is more expensive than a simple printed-on-glass antenna.

A designer who starts to build an antenna system must consider all the above issues and achieve a compromise between price and quality of reception. While professional glass suppliers can build printed-on-glass antennas, such antennas are available on the aftermarket. Figure 4.2 shows an example. The antenna is built on transparent film that can be easily glued to a car window.

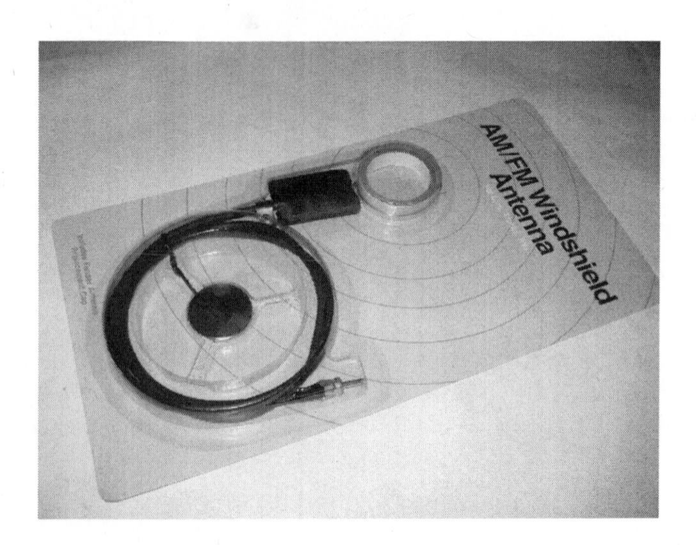

FIGURE 4.2
Aftermarket printed-on-glass antenna.

We now discuss practical design examples of a simple single AM/FM printed-on-glass antenna. The first is a passive windshield antenna and the other one is an active side-glass system.

4.2.2 Passive Windshield Antennas

The first passive antenna design is installed on the front glass window of General Motors' Sonoma pick-up truck. Metal tape is used to print the antenna prototype pattern. The tape has glue on one side for hard contact with the window. The antenna output is connected to a network analyzer that records the VSWR of the antenna in the entire FM frequency range. The designed antenna pattern is shown in Figure 4.3. Figure 4.4 is a block diagram of an antenna with a passive matching circuit that includes AM and FM branches. The matching circuit can be mounted near the antenna as shown in Figure 4.4a, inserted into the RF cable connecting the antenna to the car radio as shown in Figure 4.4b, or installed near the radio as shown in Figure 4.4c. The FM passive branch provides the maximum power conveyed from an antenna to the radio (loss is less than 1 dB). The AM branch provides the maximum voltage gain from the antenna to the radio.

4.2.2.1 FM Frequency Range

In the FM frequency range, antenna dimensions are comparable with half of wavelength. Correctly matching antenna output impedance with the radio input impedance (and RF cable) should convey maximum power from an antenna to the car radio.

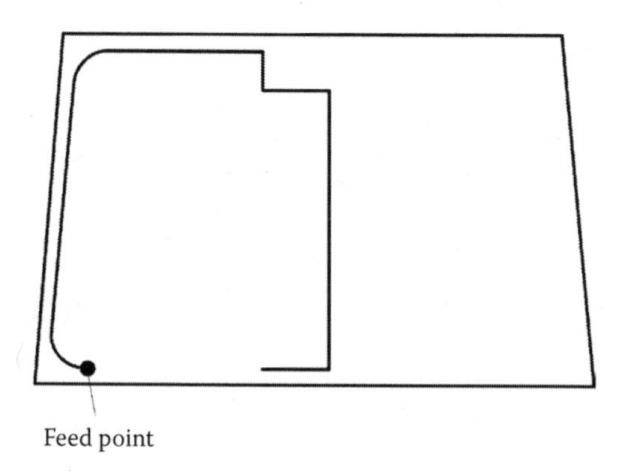

FIGURE 4.3
Antenna pattern printed on windshield.

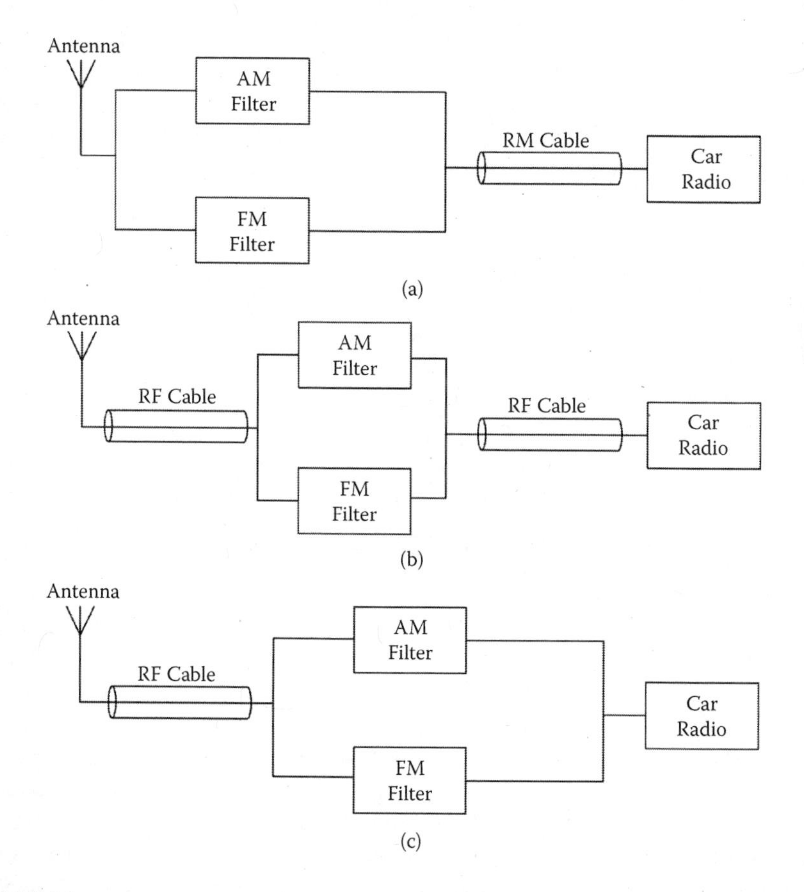

FIGURE 4.4
Block diagram of antenna system: (a) matching circuit mounted near antenna, (b) matching circuit installed into RF cable, (c) matching circuit mounted near car radio.

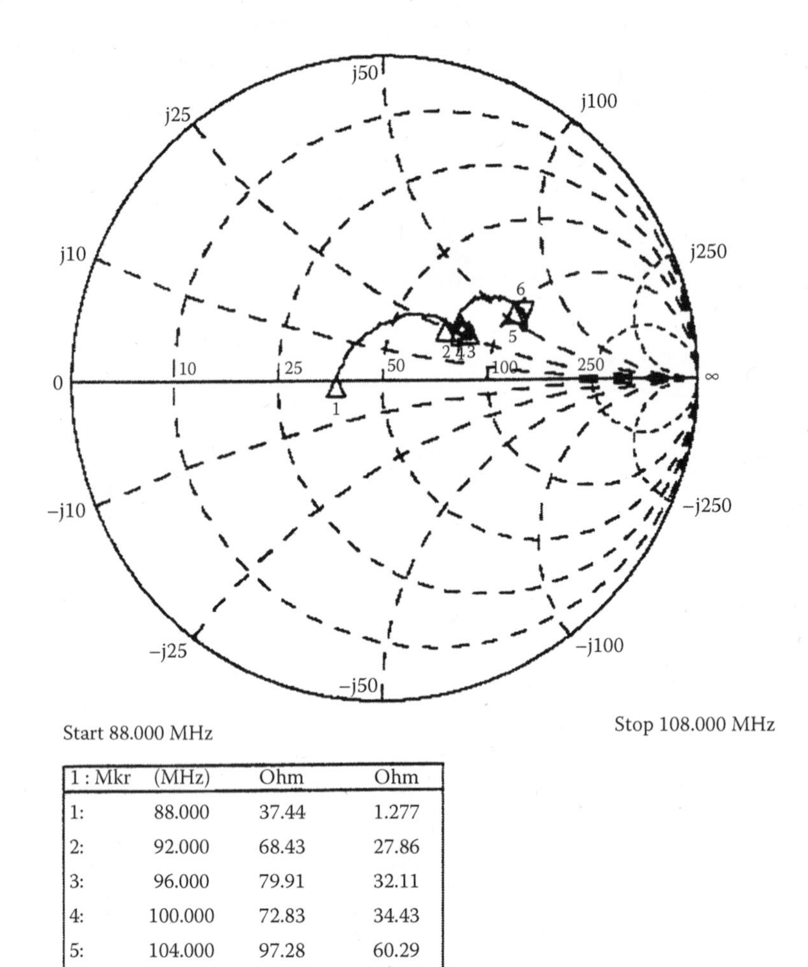

1 : Mkr	(MHz)	Ohm	Ohm
1:	88.000	37.44	1.277
2:	92.000	68.43	27.86
3:	96.000	79.91	32.11
4:	100.000	72.83	34.43
5:	104.000	97.28	60.29
6:	107.900	114.5	52.24

FIGURE 4.5
Measured output impedance of windshield printed antenna.

Figure 4.5 presents the measured impedance. The front windshield antenna has a VSWR below 2.5:1 in the entire frequency range. Radiation pattern graphs measured in the horizontal plane for right and left slant 45 degree linear polarizations of the transmitting antenna are shown in Figure 4.6 and Figure 4.7. The ratio γ that determines the dip values in the radiation pattern for the printed-on-glass antenna exceeds 25 dB for some frequency points. However, the angle widths of these dips do not exceed a few degrees. More than 90% of the measured data are within the ±10 dB range of the

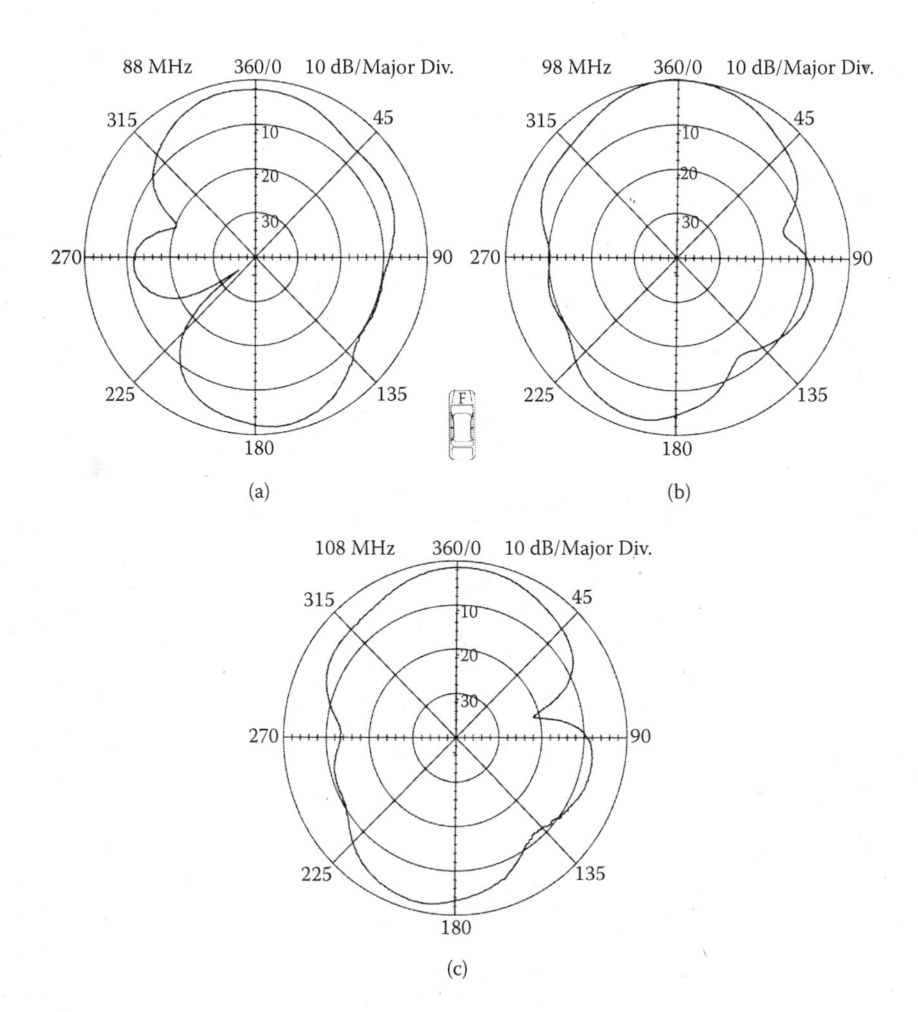

FIGURE 4.6
Measured radiation patterns for right slant polarization for windshield printed antenna: (a) 88 MHz; (b) 98 MHz; (c) 108 MHz.

average received signal value. Figure 4.8 presents the relative gain values of printed-on-glass and regular whip antennas. The average FM frequency range gain value for the printed-on-glass antenna is only 1 dB less than the whip antenna gain for both slant polarizations. The gain values presented in Figure 4.8 are averaged over 360 degrees around the car.

4.2.2.2 AM Frequency Range

Antenna length in the AM frequency range is much shorter than the wavelength. An antenna designed for this frequency band can be presented as a capacitor with a value of a few tenths of a picofarad (pF). A modern receiver

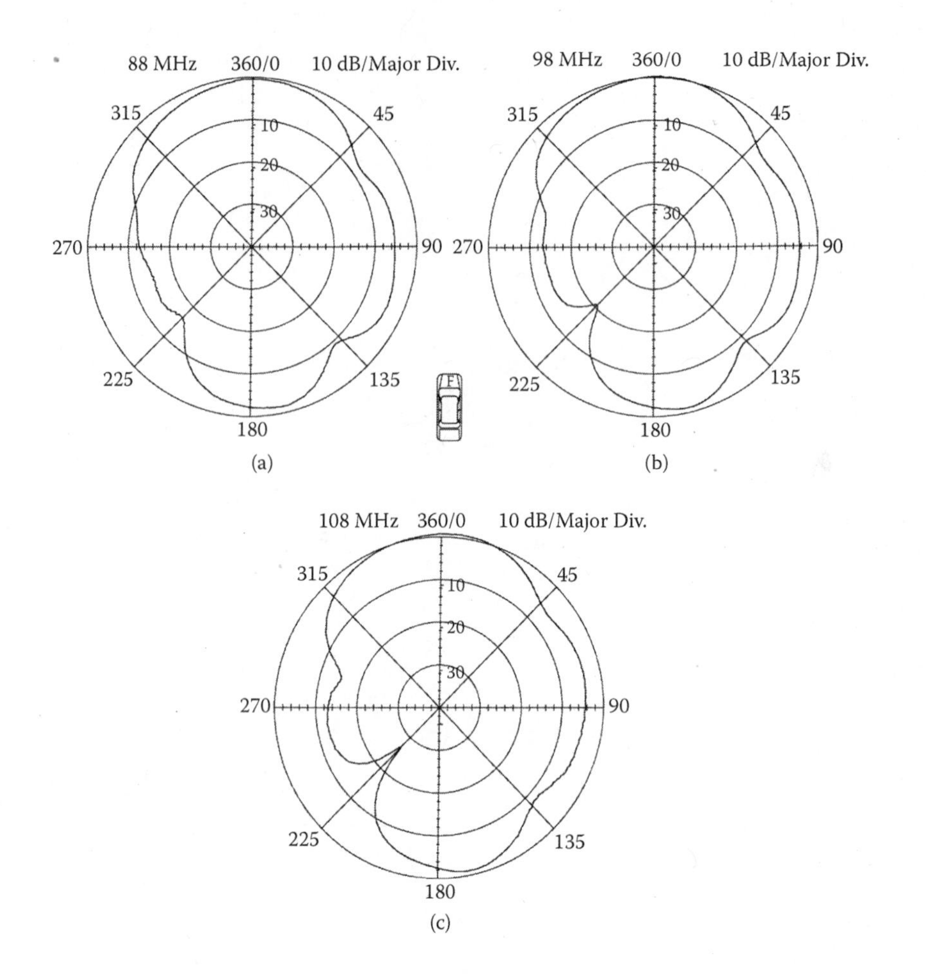

FIGURE 4.7
Measured radiation patterns for left slant polarization for windshield printed antenna: (a) 88 MHz; (b) 98 MHz; (c) 108 MHz.

in the AM frequency range uses a capacitive non-resonant input stage with an input capacitor typically on the order of 20 pF (see Reference [1], p. 294). The equivalent circuit of an antenna design for AM frequency [2] is shown in Figure 4.9a.

Antenna impedance is determined by a capacitor C_a (the radiation resistor value is far below 1 ohm) and by its voltage source (V). The capacitor $C1_{cab}$ describes the coaxial cable that connects the antenna with a passive filter inserted into the RF cable. The capacitor $C2_{cab}$ describes the coaxial cable part connecting the passive filter with the radio. The input stage of the receiver is represented by the input capacitor C_r. A passive filter circuit is placed between the antenna and RF cable to increase the voltage received by the

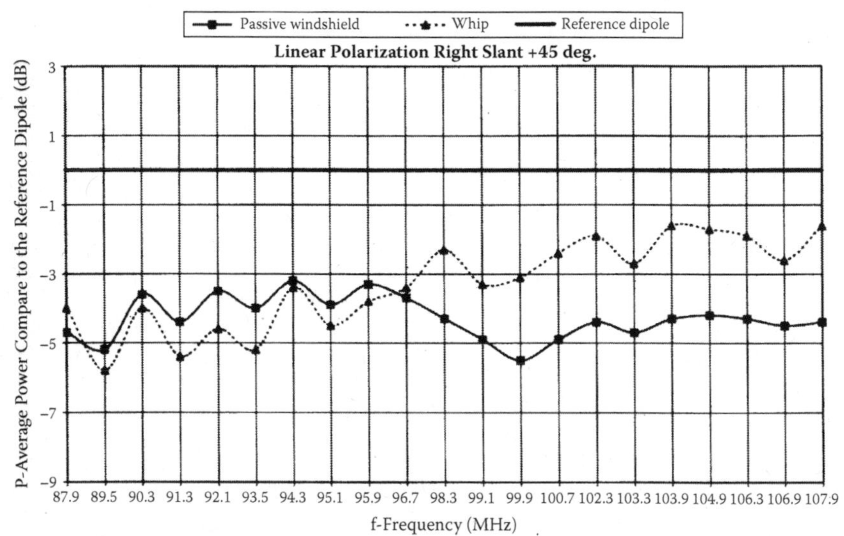

Average power for the entire frequency range: Passive Windshield –4.2 dB, Whip-3.1 dB

(a)

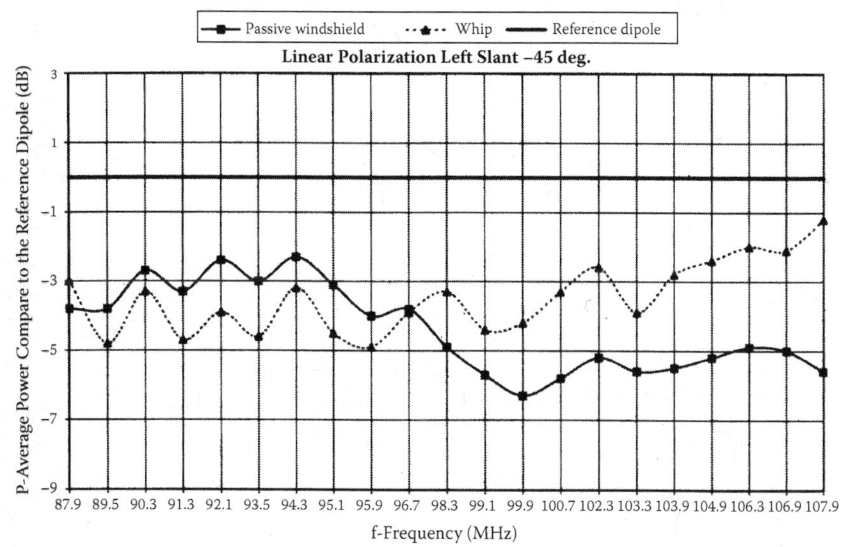

Average power for the entire frequency range: Passive Windshield –4.2 dB, Whip –3.4 dB

(b)

FIGURE 4.8
Relative gain values: (a) linear polarization, right slant; (b) linear polarization, left slant.

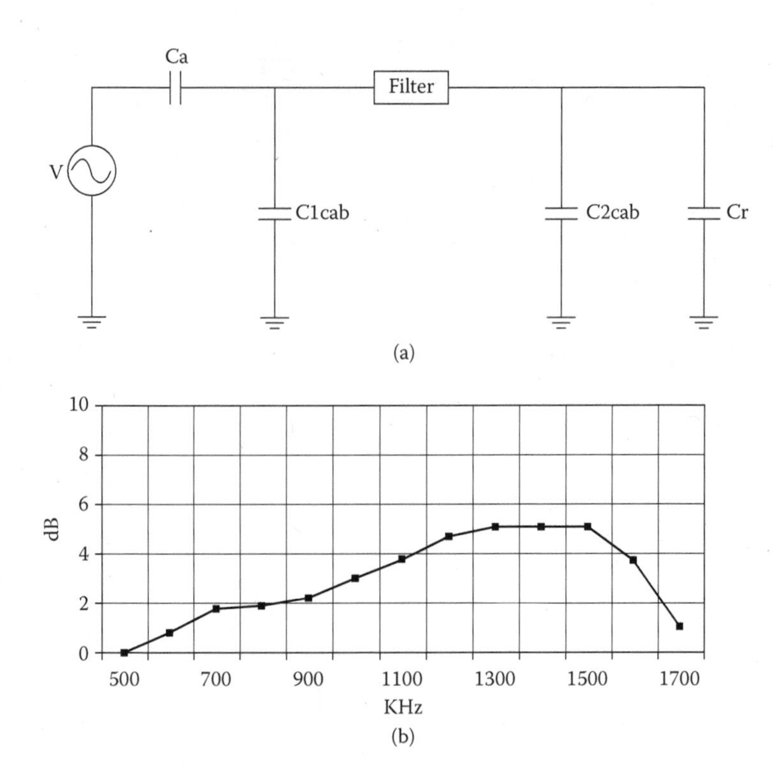

FIGURE 4.9
(a) Equivalent circuit in AM frequency band; (b) AM voltage gain versus frequency with passive matching filter.

radio. Figure 4.9b shows the voltage gain versus the AM frequencies with a passive matching filter inserted into RF cable based on Figure 4.4b.

The measurement results of the printed antenna gain compared with the reference antenna gain value and the whip antenna gain (dB scale) for different frequencies (KHz) are shown in Table 4.1. A reference 75 cm-high AM monopole antenna was mounted on the middle of a car roof. Broadcasting stations were used to determine AM gain measurements. Based on the table, on average over the entire AM frequency band, the printed antenna gain is only 0.5 dB less than the whip antenna gain. Thus, it can be concluded that

TABLE 4.1

Relative Gains for Printed and Whip Antenna Designs

Frequency	560	690	760	950	990	1310	1600
$G_{print} - G_{ref}$	–3.8	–3.3	–2.9	–3.2	–3.1	–4.6	–3.3
$G_{print} - G_{whip}$	–0.3	–0.2	+0.2	–0.5	–0.4	–0.8	–1.7

the average parameters of the designed passive antenna are almost equivalent to those of the whip antenna.

4.2.3 Simulation Results

Numerical modeling of an antenna in parallel with an experimental cut-and-try design is the shortest way to achieve performances that meet specific requirements. The commercially available FEKO software package [3] mentioned in Chapter 1 is very powerful for simulating real car body shape with an installed antenna. Such a simulation does not predict accurate parameters of the antenna mounted on/in the car but can be an excellent tool to speed the process of antenna design. For printed-on-glass or integrated-into-glass metallic strip antennas, FEKO introduces a special [4] technique described as the dielectrically coated wire method.

The method allows the changing wavelength to be taken into account when the antenna metal strip is placed on the dielectric surface. Achieving an accurate sheet metal surface (responsible for electromagnetic properties) on a vehicle is a very complex process. Some features (driver and passenger seats, interior and exterior decorations, windowpane seals, gaps, and other structures) that cause uncertainties cannot be modeled accurately. Figure 4.10 shows a simplified FEKO wire-grid model of a pick-up truck model. The first

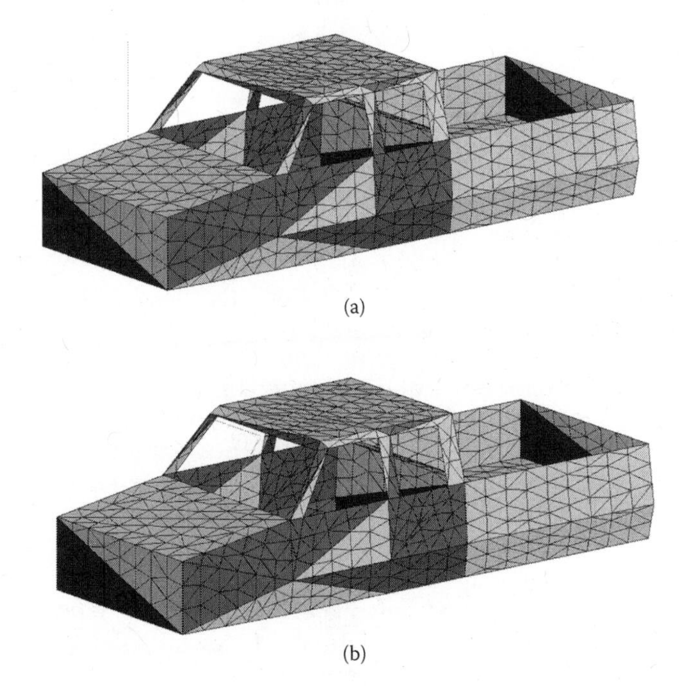

FIGURE 4.10
Wire grid of Sonoma pick-up truck with (a) whip; (b) windshield printed antenna.

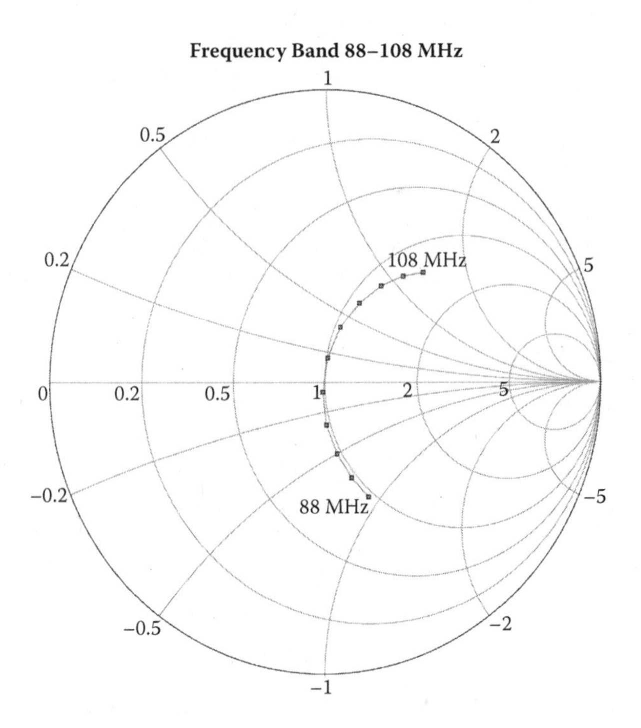

FIGURE 4.11
Simulated output impedance of whip antenna.

model (Figure 4.10a) has a regular whip antenna; the second (Figure 4.10b) has a printed-on-glass pattern.

Figure 4.11 shows the simulated output impedance of the whip antenna. Figure 4.12 shows the simulated radiation patterns for the whip antenna in the horizontal plane at a few frequency points. The left column of Figure 4.12 corresponds to 88 MHz; the middle column presents results for 98 MHz; and the right column shows simulations for 108 MHz. Results in Figure 4.12a relate to the vertical polarization of the transmitting antenna. Figure 4.12b corresponds to horizontal polarization, Figure 4.12c presents the radiation pattern for right slant orientation of the transmitting antenna, and Figure 4.12d demonstrates horizontal directionality for a left slant linear polarized transmitting antenna.

Figure 4.13 presents the output impedance of a printed-on-glass antenna, and Figure 4.14 shows the horizontal antenna directionality for different polarizations of a transmitting antenna at a few frequency points. A comparison of the measured plots for right and left slant polarizations shown in Figures 4.6 and 4.7 with the appropriate simulation curve presented in Figure 4.14 shows that the agreement between the simulation and measurement results is not perfect due to difficulties related to describing an

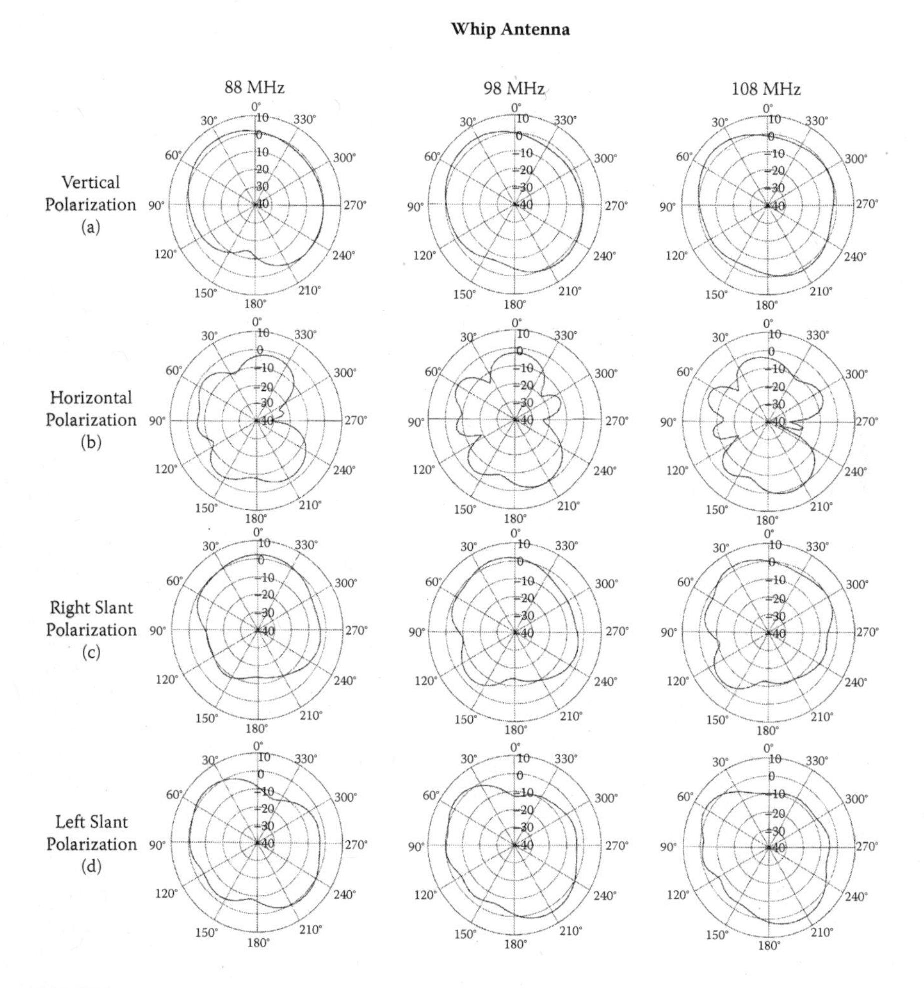

FIGURE 4.12
Simulated radiation patterns of whip antenna.

accurate car shape. However, the simulation results reproduce the area of the measured antenna impedance values and the locations of reduced radiation pattern values.

4.2.4 Side Glass Antenna Pattern Example

The second example is an active printed-on-glass antenna installed on an Escalade car model. Figure 4.15 shows the antenna topology. Figure 4.16 demonstrates the measured radiation pattern in horizontal plane for right slant transmitting antenna polarization at a few frequency points. Note that the maximum-to-minimum ratio γ is more than 15 dB for some angle positions,

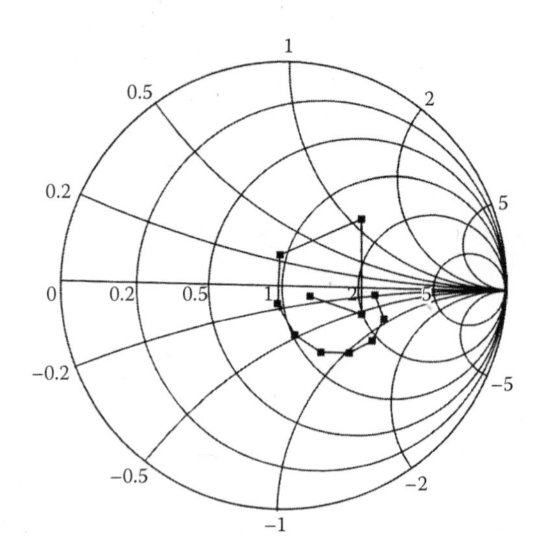

FIGURE 4.13
Simulated output impedance of windshield printed antenna.

but the width of these angle sectors is less than a few degrees. The power (right slant 45 degree linear polarization for transmitting antenna; measurements for horizontal plane) received by the passive antenna portion and active antenna system (passive portion plus amplifier) in the FM frequency range is shown in Figure 4.17.

Results are compared with the horizontally mounted reference dipole described in Section 4.1. The power for each frequency point is calculated as an average value over 360 degrees around the car. The average gain value over all frequency points in entire FM frequency band for the passive antenna portion is 8 dB less than the gain of the reference antenna. Therefore, such a system requires an amplifier to increase antenna gain without loss of the sensitivity of the antenna system.

4.2.5 Amplifier Circuit for AM/FM Antenna System

Typical requirements for an antenna amplifier operating in the AM/FM frequency range are:

1. The amplifier gain value in the AM and FM frequency ranges has to provide a stop–scan radio station regime identical to the regime for a regular whip antenna.
2. The residual noise level of an amplifier must be equal to or less than –5 dBuV with 3 dB bandwidth of 10 kHz in the AM frequency range and –5 dBuV with bandwidth of 110 kHz for the FM frequency range.

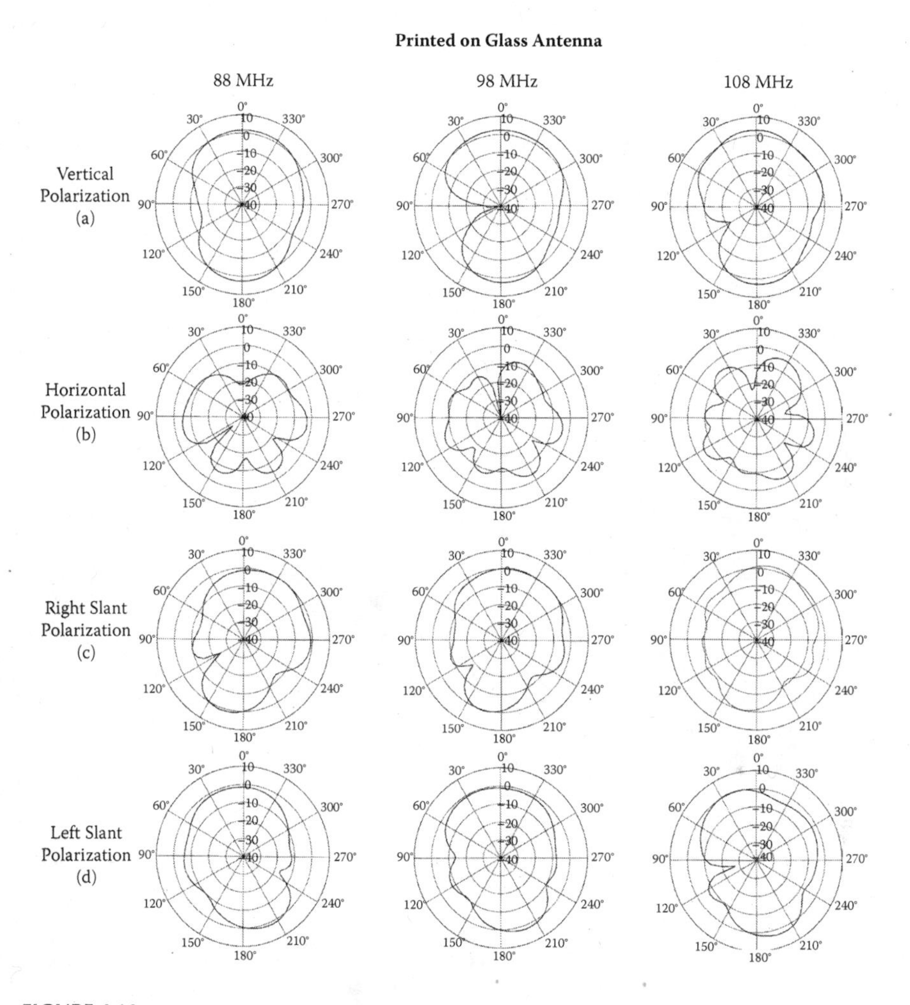

FIGURE 4.14
Simulated radiation patterns for different polarizations of windshield printed antenna: (a) vertical; (b) horizontal; (c) right slant (+45 degrees); (d) left slant (–45 degrees).

3. The amplifier noise figure in the FM frequency range must be less than 5 dB in the entire frequency range.
4. The third-order intermodulation (IM) product must be less than 60 dBuV for an output voltage of 110 dBuV in the AM and FM frequency ranges; the second-order IM for the AM frequency range must be less than 70 dBuV for an output signal voltage of 110 dBuV. In the FM frequency band it should be less than 40 dBuV for an output signal of 110 dBuV.

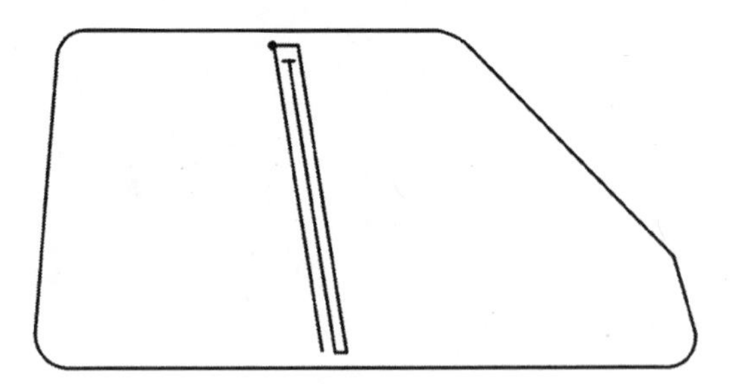

FIGURE 4.15
Antenna printed on side glass.

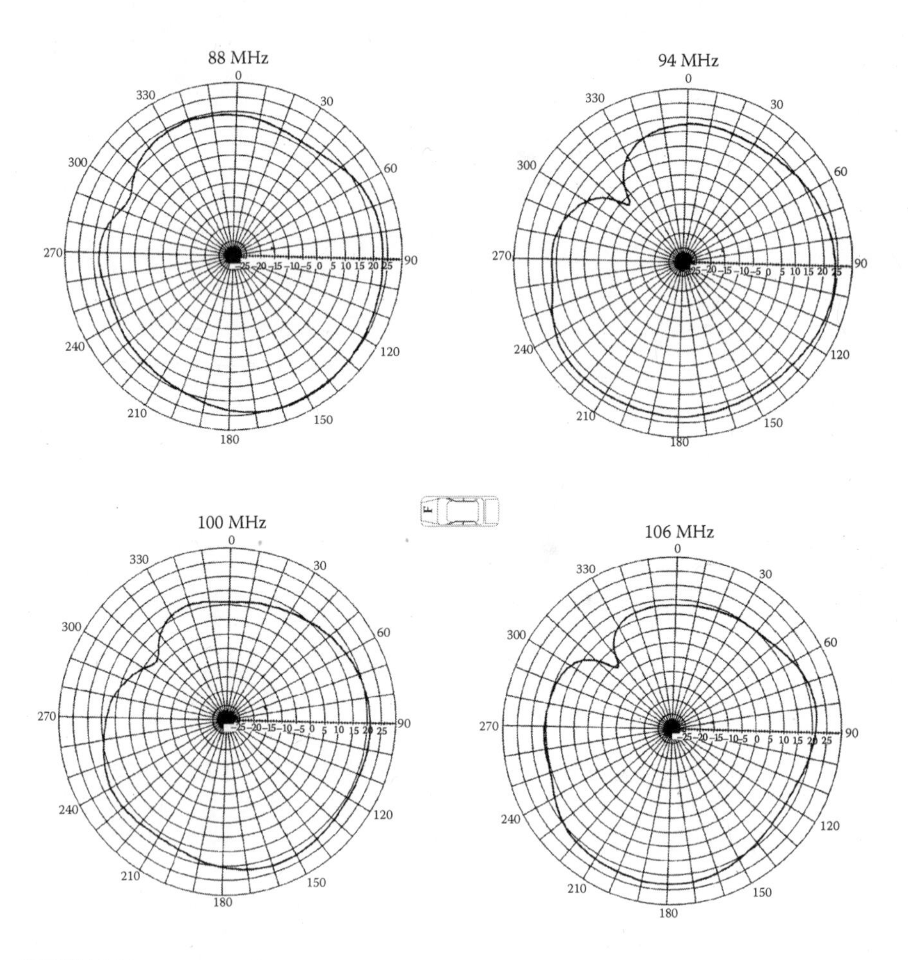

FIGURE 4.16
Radiation pattern of side glass antenna.

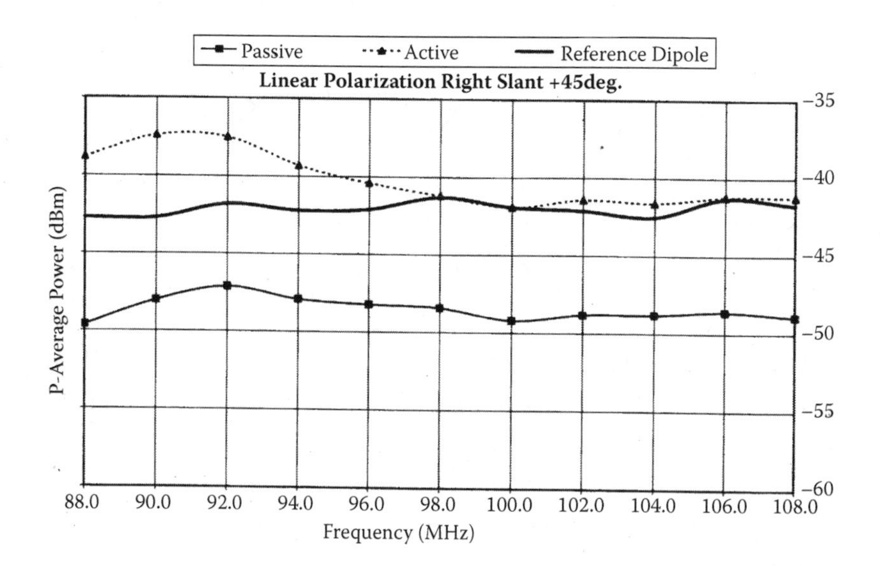

FIGURE 4.17
Relative measured power for side glass antenna.

Usually, an antenna amplifier includes two branches: one for AM amplification and another for FM amplification. Figure 4.18 is a block diagram of an amplifier example used with a single AM/FM antenna design. The AM branch consists of an input filter that passes AM signals and blocks signals with FM components, a field effect transistor (FET; active AM element) with high input and low output impedance, and an output filter.

The FM branch contains an input matching circuit to minimize amplifier noise with an input filter that passes FM signals and blocks AM signals, a transistor (FM active element) for amplifying FM signals, and an output matching circuit (with an output filter) that decouples the output AM and FM signals. The block diagram includes a circuit that protects the amplifier from high-level signals when a vehicle is driven near high-power transmission stations. An automatic control signal is applied to an electronic switch that activates a bypass of the FM transistor when the car approaches a broadcasting transmission station.

A simplified example of an electrical circuit [5] of an amplifier with a pin-diode switch to bypass an amplifier in the vicinity of a strong FM broadcasting station is presented in Figure 4.19. The circuit consists of an AM signal branch with FET element Q1, input and output filters, an FM signal branch with transistor Q2, input noise matching and filter elements, and output matching and filter elements. The system also includes a control circuit with a few pin diodes and a bypassing branch. When the car nears a strong FM transmission station, the detected voltage exceeds the predetermined value, the comparator COM generates DC voltage, which controls pin-diode

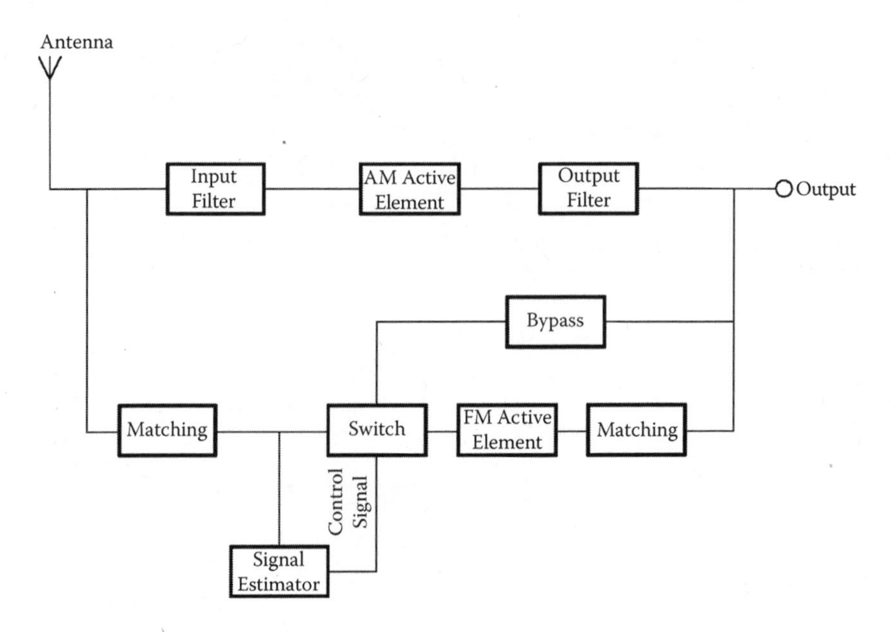

FIGURE 4.18
AM/FM antenna amplifier topology.

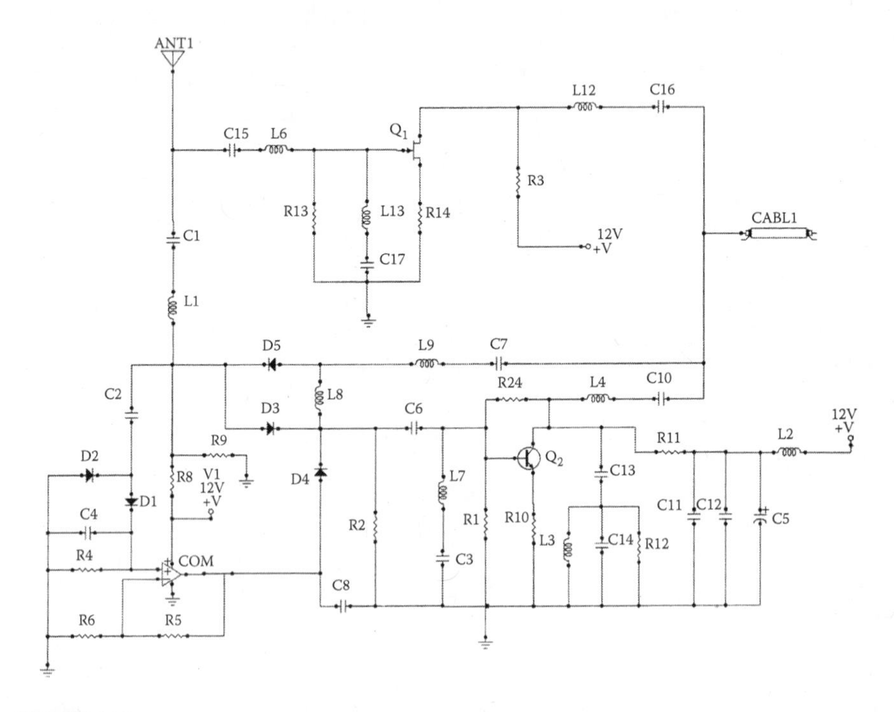

FIGURE 4.19
Example of electrical circuit of amplifier with overload protection branch.

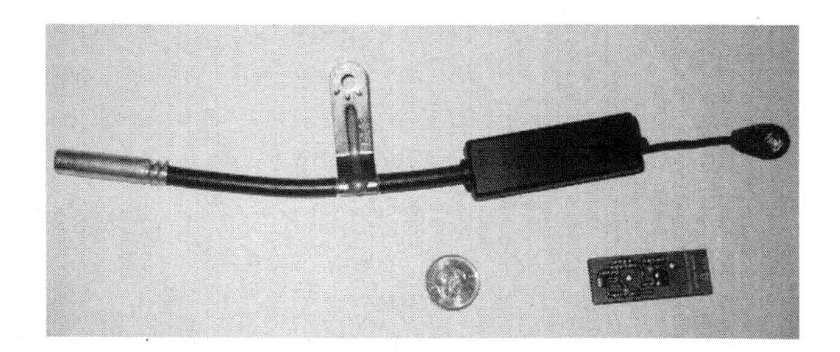

FIGURE 4.20
Dimensions of AM/FM amplifier.

switches D3, D4, and D5, in such a way that the FM signal goes through pin diode D5, the inductor L9, and capacitor C7, and does not go through the transistor. If the voltage detected by diodes D1 and D2 is less than the reference voltage level, then pin diode D5 is closed, and the FM signal goes through pin diode D3, capacitor C6, transistor Q2, inductor L4, and capacitor C10. The example of an amplifier built on an FR-4 printed board according to the Figure 4.19 is presented in Figure 4.20. The amplifier dimensions are $60 \times 20 \times 12$ mm^3.

The measured noise figure of an amplifier does not exceed 4 dB in the entire FM frequency range. The gain of the amplifier in AM band does not exceed 6 dB in the entire frequency range; the gain in the FM band is about 6 to 11 dB, depending on the frequency point. A bypassing circuit starts to operate when the signal applied to the amplifier exceeds 0 dBm. Figure 4.21 shows the output response for two applied input signals of 95 and 100 MHz (each has 15 dBm input power) applied to an amplifier built with options for a no-protection circuit and a protection circuit, as shown in Figure 4.19. The third intermodulation product is about –25 dB compared to the main maximum signal value for a circuit without protection and –60 dB for a circuit with protection elements.

Usually, measurements of an active car antenna are finalized by driving tests of:

- AM reception
- AM overload
- AM fringe
- FM reception
- FM overload
- FM fringe
- FM multipath

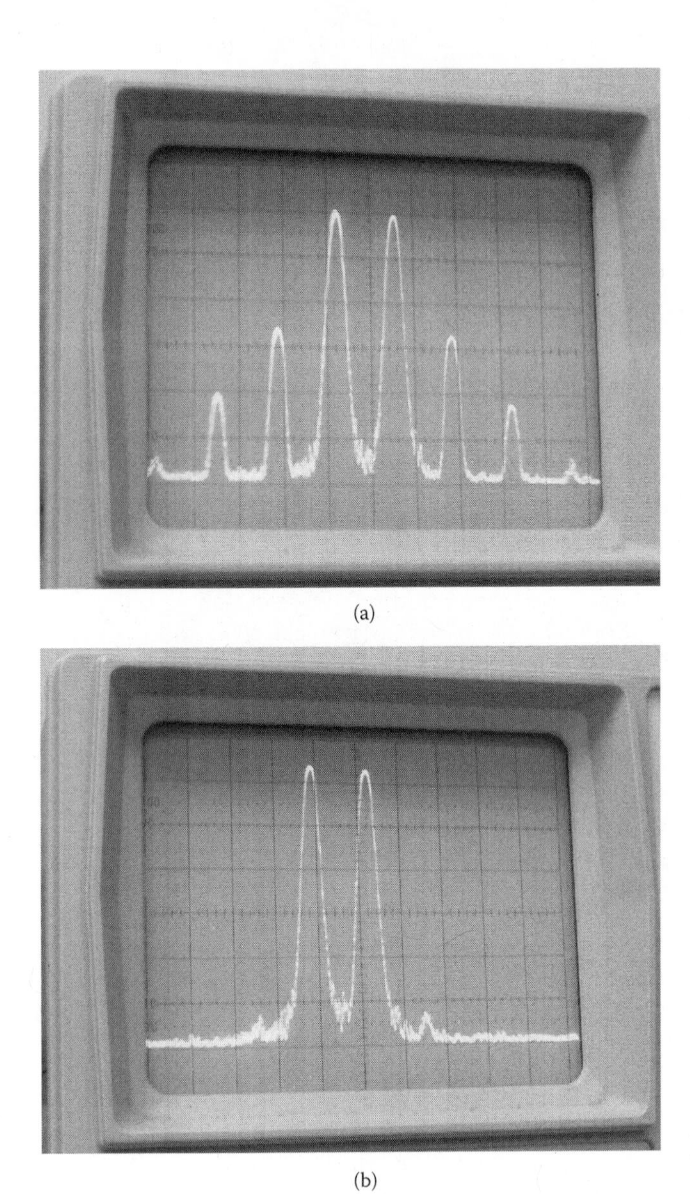

(a)

(b)

FIGURE 4.21
Third-order intermodulation product for amplifier shown in Figure 4.20: (a) amplifier without protection circuit; (b) amplifier with protection circuit.

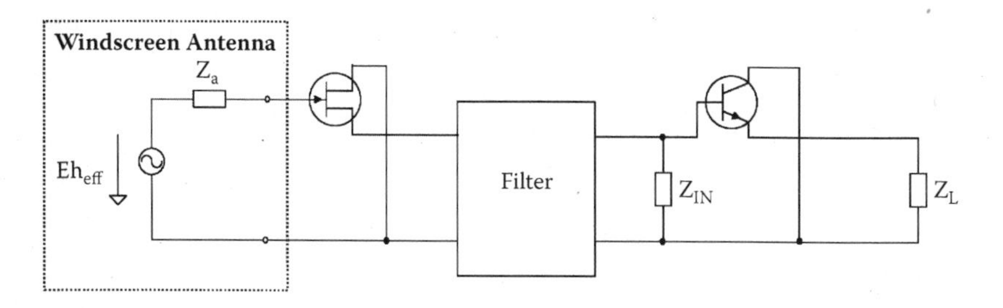

FIGURE 4.22
Block diagram of amplifier with high input impedance.

4.2.6 Amplifier with High Input Impedance

When the allowable window space for an antenna is very small, power matching between an antenna and an amplifier may become a problem. A German team from the Institute of High Frequency Technology and Mobile Communication under H. Lindenmeier investigated an active printed-on-glass antenna system with an amplifier that has high input impedance for FM reception [6,7]. They described the design of a broadband amplifier with high input impedance that can provide a low level of residual noise and reasonable transmission amplifier gain. The authors examined a few active elements: the GaAs FET, junction gate field effect transistor (JFET), and high electron mobility transistor (HEMT).

A typical active antenna block diagram with two active transistor elements [8] is shown in Figure 4.22. The system includes a windscreen antenna pattern, high input impedance FET transistor, lossless passive filter, and bipolar transistor stage. A detailed investigation of an active antenna for the FM frequency range with high input impedance low noise and highly linear amplifiers is presented in a doctoral thesis [8]. The use of the high input impedance stage eliminates the problem of designing a specific matching circuit in front of a bipolar transistor amplifier minimizing noise. Residual amplifier noise is independent from the imaginary part of the antenna impedance (only the real part must be more than 10 ohms). The amplifier exhibits low noise, high linearity, and good third-order intermediation distortion (–65 dB with amplifier input levels up to 130 dBuV). The author presents the simulation and gain and noise measurement results for the active antenna circuit. Analysis is carried out for a number of curves of antenna output impedance shown in Figure 4.23.

For impedance curves A, B, and C, the gain value over the entire FM frequency range varies from 6 to 7 dB and the noise does not exceed 3 dB. The poor noise figure for impedance curve D as explained in the thesis is due to the real part of impedance measurement lower than 10 ohms. Such a design concept allows the use of an amplifier with any pattern, which is very convenient from a commercial view. The design can be used in a space diversity

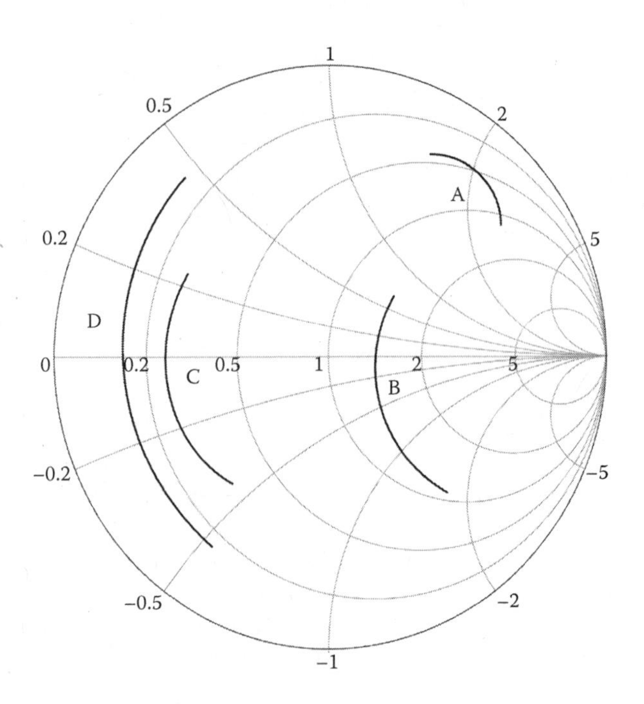

FIGURE 4.23
Impedance curves of antenna structure in impedance plane in FM frequency range.

system that uses few compact FM printed-on-glass antennas for which impedance noise matching may be a serious problem.

4.3 Short Roof AM/FM Antenna

4.3.1 Helix Antenna Radiation Modes

A helix antenna has two major radiation modes [9]: normal and axial. The axial mode provides maximum radiation along the helix axis and corresponds to the helix circumference of the order of the wavelength. The normal mode supplies radiation perpendicular to the helix axis, when the helix diameter is small compared to wavelength and uniform current distribution around each turn of the helix is assumed. The normal mode is the main regime for automotive applications.

A roughly helical antenna can be presented as a combination of the elementary loop and monopole elements as shown in Figure 4.24. A practical

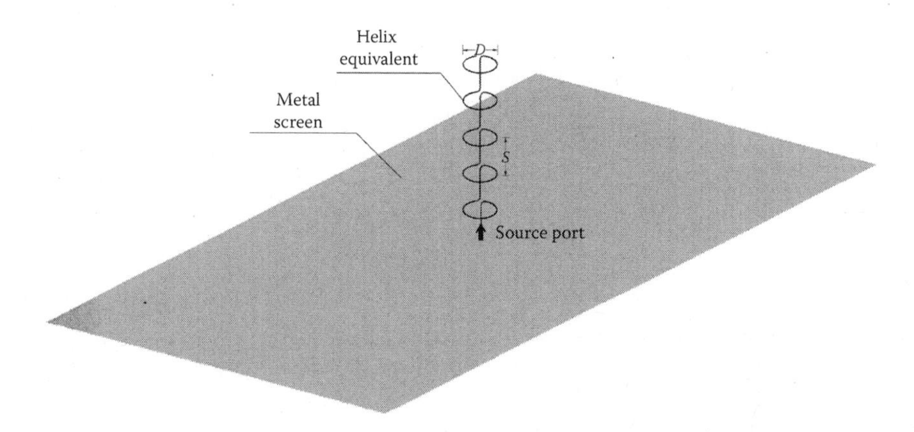

FIGURE 4.24
Helix antenna combining loop and monopole.

automotive version has a loop diameter that is much less than the wavelength in the FM frequency band. Each elementary loop radiates only the horizontally polarized electric field component $E_{\varphi}(\theta,\varphi)=E_{\varphi}(\theta)$ that does not depend on the φ angle value. Each elementary monopole component of the helix radiates only the θ component of the electric field $E_{\theta}(\theta,\varphi)=E_{\theta}(\theta)$. Small loop and monopole antenna components do not radiate energy along the helix axis. The total (from all helical loops and monopole components) electric field angle orientation depends on the ratio *AR* between the values of these two components (Reference [9], pp. 505–509):

$$AR \approx \frac{2\cdot\lambda\cdot S}{(\pi\cdot D)^2} \tag{4.2}$$

where S = the spacing between adjacent turns, D = helix diameter, and λ = wavelength. The ratio $AR \ll 1$ corresponds to the horizontally polarized radio wave. The helix antenna for automotive use has a geometry that satisfies the condition $AR \gg 1$. It corresponds to a radiation pattern equivalent to the short monopole antenna with the same helix height that equals H.

Under resonance conditions, a short helical antenna with height less than wavelength exhibits resistive radiation impedance and therefore does not require additional series inductor tuning. However, the impedance of a short monopole has a high capacitive value that is changed insignificantly in the frequency band. Compensation for such capacitance with inductive reactance that usually has loss resistor R_{loss} (quality factor $Q=2\cdot\pi\cdot f\cdot L/R_{loss}$) can dramatically reduce the efficiency of an antenna [10] and, as a result, reduce gain. Therefore, short helix antennas are more preferable than short monopoles with matching inductors.

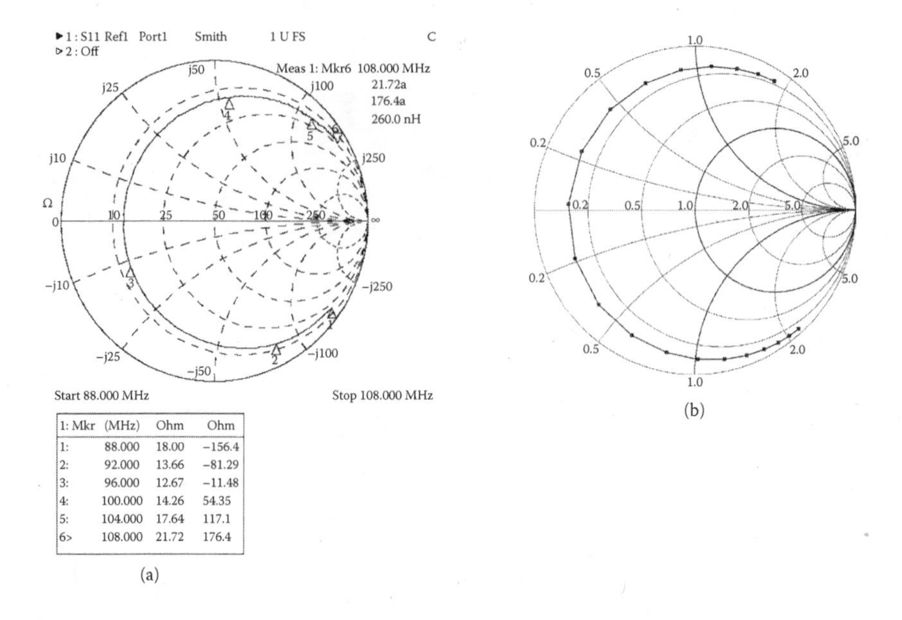

FIGURE 4.25
Impedance of car roof helix antenna: (a) measurement results; (b) simulation.

4.3.2 Helix Antenna for FM Frequency Band

Figure 4.25a shows the measured impedance in the FM frequency band of the helix car roof antenna (passive portion) with the following geometry parameters: $H = 16$ cm ($H \approx 0.053 \cdot \lambda$); $D = 0.7$ cm; wire radius = 0.05 cm, and the number of turns $N \approx 100$. The antenna has a straight base with a length around 3 cm. The helical coil is wound on a flexible insulation fiberglass rod. An insulating resin pipe is used to cover the wound coil, and the coil is fixed on the rod. Impedance is measured for the passive portion antenna placed on a circular metal plate with a diameter of 1 m. The same geometry simulated with the NEC-PRO software demonstrates the impedance curve shown in Figure 4.25b.

We do not see perfect agreement of the two graphs, but they tend to show good agreement between simulated and measurement results. The difference between the measured and simulated results can be explained by the fact that the simulation does not take into account the loss resistor of the antenna. The network analyzer measures input impedance—the sum of the radiating impedance and loss resistance.

The experimental received power results (horizontal plane, vertical polarization) of an antenna with amplifier are shown in Table 4.2. Row 1 corresponds to the reference quarter wave mast antenna mounted in the middle of the car roof, row 2 corresponds to the active short helical antenna mounted in the middle of the roof and Δ is a difference between them. The average difference between the

TABLE 4.2

Power Measurements for Whip and Active Helix Antennas. Stop-scan radio station regime is identical for both antennas

Frequency (MHz)	90	92	95	97	100	102	104	106	108	Average
#1	–40.2	–40.7	–39.9	–41	–42.7	–42.7	–42.3	–44.1	–42.1	–41.7
#2	–40	–41.4	–41.7	–40.6	–40.2	–40.4	–44.4	–46.6	–45.6	–42.3
Δ dB	–0.2	0.7	1.8	0.4	–2.5	–2.3	2.1	2.5	3.5	0.6

reference antenna and the active helix does not exceed 0.6 dB. The residual noise measured at 110 KHz bandwidth is about –110 dBm (50 ohm impedance load).

4.4 Short Meander Antenna

A short meander line monopole antenna [11,12] is shown in Figure 4.26. The prototype dimensions [11] are height = 106 mm, base gap from ground plane = 9 mm, copper wire diameter = 1.7 mm, wire spacing = 10 mm, and ground plane = 1×1 m^2. The simulated meander antenna resistance is around 2 ohms. Both simulated and measured meander reactance is within –190 to –170 ohms

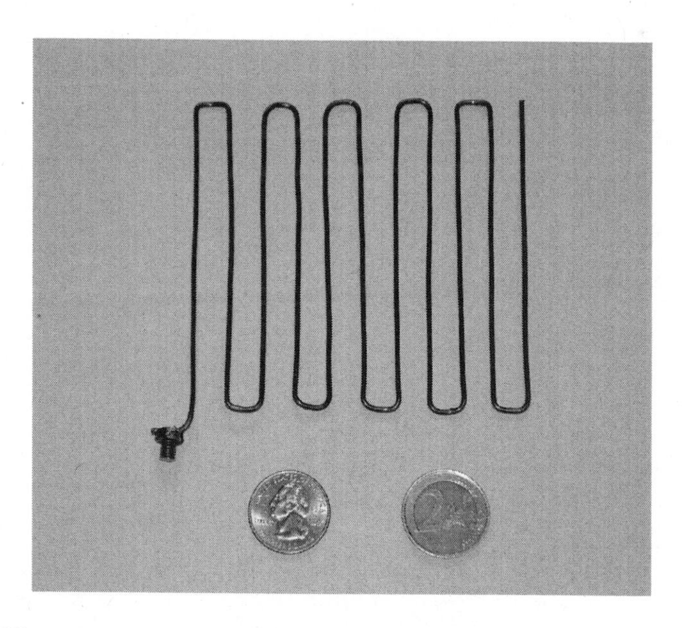

FIGURE 4.26
Meander line antenna for FM broadcasting application. (From E. Perri et al., IEEE Antennas and Propagation Society International Symphosium. 2008. Copyright 2008 IEEE. With permission.)

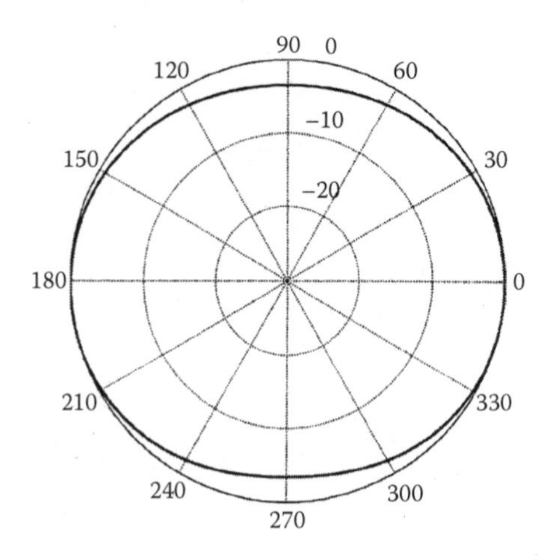

FIGURE 4.27
Radiation pattern in horizontal plane for vertical polarization. (From E. Perri et al., IEEE Antennas and Propagation Society International Symphosium. 2008. Copyright 2008 IEEE. With permission.)

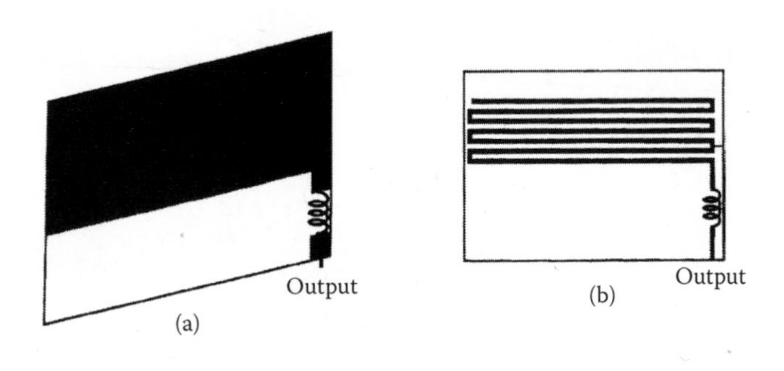

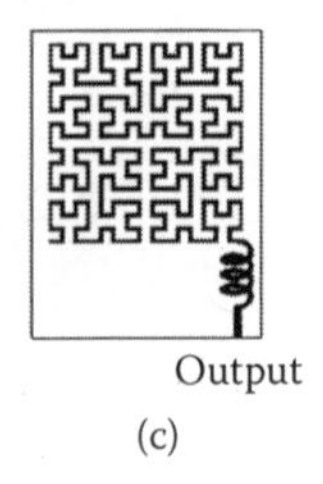

FIGURE 4.28
Different options of meander line antenna geometry for FM band.

over the entire FM band. The meander radiation pattern at the horizontal plane for a vertically polarized transmitting antenna is illustrated in Figure 4.27. The radiation efficiency of this antenna is higher than the efficiency of a simple monopole with the same height value, assuming that both antennas are matched with passive elements to compensate for reactance values.

A patent application [12] examines a circuit board antenna for AM/FM broadcasting applications. Three options are shown in Figure 4.28. All are used with amplifiers. The height of antenna pattern (Figure 4.28a) from the car roof is about 7 mm. The antenna height is about 90 mm. The antenna resonates in the FM band with an inserted series inductor value of about 1000 nH. As shown, the received AM voltage is about –15 dB in comparison with that of a metal rod antenna about 400 mm tall. In the FM frequency band, the antenna has maximum gain (vertical polarization, horizontal plane) about 8 dB less than a 180 mm rod antenna. In the AM band, the voltage gain is 10 dB less than the gain of the 180 mm rod antenna when the distance between the ground and the bottom part of the meander is about 25 mm and the height of antenna is 60 mm. The antenna can be used as an FM diversity component with a main FM printed-on-glass antenna.

4.5 Diversity FM Antennas

4.5.1 Two Antenna Diversity Elements

Many diversity printed-on-glass antenna designs for AM/FM auto applications have been published [13–16]. Figure 4.29 shows examples of two FM antennas mounted on the rear window of a car: one is located above the

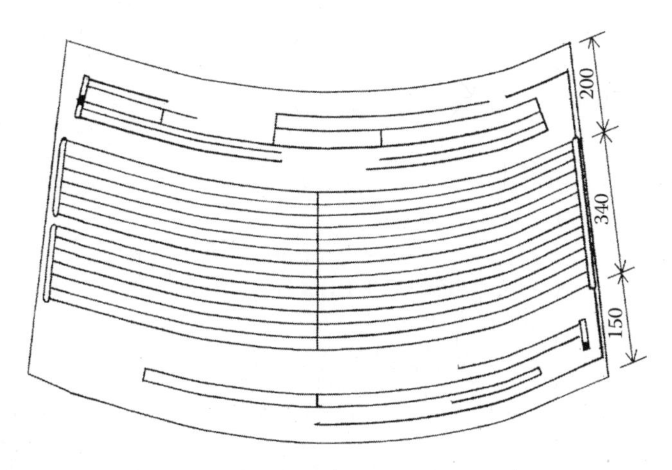

FIGURE 4.29
Patterns of two diversity rear glass FM antennas.

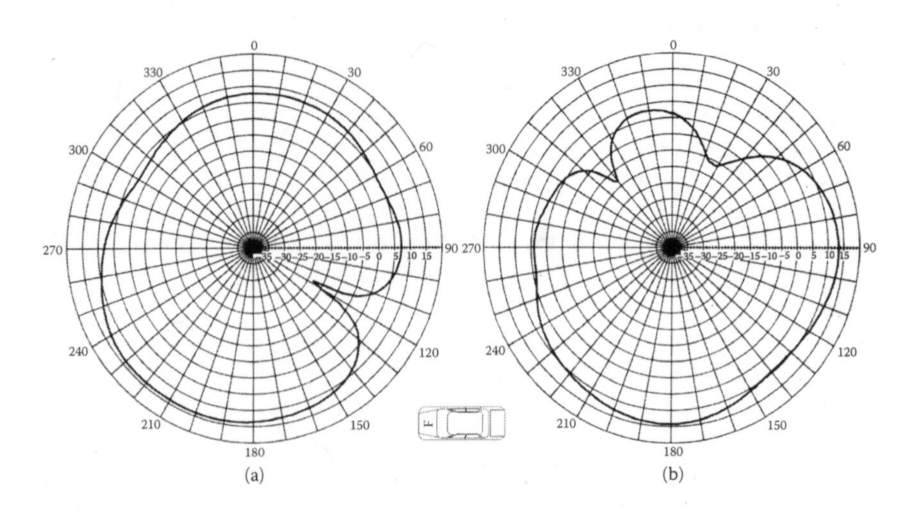

FIGURE 4.30
Radiation patterns of two diversity antenna elements: (a) top antenna; (b) bottom antenna.

defogger and the other below the defogger. Both have output VSWRs over entire broadcasting frequency bands below 4:1 for a 50 ohm load. The radiation patterns for the first and second antennas in the horizontal plane for right slant linear polarization of a transmitting antenna at 100 MHz are presented in Figure 4.30a and Figure 4.30b, respectively.

The space between the antennas is about a quarter wave at 100 MHz. The figure illustrates that at the angles where directionality of the first antenna has reduced gain value, the directionality of the second antenna exhibits increased gain values and vice versa. Thus, the "overlap" radiation pattern of this system is more omnidirectional in comparison with curves (a) or (b). Based on a simplified estimation (Equation 3.38), the correlation coefficient between the signals received by the first and second antennas is about 0.2. Assuming that the outage is 0.1, the diversity gain estimation value is 5 to 7 dB (Figure 3.5), depending on the diversity gain algorithm. The simplest diversity algorithm is based on the switched scanning technique (Reference [17], pp. 18,19,357–360).

As noted in Chapter 3, when the signal received by the first antenna drops below a predetermined but floating threshold (determined by the averaged signal level in the driving area), the control logic switches the system to the second antenna. When the signal received by the second antenna drops below a certain level, the control logic connects both antennas (in phase) to the car radio. When the signal from both antennas decreases below the threshold level, the control logic connects the two antennas (out of phase) to the radio input. The switching procedure continues for the duration of the drive.

4.5.2 Single Reconfigurable Antenna Element for Space Diversity Applications

The space diversity system requires at least two antenna patterns for improving reception quality. Two physically separated antennas require two feeding points, two RF cables, and two amplifiers—an expensive option for the auto industry. It is a known application of reconfigurable antennas in communication systems [20].

A reconfigurable antenna is capable of dynamic (electronic or mechanical) modification of the radiation pattern. Figure 4.31 shows two reconfigurable printed-on-glass options in the FM frequency band that can be used in a space diversity system. Both designs use an electronic control switch (for example, a pin diode) for reconfiguring the antenna pattern to eventually achieve a single feeding point. In the first design, the open pin diode position corresponds to the antenna pattern with the horizontal portion, and the short

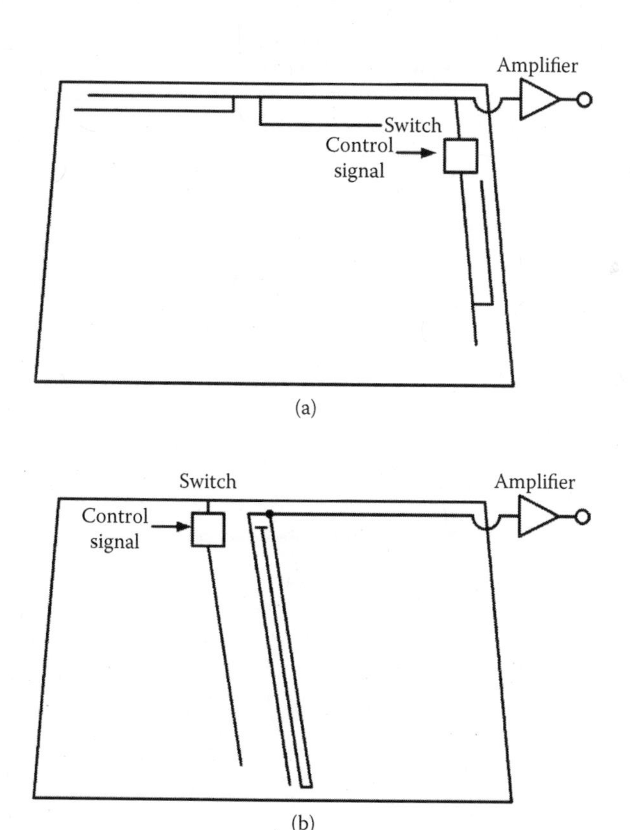

FIGURE 4.31
Reconfigurable designs: (a) antenna with controllable vertical and horizontal portions; (b) antenna with controllable parasitic element.

pin diode corresponds to the antenna with horizontal and vertical portions. The pin diode can be mounted in a case with the main diversity circuit. The second design [21] is based on a parasitic element that changes the antenna directionality when the pin diode is on or off.

4.5.3 Correlation Signal Analysis for Car Roof Spaced Antennas

A car roof antenna has an almost omnidirectional radiation pattern. However, as noted earlier, short-term fading still affects reception quality in the FM range for a mast antenna mounted on the car roof.

Shimizu and Kuwahara [22] estimate the correlation coefficient for two spaced roof mast antennas designed for space diversity. Both diversity quarter wave monopoles are mounted on the car roof and modeled by NEC2 software as shown in Figure 4.32. The monopole is tilted 45 degrees to the plane of the car roof. Two receivers measure a received electric field while the car is running. The measurement interval is 10 cm and the received frequency is 82.5 MHz. The measurement course equals a 4 km section.

The space between antennas is changed and results of the simulation and measurement of the correlation coefficient between received signals are shown in Figure 4.33. As we can see, the tendency of changing the correlation coefficient as a function of the space shows good agreement between the calculation and measurement results. The roughly estimated space between the two FM antennas that produces independent signals received by the first and second antennas is about 0.3 to 0.5 m (correlation coefficient = 0.2 to 0.1). Based on test results, the 0.3 to 0.5 m spacing between the antennas is enough to provide reasonable quality of the diversity system.

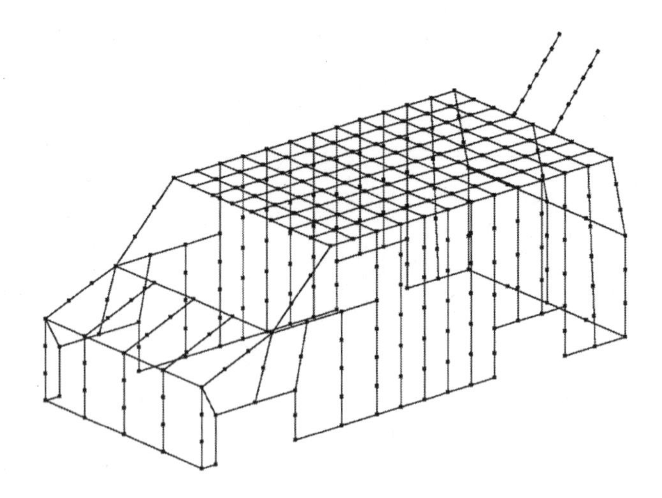

FIGURE 4.32
Simulation model for diversity monopole antennas mounted on car roof. (From M. Shimizu and Y. Kuwahara, *IEEE Antennas and Propagation International Symposium*. 2007. Copyright 2007 IEEE. With permission.)

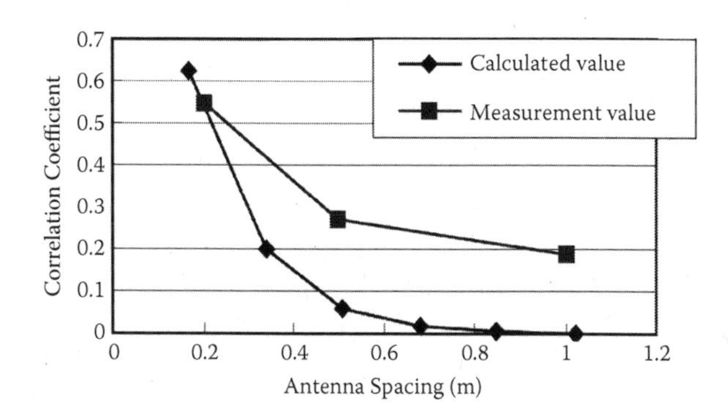

FIGURE 4.33
Correlation coefficient for space diversity antennas. (From M. Shimizu and Y. Kuwahara, *IEEE Antennas and Propagation International Symposium*. 2007. Copyright 2007 IEEE. With permission.)

4.6 Antennas for Digital Audio Broadcasting

Digital audio broadcasting (DAB) is a technology used by radio stations in different countries. DAB radio operates in Band III (174–240 MHz; known as Eureka 147) and in the L frequency band (1452–1492 MHz). As of 2006, approximately 1,000 stations worldwide broadcast in DAB format. Europe uses the 174 to 240 MHz band and Canada uses the L band terrestrial network. A DAB tuner installed in a car provides improved reception of AM and FM stations. DAB is more bandwidth efficient than the analogue technique for national radio stations, allowing placement of more stations into a smaller section of the spectrum.

4.6.1 L Band Antenna Geometry

The L band antenna [23] is a monopole printed on a thin dielectric substrate. Surface-mounted low-profile omnidirectional monopoles with linear polarization performance are very popular [24–26]. The proposed antenna is fabricated on a low cost RF material FR-4 substrate with a dielectric constant ε_r = 4.5 and a substrate thickness of 1.6 mm. Figures 4.34 and Figure 4.35 show two proposals. The quarter wavelength antenna shown in Figure 4.34 is preferred for on-vehicle applications and can be used for roof or trunk mounting. The amplifier circuit is arranged on the bottom side of the ground plane.

The bow tie-shaped antenna in Figure 4.35, is preferred for in-vehicle applications and can be attached to a window. The extended ground is used for 50 ohm impedance matching. The antenna patch and amplifier are placed on one side of a substrate, while the ground plane with extended ground parts is arranged on the other side. This antenna was designed and optimized for the L-band frequency range by using a FEKO electromagnetic simulator.

Figure 4.36 and Figure 4.37 show the measured input impedance and VSWR values for antennas without amplifiers. The amplifiers for both antennas were

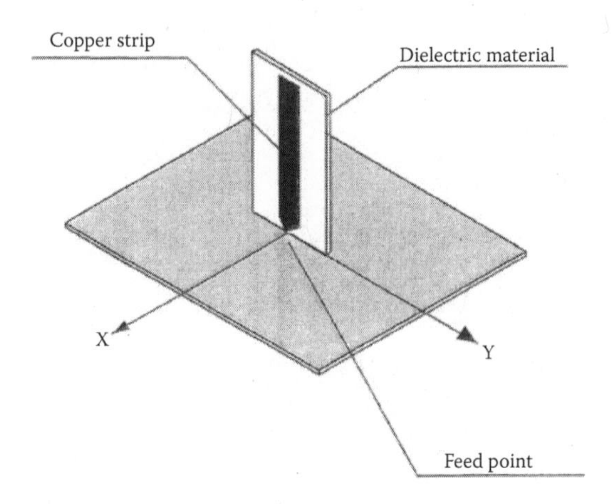

FIGURE 4.34
Geometry of quarter wave L band antenna for roof application. (From V. Rabinovich, *Microwave and Optical Technology Letters*, Vol. 39, 2003. Copyright 2003 Microwave and Optical Technology Letters. Reprinted with permission of Wiley-Blackwell, Inc.)

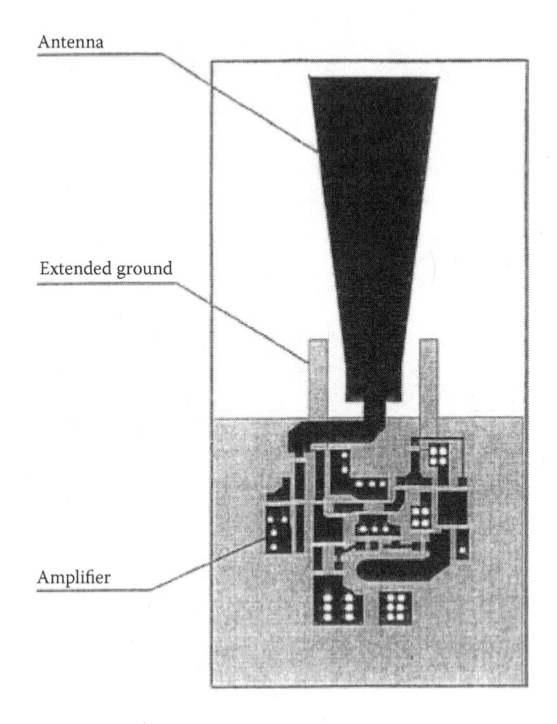

FIGURE 4.35
Geometry of bow tie L band antenna. (From V. Rabinovich, *Microwave and Optical Technology Letters*, Vol. 39, 2003. Copyright 2003 Microwave and Optical Technology Letters. Reprinted with permission of Wiley-Blackwell, Inc.)

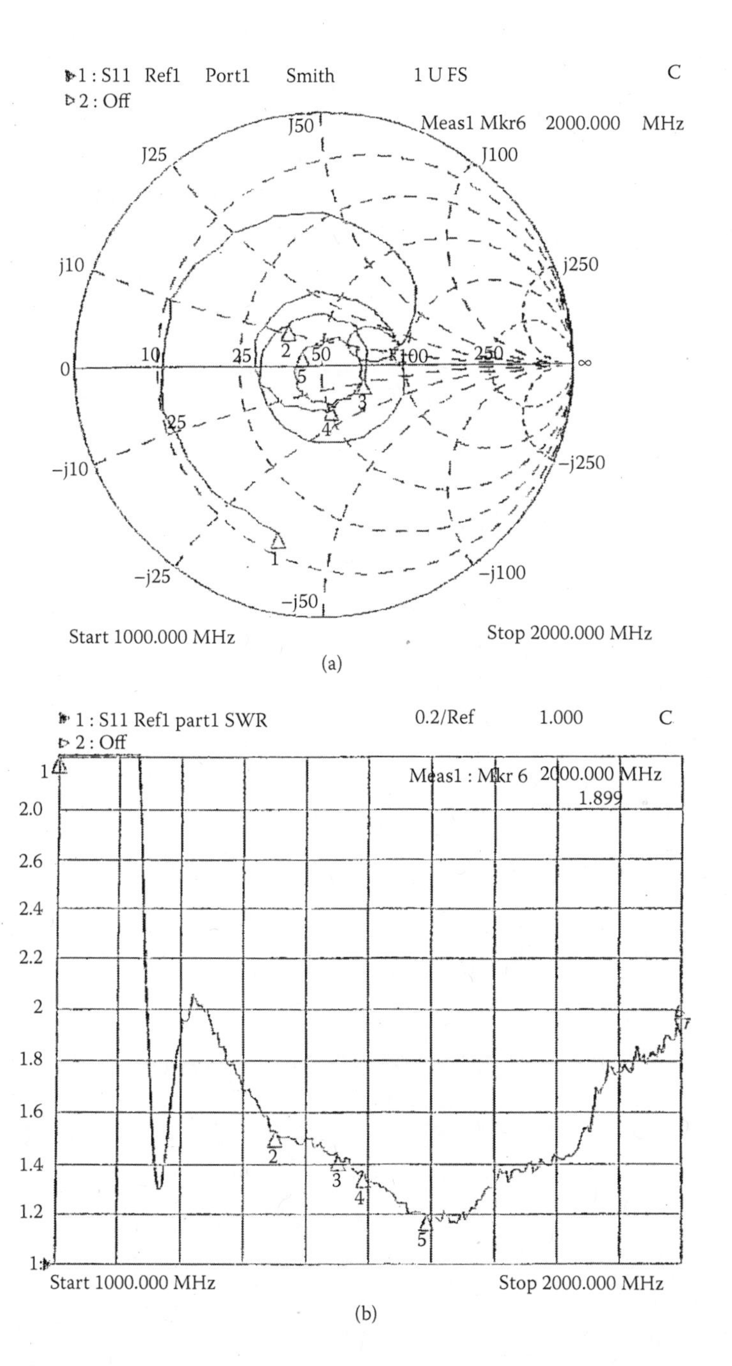

FIGURE 4.36
L band quarter wave antenna on metal plate: (a) Smith chart; (b) VSWR. (From V. Rabinovich, *Microwave and Optical Technology Letters*, Vol. 39, 2003. Copyright 2003 *Microwave and Optical Technology Letters*. Reprinted with permission of Wiley-Blackwell, Inc.)

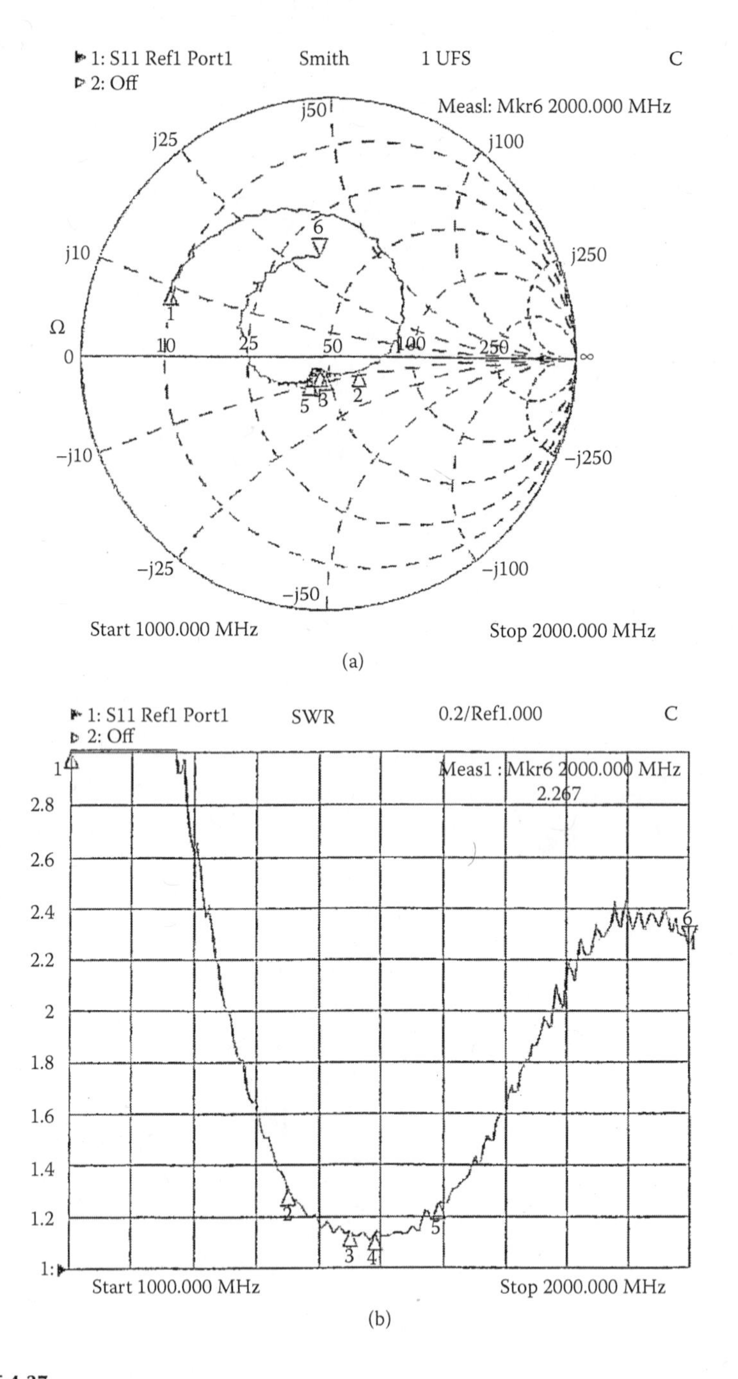

FIGURE 4.37
Bow tie-shaped antenna: (a) Smith chart; (b) VSWR. (From V. Rabinovich, *Microwave and Optical Technology Letters*, Vol. 39, 2003. Copyright 2003 Microwave and Optical Technology Letters. Reprinted with permission of Wiley-Blackwell, Inc.)

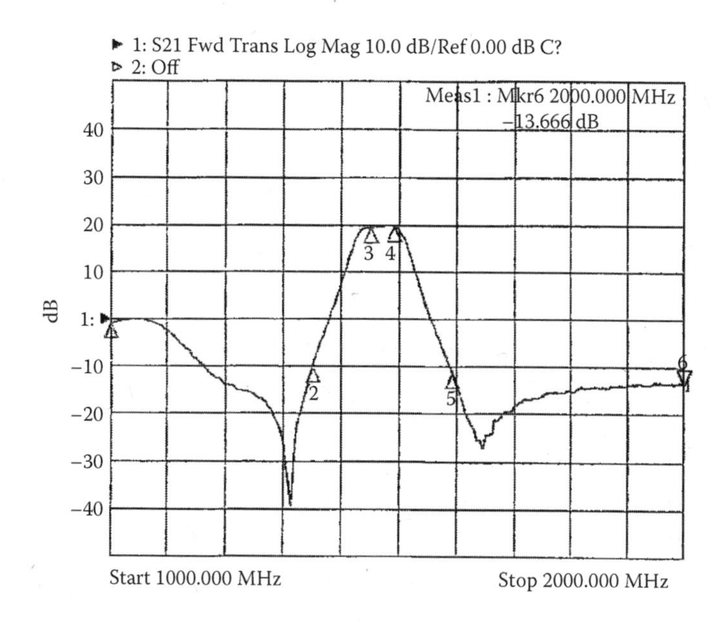

FIGURE 4.38
Amplifier gain. (From *Microwave and Optical Technology Letters,* Vol. 39, 2003. Copyright 2003 Microwave and Optical Technology Letters. Reprinted with permission of Wiley-Blackwell, Inc.)

designed using computer software by the Eagleware Corporation (now part of Agilent Technologies) [27]. Each amplifier has two transistor stages and a dielectric L band filter. Both transistors are ATF 34143 models from Avago Technologies. The dielectric L band filter is from TOKO. The power gain of the amplifier in the L band frequency band is shown in Figure 4.38. The noise figure of the designed amplifier is about 1.3 dB. Noise measurements were made with the aid of N8973A noise meter from Agilent Technologies. The measured noise does not include losses of the monopole.

4.6.2 Radiation Pattern Measurements for L Band Antenna

Horizontal plane antenna radiation patterns (without vehicle) measured in an automated anechoic chamber are omnidirectional. The gain of the first antenna on the metal plane with linear size $d/\lambda = 1$ m is approximately the same as the gain of a wire monopole with the same ground plane. The gain of the second antenna is less than the gain of the free-space half-wave dipole by 0.5 dB.

A Pontiac Sunfire served as the base car for the measurements of the antennas mounted on and in a vehicle. The measurement site uses a turntable that rotates (in horizontal plane) the automobile with the antenna to be measured. The turntable is placed on a hill to block reflections from the surrounding environment. The transmitting antenna was vertically polarized during the antenna measurements.

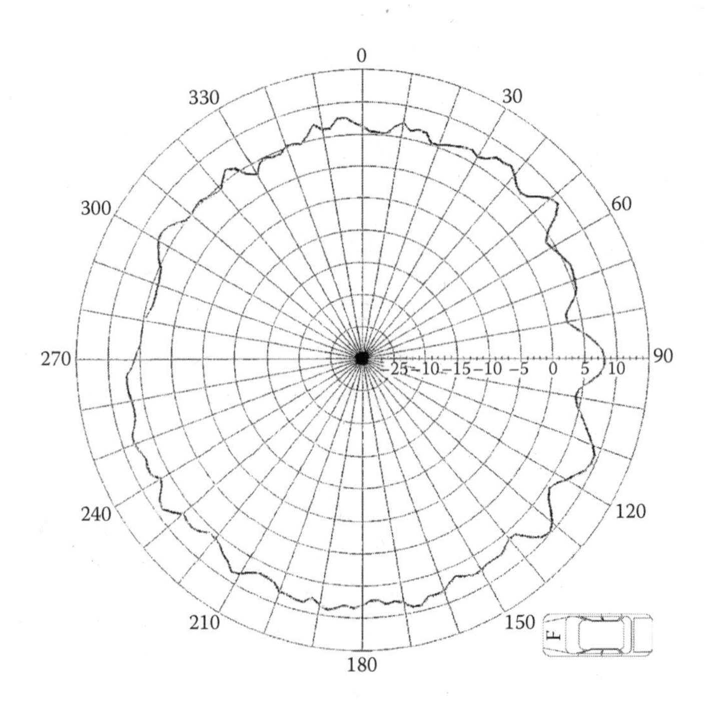

FIGURE 4.39
Radiation pattern of roof antenna. (From V. Rabinovich, *Microwave and Optical Technology Letters*, Vol. 39, 2003. Copyright 2003 Microwave and Optical Technology Letters. Reprinted with permission of Wiley-Blackwell, Inc.)

Two on-vehicle (quarter wave monopole) radiation patterns measured on automobile turntable are shown in Figure 4.39 (roof mount) and Figure 4.40 (trunk mount). The average gain (over 360 degrees in azimuth) for both antenna locations is almost the same, but the deviation of a radiation pattern for the trunk location is more than the deviation for the roof antenna. The trunk location is reasonable if the roof is restricted or not available for antenna placement. However, because roof and trunk antennas protrude from exterior surfaces, they are exposed to destructive impacts and create aerodynamic disturbances.

For this reason, we investigated a few in-vehicle (bow-tie shape) antenna positions, all of which corresponded to the vertical orientation of the antenna in Figure 4.35. Three in-vehicle radiation patterns are shown in Figures 4.41 to 4.43. The average gain of the side glass antenna is less than the gain of the roof antenna by 2.7 dB. The average gains of the front and back glass antennas are less than the gain of the roof antenna by 6.5 dB. This low gain level for the front and back window antennas can be explained by the inclined window planes, compared to the vertical plane. A comparison of three in-vehicle positions shows that the side location is preferred.

Results show significant directionality fluctuations for in-vehicle antenna positions. As noted earlier, a space diversity system [17] can be used to

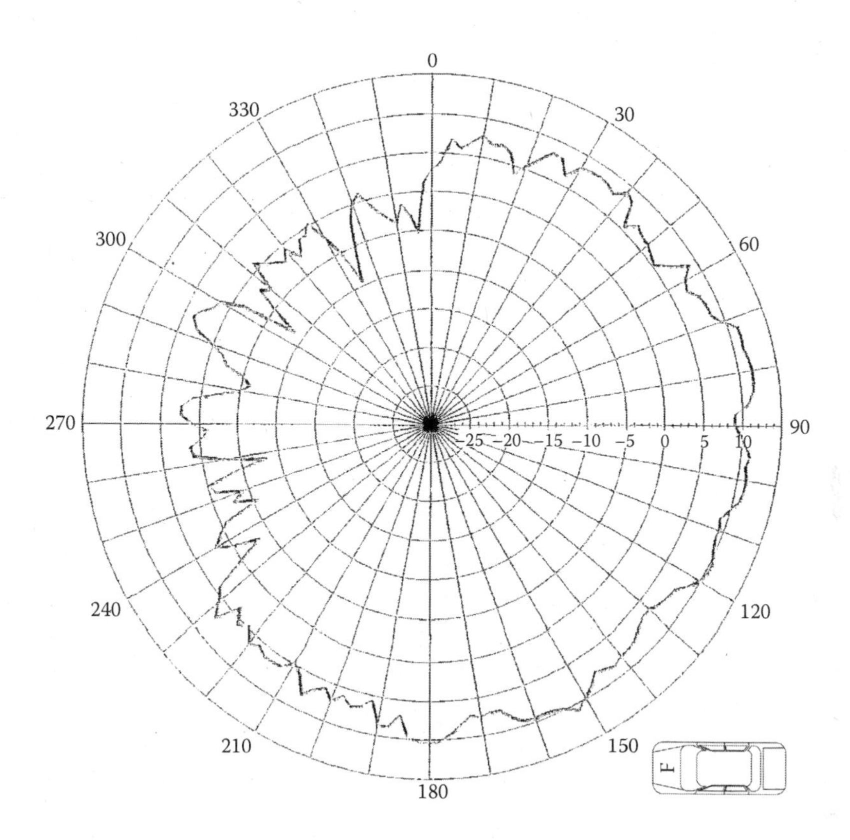

FIGURE 4.40
Radiation pattern of trunk antenna. (From *Microwave and Optical Technology Letters*, Vol. 39, 2003. Copyright 2003 Microwave and Optical Technology Letters. Reprinted with permission of Wiley-Blackwell, Inc.)

improve the reception of an in-vehicle antenna. Space diversity requires two or more antennas. The greatest fading reduction effect is expected when the correlation of signals of the two antennas is near zero. The half wavelength distance between the two antennas provides small correlation in receiving signals [29] and improves system performance.

4.6.3 Antennas for Band III

A simple quarter wave omnidirectional monopole about 35 cm long and mounted on a car roof can be used at 170 to 240 MHz (Band III). The impedance curve of such a monopole mounted at the middle of a metal screen 1 m long and 1 m wide in the entire band and simulated with FEKO software [3] is shown in Figure 4.44a.

To meet the miniaturization requirements of modern car antenna systems, the designs of compact monopole antennas are very important. One of the

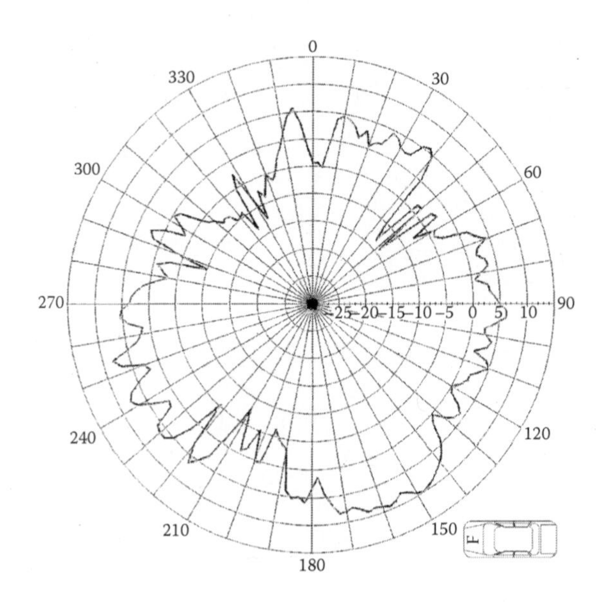

FIGURE 4.41
Radiation pattern of antenna attached to side glass. (From V. Rabinovich, *Microwave and Optical Technology Letters*, Vol. 39, 2003. Copyright 2003 Microwave and Optical Technology Letters. Reprinted with permission of Wiley-Blackwell, Inc.)

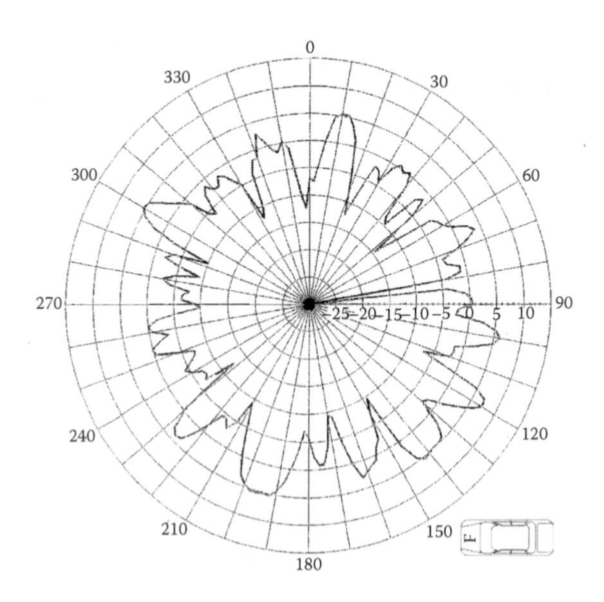

FIGURE 4.42
Radiation pattern of antenna attached to back glass. (From V. Rabinovich, *Microwave and Optical Technology Letters*, Vol. 39, 2003. Copyright 2003 Microwave and Optical Technology Letters. Reprinted with permission of Wiley-Blackwell, Inc.)

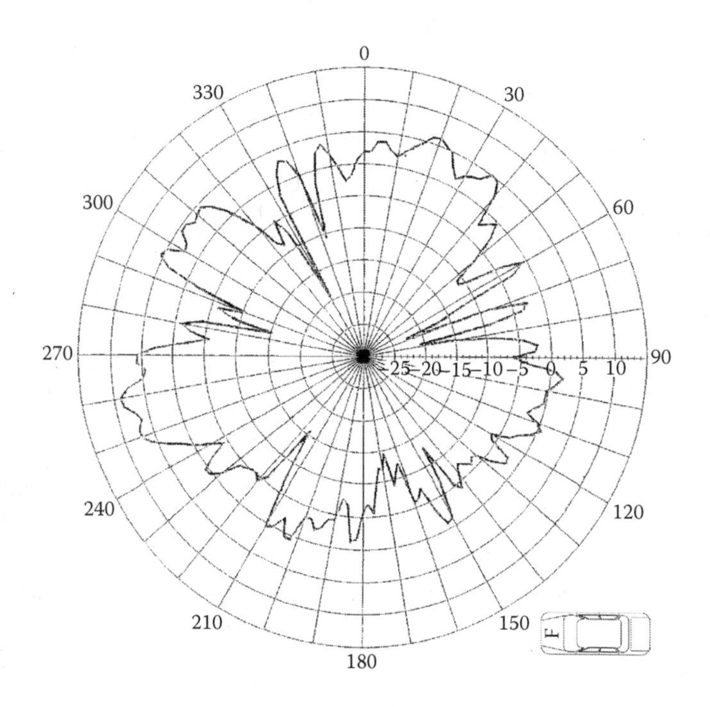

FIGURE 4.43
Radiation pattern of antenna attached to front glass. (From V. Rabinovich, *Microwave and Optical Technology Letters*, Vol. 39, 2003. Copyright 2003 Microwave and Optical Technology Letters. Reprinted with permission of Wiley-Blackwell, Inc.)

simplest methods of reducing length uses a helical monopole similar to the AM/FM helical antenna. The simulated curve in Figure 4.44b shows impedance of a helical antenna with a total length of about 19 cm, 30 turns, and a turn radius of 0.4 cm. An amplifier inserted between the antenna and car radio provides a reasonable SNR and impedance matching of the antenna and receiver.

A low profile broadband printed monopole antenna [30] for vehicle application is shown in Figure 4.45. The proposed antenna is composed of four metal pattern sections printed on a flexible printed circuit board (FPCB) 80×106 mm^2. The sections are folded onto a square-cylindrical form base with side length of 20 mm and height of 109 mm. The sections have slits 10 mm long and 2 mm wide that allow a decrease of monopole size. One of the four sections is a simple rectangular strip without meandering slits. Experimental results show a return loss of less than –10 dB and a gain of more than 0 dBi in the band from 174 to 218 MHz.

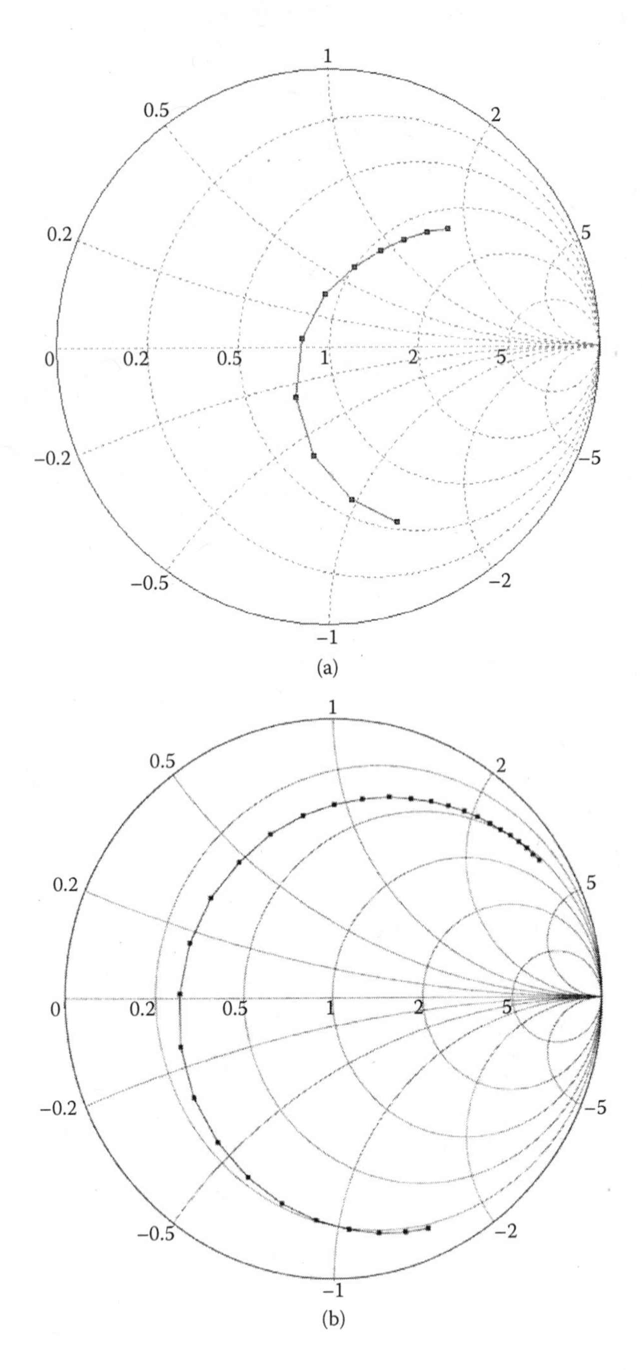

FIGURE 4.44
DAB III antenna parameters: (a) monopole impedance; (b) helix antenna impedance.

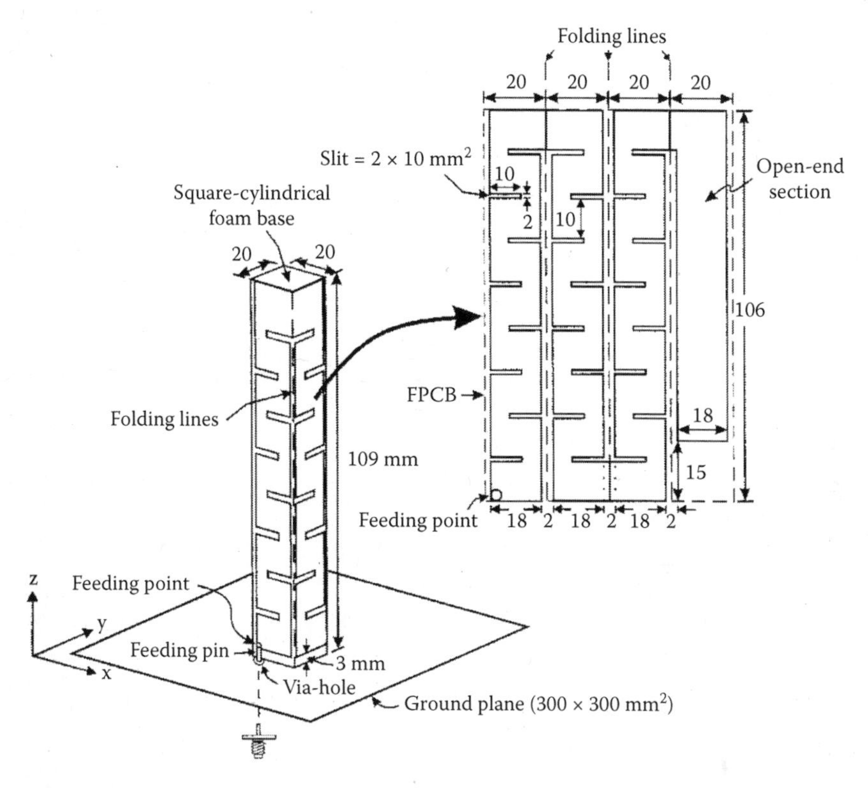

FIGURE 4.45
Geometry of DAB III monopole antenna printed on flexible printed circuit board. (From S. Su et al. *Microwave and Optical Technology Letters,* Vol. 42, 2004. Copyright 2004 Microwave and Optical Technology Letters. Reprinted with permission of Wiley_Blackwell, Inc.)

References

1. K. Fujimoto and J.R. James, *Mobile Antenna Systems Handbook,* Boston: Artech, 1994.
2. B.M. Al-Khateeb, Hidden Planar Antennas with Extended Communication Range for Remote Keyless Entry Automotive Applications, PhD Dissertation, Oakland University, Detroit, 2006.
3. FEKO Comprehensive EM Solutions. Product of EMSS-SA (Pty) Ltd., Stellenbosch, South Africa, www.feko.co.za.
4. U. Jakobus, N. Berger, and F. Landstorfer, Efficient techniques for modeling integrated windscreen antennas within the method of moments, Millennium Conference on Antennas and Propagation, Davos, Switzerland, 2000.
5. V. Rabinovich et al., Active window glass antenna system with automatic overload protection circuit, U.S. Patent 6,553,214, April 2003.

6. L. Reiter et al., Compact antenna with novel high impedance amplifier diversity module for common integration into narrow dielectric parts of a car skin, Institute of High Frequency Technology and Mobile Communication, University of Bundeswëhr Munich.
7. H. Lindenmeier, J. Hopf, and L. Reiter, Active Broad-Band Reception Antenna, U.S. Patent 6,603,435, August 2003.
8. A. Ramadan, Active Antennas with High Input Impedance Low Noise and Highly Linear Amplifiers, PhD Dissertation, University of Bundeswëhr Munich, 2005.
9. C. Balanis, *Antenna Theory, Analysis, and Design*, New York: John Wiley & Sons, 1997.
10. K. Fujimoto et al. *Small Antennas*, New York: John Wiley & Sons, 1993.
11. E. Perri et al., Very short meander monopole antennas, *IEEE Antennas and Propagation Society International Symposium*, pp. 1–4, 2008.
12. M. Ikeda et al., Antenna Apparatus, U.S. Patent Application 20080117111, 2008.
13. H. Oka, Glass Antenna and Glass Antenna System Using the Same, U.S. Patent 6906671, 2005.
14. H. Lindenmeier et al., Diversity Antenna on a Dielectric Surface in a Motor Vehicle Body, U.S. Patent 6,603,434, August 2003.
15. A. Fuchs et al., Vehicle Window Antenna System, U.S. Patent 6,239,758, May 2001.
16. H. Murakami et al., Window Glass Antenna for a Motor Vehicle, U.S. Patent 5,231,410, July 1993.
17. W.C.Y. Lee, *Mobile Communication Engineering*, 2nd Ed., New York: John Wiley & Sons, 1998, pp. 357–360.
18. A.J. Rustako et al., Performance of feedback and switch space diversity 900 MHz FM mobile radio systems with Rayleigh fading, *IEEE Transactions on Vehicular Technology*, 22, 173–184, 1973.
19, H. Lindenmeier et al., Scanning antenna diversity system for FM radio for vehicles, U.S. Patent 7,127,218, October 2006.
20. R.G. Vaughan, Switched parasitic elements for antenna diversity, *IEEE Transactions on Antennas and Propagation*, 47, 399–405, 1999.
21. L. Low and R.J. Langley, Single feed antenna with radiation pattern, *Electronics Letters*, 40, 975–976, 2004.
22. M. Shimizu and Y. Kuwahara, Analysis of a diversity antenna mounted on the vehicle for FM radio, *IEEE Antennas and Propagation International Symposium*, pp. 1068–1071, 2007.
23. V. Rabinovich, L-band active receiving antenna for automotive applications, *Microwave and Optical Technology Letters*, 39, 319–323, 2003.
24. K.F. Tong et al., A miniature monopole antenna for mobile communications, *Microwave and Optical Technology Letters*, 27, 262–263, 2000.
25. H.M. Chen, Microstrip-fed dual-frequency printed triangular monopole, *Electronics Letters*, 38, 619–620, 2002.
26. J. Lee, S. Park, and S. Lee, Bow-tie wide-band monopole antenna with the novel impedance-matching technique, *Microwave and Optical Technology Letters*, 33, 448–452, 2002.

27. Eagleware RF and Microwave Design software, Eagleware Corporation, www.eagleware.com
28. M. Daginnus et al., SDARS antennas: environmental influences, measurement, vehicle application investigations, and field experiences, Society of Automotive Engineers 2002 World Congress, Detroit, 2002.
29. J.D. Parsons, *The Mobile Radio Propagation Channel*, New York: John Wiley & Sons, 1996, pp. 137–140.
30. S. Su, et al., Low-profile broadband printed VHF monopole antenna for vehicular applications, *Microwave and Optical Technology Letters*, 42, 349–350, 2004.

5

Cellular Antennas

5.1 Single Band Monopole on Roof

The quarter wave monopole antenna is the simplest design appropriate for vehicle cellular phone applications in any frequency range. Such a vertically polarized antenna has an omnidirectional radiation pattern in the horizontal plane. Based on Reference [1], p. 155, the resonance impedance of a quarter monopole antenna mounted on an infinite metal screen is 36 ohms and the maximum gain is twice (3 dB over) that for a half wave dipole, that is, 5.14 dBi.

Figure 5.1 plots the radiation pattern in the elevation plane for a component designated $F_\theta(\theta, \varphi = \text{const})$. The radiation pattern $F_\theta(\theta,\varphi)$ as a function of φ for any θ value is omnidirectional; component $F_\varphi(\theta,\varphi)$ for any θ versus φ equals 0.

It is interesting to investigate the effects of a finite metal screen (representing a car roof) on monopole antenna parameters. The simulation results of a monopole used in the 824 to 894 MHz frequency band are demonstrated. Figure 5.2a shows a monopole mounted in the middle of a metal roof (dimensions 2 m × 1 m). Figure 5.2b presents the antenna impedance calculated with FEKO software. An antenna height of 8.16 cm was chosen from the resonance condition for the middle frequency point of 854 MHz. Antenna impedance for a resonant frequency point differs insignificantly from a value equal to 36 ohm. Figure 5.3 shows radiation patterns for the 854 MHz antenna in the horizontal plane for different polarization orientations of the transmitting antenna: (a) vertical, (b) horizontal, (c) right slant (+45 degrees), and (d) left slant (–45 degrees).

To achieve linear slant polarization, the linear polarized transmitting antenna is inclined to the horizon at a predetermined angle value. As follows from the simulations, the maximum gain for vertical polarization in horizontal plane $F_\theta(\theta = 90°, \varphi)$ is about –1 dBi, and parameter γ, which determines the maximum to minimum ratio of the directionality over the 360° azimuth plane, is equal to only 0.2 dB. This means that this radiation pattern for such screen size is omnidirectional. For a horizontal polarization component $F_\varphi(\theta = 90 \text{ degrees}, \varphi)$, see Figure 5.3b, the maximum directionality is about –9 dB compared to the vertical polarization component. The cross

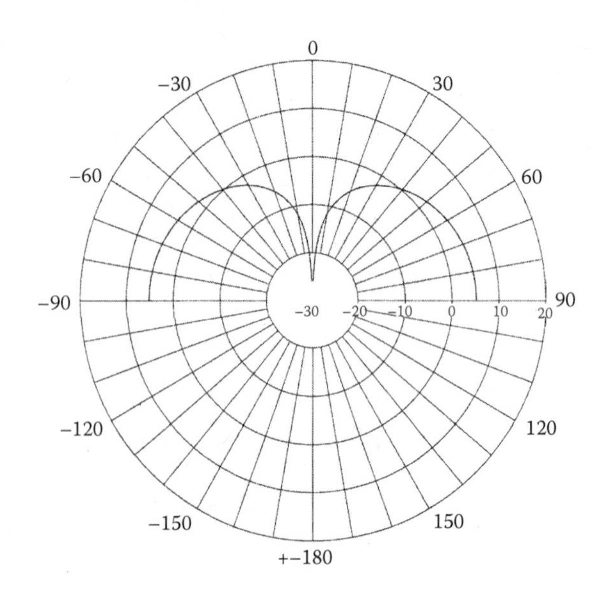

FIGURE 5.1
Radiation pattern of monopole mounted on infinite metal screen in elevation plane, θ-polarization.

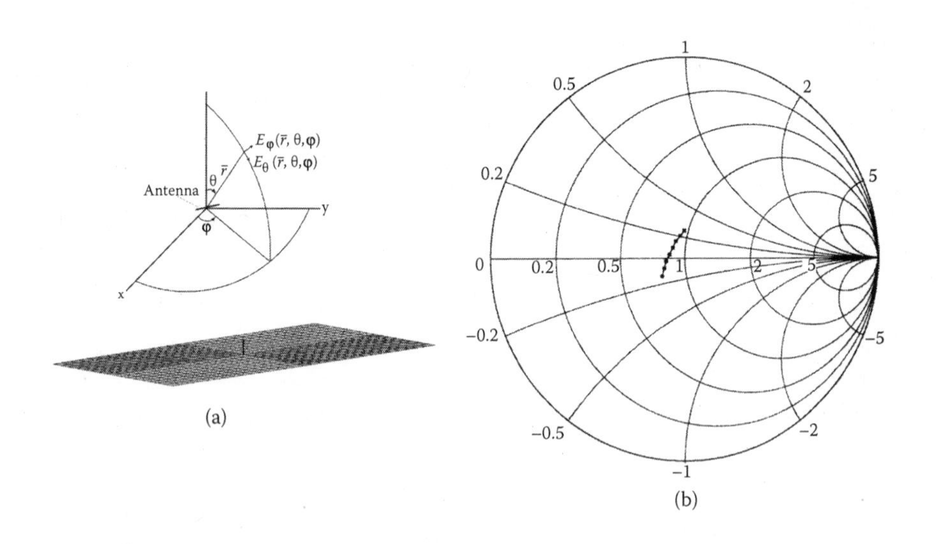

FIGURE 5.2
(a) Geometry of simple quarter wave monopole mounted at middle of 2 m × 1 m metal screen. (b) Antenna impedance in 824 to 894 MHz frequency band.

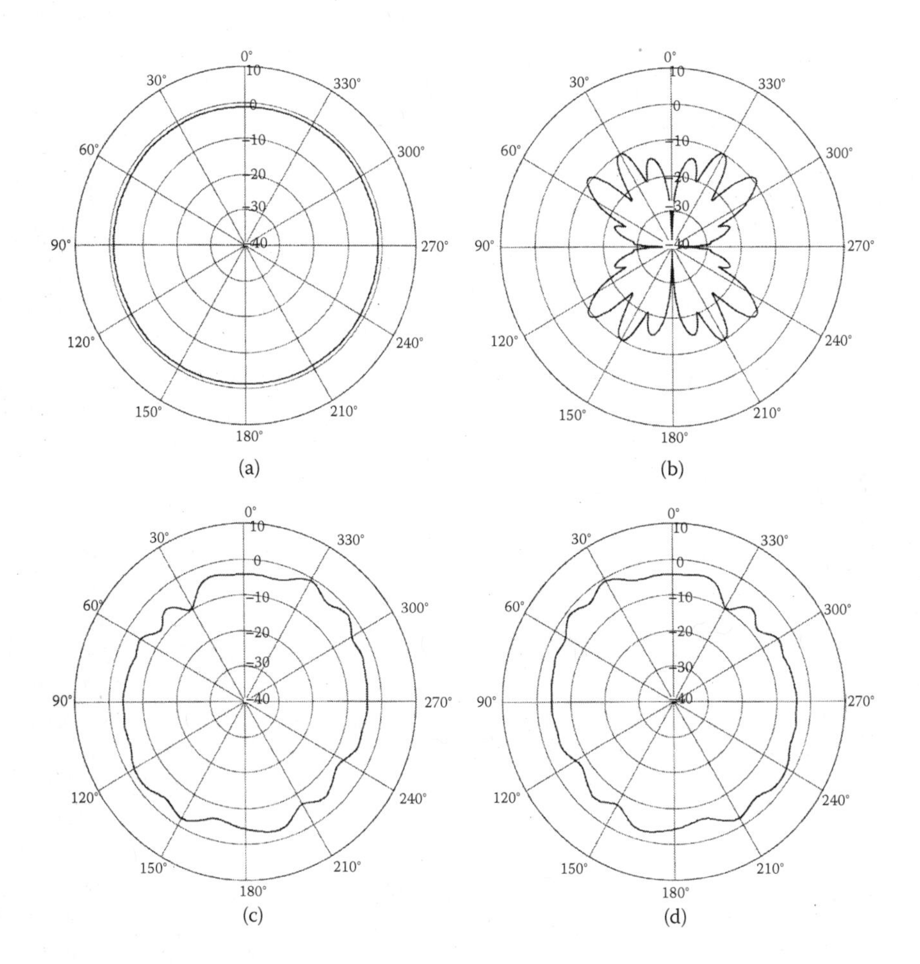

FIGURE 5.3
Radiation patterns of middle mounted monopoles in horizontal plane: (a) vertical polarization; (b) horizontal polarization; (c) right slant + 45 degrees polarization; (d) left slant −45 degrees polarization.

component arises from the current flowing along the edges of the metal plate in the horizontal direction. The radiation patterns for the right and left slant components have maximum values approximately equal to those obtained for vertical polarization.

Figure 5.4a and Figure 5.4b show the radiation patterns $F_\theta(\theta, \varphi = 0$ degrees) and $F_\theta(\theta, \varphi = 90$ degrees) of a monopole in the elevation plane. The maximum gain value (about 4 dBi) for both radiation patterns corresponds to an elevation angle about 30 degrees above the horizon.

Usually, a cellular phone antenna is mounted near the rear roof rim, not in the middle of the roof. Figures 5.5 through 5.7 show antenna geometry, simulated antenna impedance, and radiation patterns (854 MHz) for different

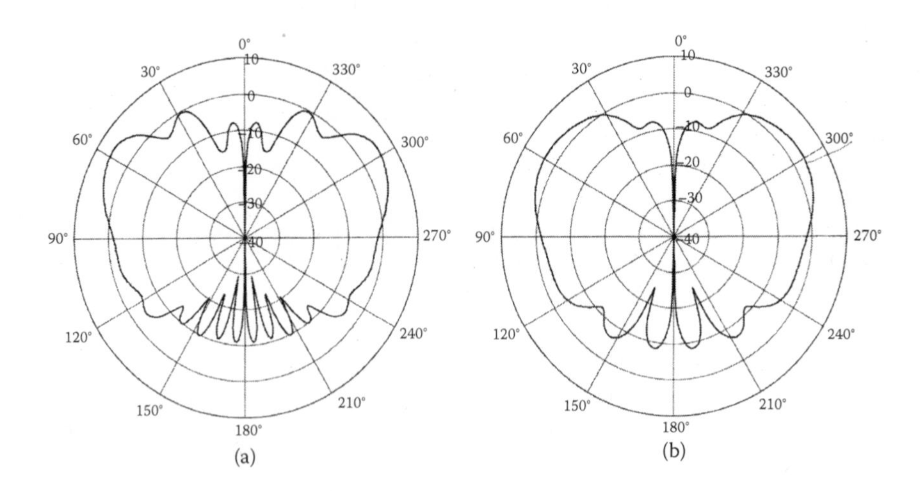

FIGURE 5.4
Radiation pattern of middle mounted monopole in elevation plane for θ-polarization: (a) $\varphi = 0$ degrees; (b) $\varphi = 90$ degrees.

transmitting antenna polarizations in horizontal and vertical planes for a monopole 8.16 cm long, mounted at the rear edge (5 cm from the rear rim) of a car roof. The impedance and radiation pattern F_θ ($\theta = 90$ degrees, φ) have almost the same plots as antenna placement at the center of the roof. The orthogonal component of the antenna directionality F_φ ($\theta = 90$ degrees, φ)

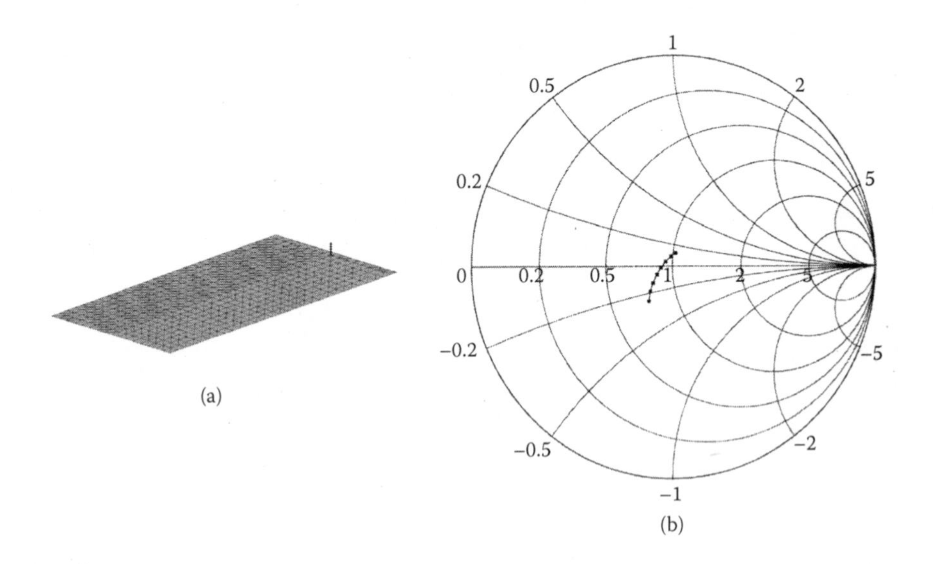

FIGURE 5.5
(a) Geometry of monopole mounted at edge of metal screen 5 cm from rear rim. (b) Antenna impedance in 824 to 894 MHz band.

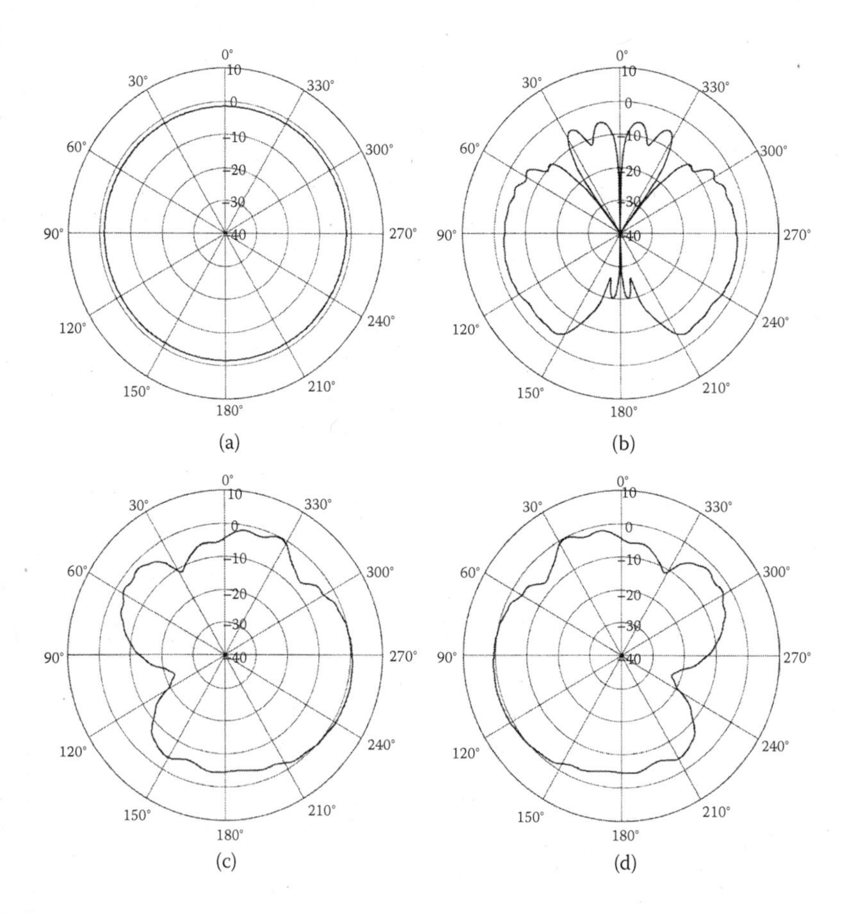

FIGURE 5.6
Radiation patterns of monopole mounted at edge of metal screen in horizontal plane: (a) Vertical polarization; (b) horizontal polarization; (c) right slant +45 degrees polarization; (d) left slant −45 degrees polarization.

becomes 7 dB higher in comparison with a similar value for a middle-of-roof mounting. The radiation pattern F_θ ($\theta,\varphi = 0$ degrees) in the elevation plane is significantly asymmetrical because the antenna is shifted from the roof center along the X axis. However, the radiation pattern F_θ ($\theta,\varphi = 90$ degrees) is symmetrical and is similar to that shown in Figure 5.4b.

Sometimes, monopole antennas of half a wavelength or three quarters of a wavelength are used for automotive cellular phone applications (Reference [1], p. 159). A three quarter wavelength-high antenna has a coiled conductor portion placed at a distance from the antenna base equal to a quarter of the wavelength, as shown in Figure 5.8. The coil is a load to decrease the out-of-phase components of the current along the antenna [2] and functions as a

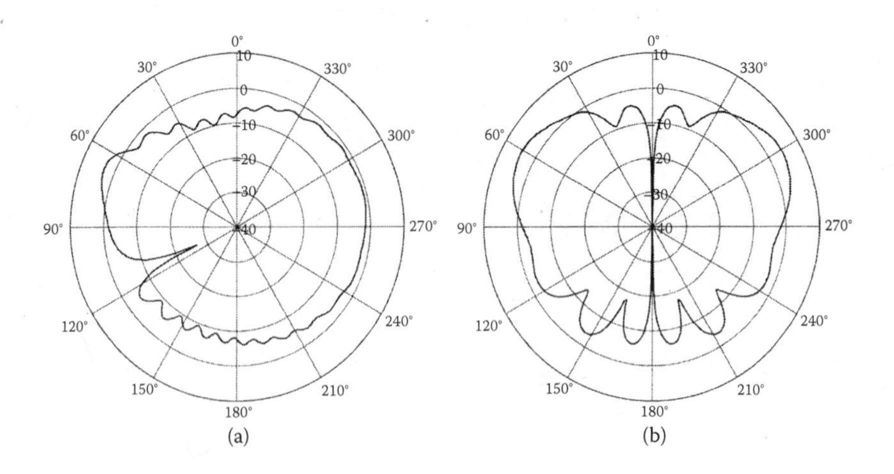

FIGURE 5.7
Radiation patterns of edge mounted monopole in elevation plane for θ-polarization: (a) $\varphi = 0$ degrees; (b) $\varphi = 90$ degrees.

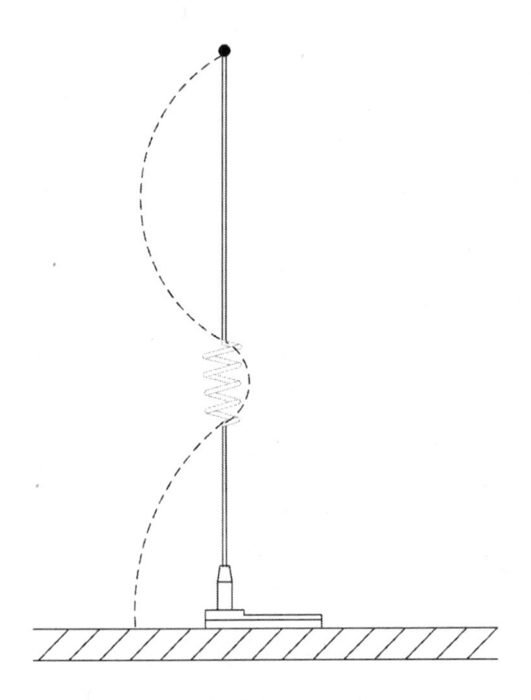

FIGURE 5.8
Three quarter wavelength monopole antenna geometry.

phase compensation element. The current distribution along the monopole with the coil is represented by a dotted line.

5.2 Glass Mounted Monopole

A simple monopole antenna mounted on rear car window [3,4] is shown in Figure 5.9. The antenna operates in the 820 to 890 MHz band and includes an external portion attached to the exterior surface of the window and an internal portion affixed to an interior surface of the window opposite the

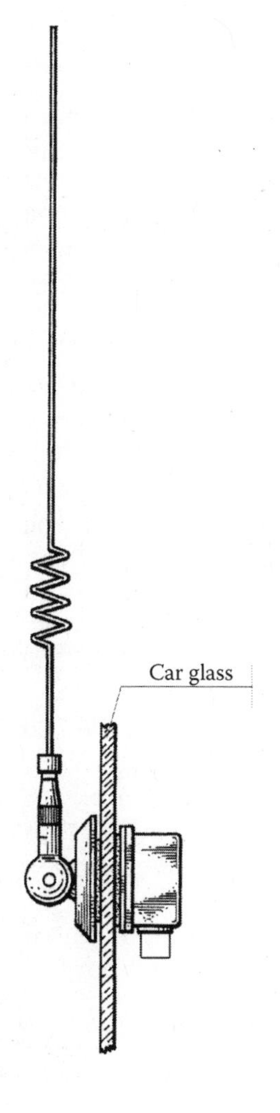

FIGURE 5.9
Monopole antenna mounted on rear glass.

exterior portion. The interior portion is electrically connected to the RF cable connected to the cellular transceiver.

Typically, the coupling capacitor of the external plate and internal plate along with the glass has a value of a few picofarads (glass dielectric constant $\varepsilon = 6.5$; glass thickness is 5 mm). The system may include a circuit for matching the impedance of the antenna to the impedance of the RF cable connected to the transceiver. As a rule, glass insertion losses do not exceed 1.5 to 2 dB, depending on the coupling design circuit over the entire design frequency band and VSWR less than 2:1 for a 50 ohm load as shown in the publication [4]. The antenna assembly contains a conductive swivel mechanism so that the angle of the antenna can be re-adjusted during operations in different communication areas.

5.3 Dual-Band Monopole

As noted earlier, modern cellular phone equipment usually uses frequencies from 800 to 900 MHz and from 1700 to 1900 MHz. Therefore, an antenna must operate in both frequency bands. Many companies now manufacture dual-frequency band antennas for cellular applications. Antenna length can vary from 28 cm to 36 cm, depending on the manufacturer.

One such antenna presented in a patent [5] and shown in Figure 5.10 has two tuning elements: a phasing inductor coil and a sleeve. The figure illustrates the distribution of currents along the antenna: a solid curve corresponds to 800 MHz and a dotted curve to 1900 MHz. The phasing coil per the patent provides an in-phase condition between the lower and upper parts of an antenna at both frequency ranges, eventually increasing the gain. The upper end of the sleeve choke is shortened to the antenna and the lower end is open. The sleeve choke forms a shortened transmission line that has an effective electrical length of a quarter wavelength at 1900 MHz. The sleeve eliminates current flows above the choke at 1900 MHz. At 800 MHz, the choke has an insignificant effect.

Usually, such antennas have gain values around 3 to 5 dBi in the 800 MHz band and about 2 to 4 dBi at 1900 MHz frequency depending on the base-mounting technique. Antennas may be mounted magnetically or through glass. Some antennas may include additional passive circuits for providing the best 50 ohm match between an antenna and receiver/transmitter with an RF cable. The matching circuit should be mounted at the antenna base. A typical VSWR of a dual band through a glass-mounted antenna [6] is shown in Figure 5.11. The VSWR is less than 2:1 in both frequency ranges.

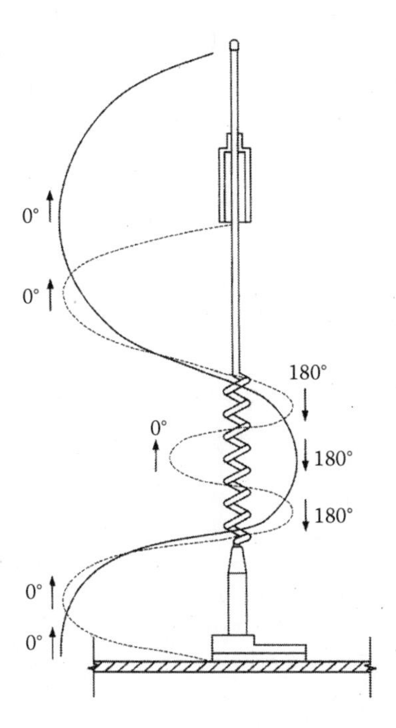

FIGURE 5.10
Dual-band monopole antenna on car glass. Solid curve shows current distribution for lower 800 MHz band; dotted curve shows current distribution for upper 1900 MHz band.

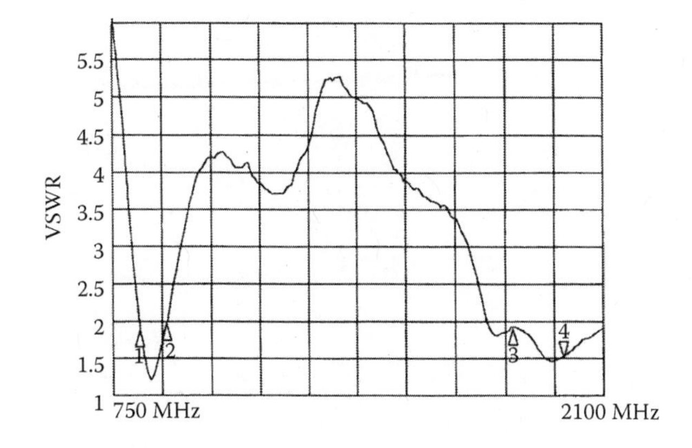

FIGURE 5.11
Typical VSWR of dual-band monopole.

5.4 Helix

5.4.1 Single Band Simulation Results

Usually, the limited height space for automotive applications dictates the use of small antennas, particularly for the cellular frequency range. A helix antenna is a good candidate for reduced size application. Figure 5.12 shows the topology chosen for the simulation using NEC-Vin Professional software. The antenna has a straight base portion with length of 5 mm and a helix portion with a diameter of 8 mm; the wire diameter is 0.4 mm. The antenna is mounted at the center of an infinite metal screen. Figure 5.13 shows the simulated impedance for a few height values H_{hel} of the helix portion. The number of turns for each height value is chosen from the resonance condition at about 854 MHz. This is the middle frequency point in the AMPS band. Antennas are modeled as lossless so that radiation resistance may be directly determined.

Table 5.1 shows the simulation results: shortening factor f_{sh} determined by the ratio of the quarter wave monopole to the total height of the simulated antenna, H_{tot} (helix portion together with straight portion), radiation resistor R_{rad}, and bandwidth *BANDW* corresponding to 3 dB power attenuation. The antenna bandwidth was estimated by matching 50 ohm at the frequency point equal to 850 MHz. A simple circuit used for 50 ohm matching is shown in Figure 5.14 and circuit parameters (capacitor and inductor values) are

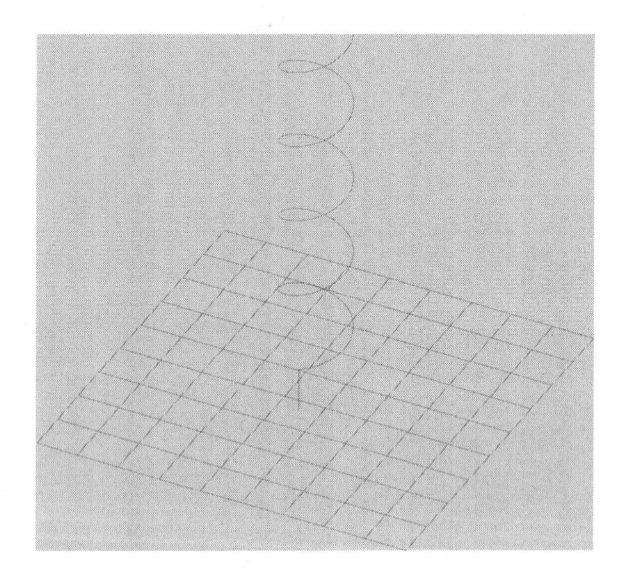

FIGURE 5.12
Example of single band helix cellular phone antenna on infinite metal screen.

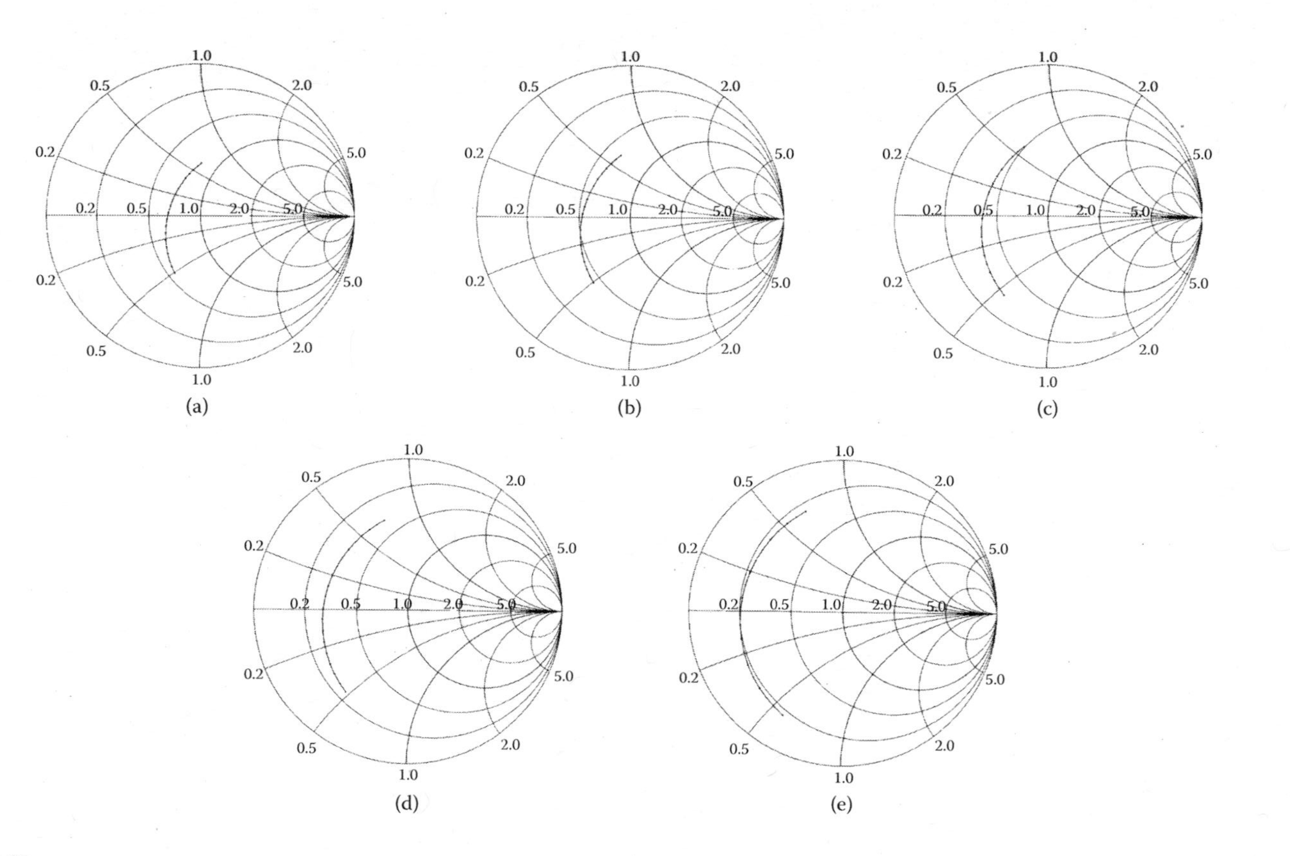

FIGURE 5.13
Simulated antenna impedance for different dimensions H_{hel} of helix monopole: (a) = 75 mm; (b) = 65 mm; (c) = 55 mm; (d) = 45 mm; (e) = 35 mm.

TABLE 5.1

Resonance Properties of Helical Monopoles of Different Heights

H_{tot}(mm)	81.6 Straight Wire	75 Helix + 5 mm Straight Wire at Base	65 Helix + 5 mm Straight Wire at Base	55 Helix + 5 mm Straight Wire at Base	45 Helix + 5 mm Straight Wire at Base	35 Helix + 5 mm Straight Wire at Base
f_{sh}	1	1.1	1.26	1.48	1.8	2.3
R_{rad}(Ohm)	36	30	26	20	14.6	10
BANDW	11.7%	8.9%	7.5%	5.9%	4%	2.7%
$C(pF)$		6	6	6	6	7.8
$L(nH)$		12	10	8	6	4.8

presented in Table 5.1. Note that the radiation resistance decreases and the bandwidth narrows as the helix height is shortened. This analysis assumed that the antenna had a 0.4 mm fixed wire diameter. A decrease in wire diameter led to an increase in loss resistance (shown in Equations (2.11) and (2.13) in Chapter 2) and a decrease in radiation efficiency [7].

5.4.2 Dual-Band Helix

We discussed the application of dual-band cellular antennas of more than a quarter wavelength long. Original equipment manufacturers (OEMs) use small cellular dual- or tri-band antennas in a single package intended for GPS, SDARS, or AM/FM application. Figure 5.15 shows the topology of a multiband cellular antenna that operates in the GSM 850, GSM 1800, and GSM 1900 bands. This antenna presents a helical–monopole combination similar to the geometry [8] cited for handset cellular phone operation. The combination incorporates a helix for a lower frequency band (824 to 894 MHz) and a

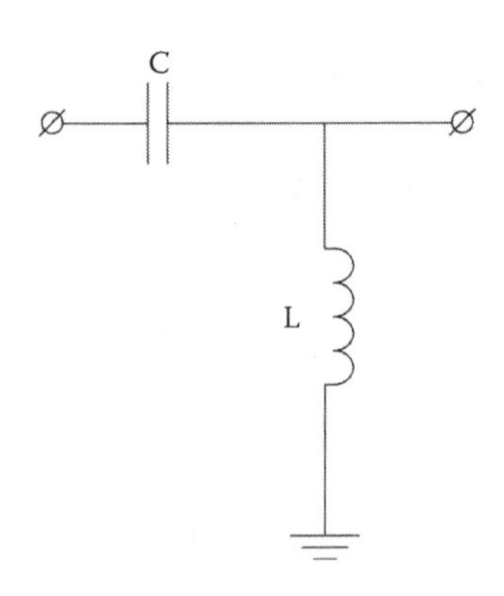

FIGURE 5.14
Matching circuit for helical design.

FIGURE 5.15
Dual-band helical monopole combination.

monopole inside the helix for high (1800 to 1900 MHz) band. A helical antenna is wound to reduce total length. The simulated antenna is placed on an infinite conductive plate.

The dimensions of the antenna are the following: total helix height = 45 mm; length of the straight offset portion = 5 mm, wire radius = 0.2 mm, helix diameter = 8 mm, and space between helical turns ~13 mm. The antenna has an additional simple matching circuit with topology similar to that shown in Figure 5.14.

Figure 5.16 shows the impedance of the antenna in both frequency ranges. Figure 5.17 shows the simulation losses in dB scale (using the Eagleware program [9]) between an antenna and 50 ohm input impedance receiver at a lower (GSM 850) operating frequency band without (curve a) and with (curve b) a matching circuit. The matching components are C = 10 pF and L = 19 nH. Similar plots for the upper band (GSM 1800 and GSM 1900) are shown in Figure 5.18. The circuit improves the matching to 50 ohm in the lower frequency band without worsening the impedance in the upper band. The shortening factor for the helical antenna is ~1.8. The simulated vertically polarized radiation antenna pattern at the frequency 855 MHz in the horizontal plane is omnidirectional and similar to the directionality for the monopole above the metal screen. The elevated radiation pattern for the vertically polarized wave is equivalent to the directionality for the monopole for the same plane. The maximum gain of the dual-band antenna

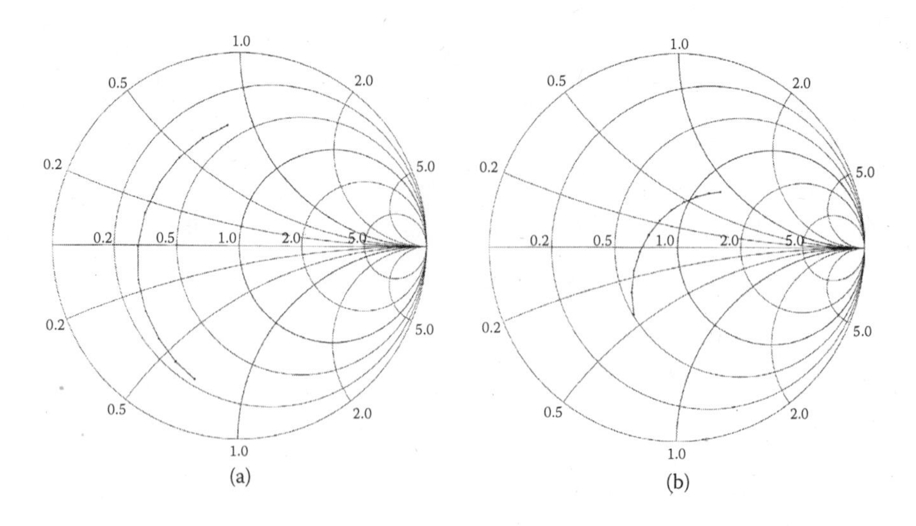

FIGURE 5.16
Antenna impedance: (a) lower frequency 800 to 900 MHz band; (b) upper frequency 1700 to 1900 MHz band.

at the lower frequency is estimated as 0 dBi. The peak gain at the 1800 MHz frequency range is estimated as 2.5 dBi.

Figure 5.19 shows another dual-band helical antenna [10] that has nonuniform pitch and can be implemented in two options: decreasing pitch angle or increasing pitch angle. Figure 5.20 shows the return losses for these structures. For both options, the first frequency (AMPS) resonance location is the same if the total wire length is constant. The second resonance can be shifted high or low, depending on how the pitch is varied. A helix with a decreasing pitch angle will move the second resonance closer to the first. Conversely, a helix with an increasing pitch angle will move the second resonance away from the first. When choosing a pitch angle shift value, it is possible to tune the operating frequency bands of the designed antenna. This structure can be used for AMPS, GSM, and Bluetooth applications. A 10 dB return loss can be achieved for both frequency bands without a matching circuit. Reference [10] notes that bandwidth and return losses may be improved further by using a proper matching circuit.

5.5 Compact Printed Circuit Board Antennas

5.5.1 Single Band Meander Line Design

Figure 5.21 illustrates a simple single-band printed antenna design. The element is printed on a thin (1.6 mm) dielectric substrate FR-4 board. The antenna is attached and grounded at the bottom on a base connected to the vehicle

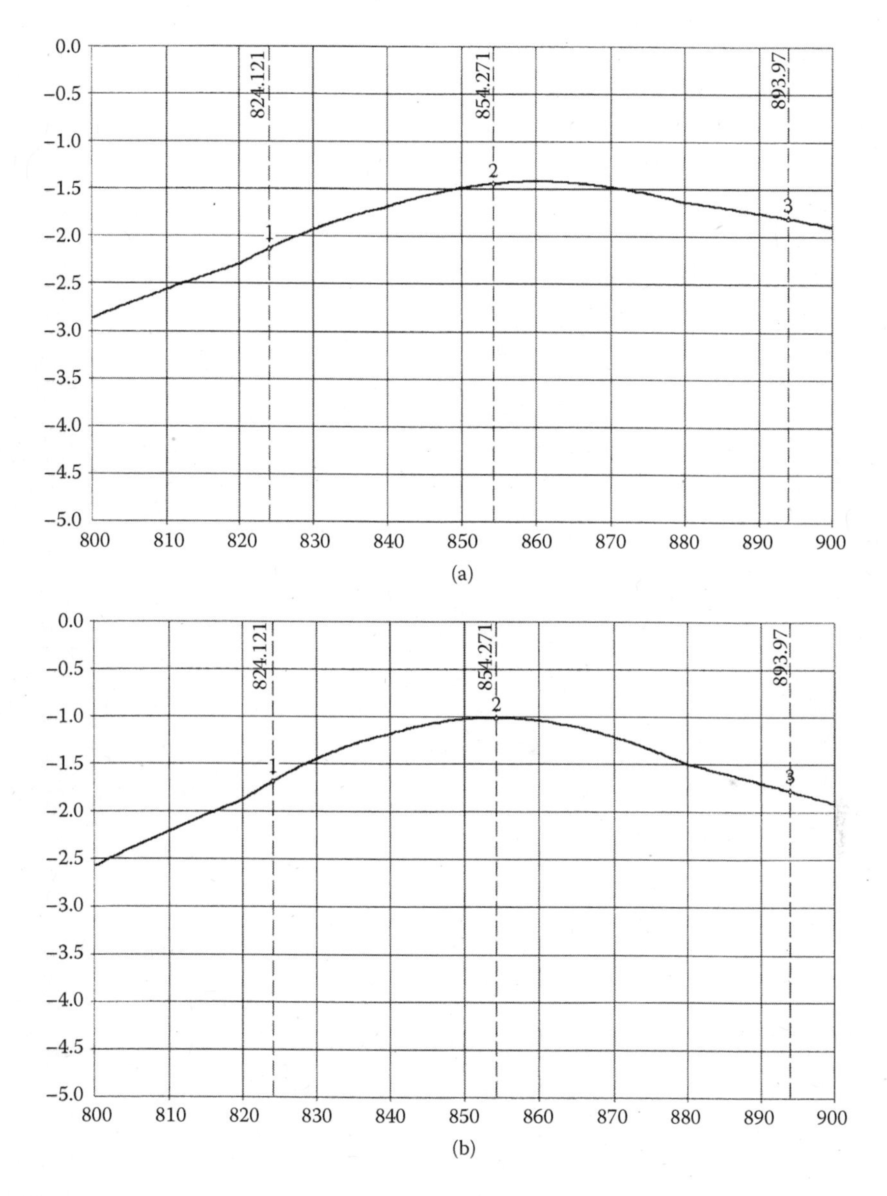

FIGURE 5.17
Insertion losses between antenna and receiver. (a) Lower band before matching. (b) Lower band after matching.

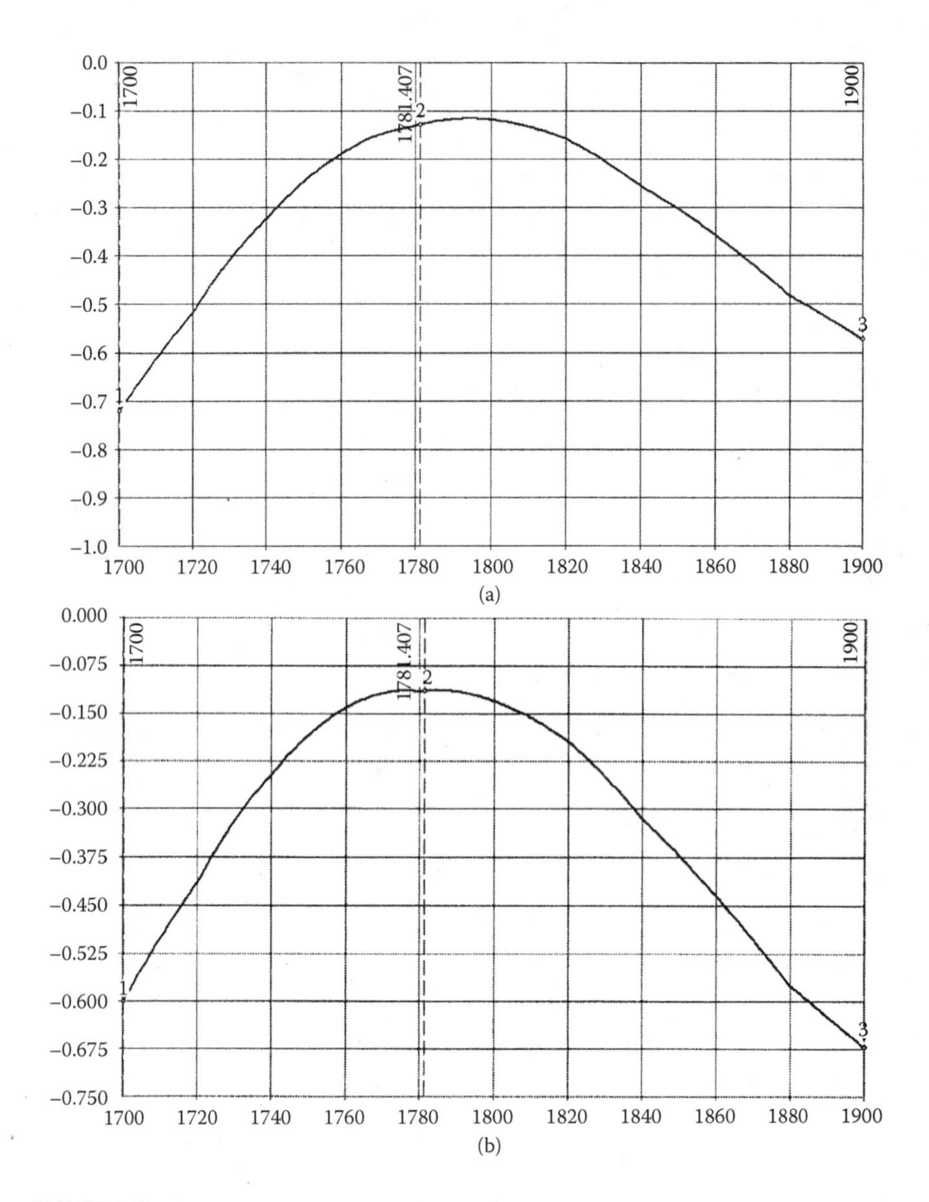

FIGURE 5.18
Insertion losses between antenna and receiver: (a) Upper band before matching; (b) Upper band after matching.

roof. The antenna is fed using a coaxial cable to the receiver. Height is 55 mm and width is 17mm. The Smith chart and VSWR of an antenna mounted on a 1 m squared metal screen are presented in Figure 5.22. The antenna has a gain in the 860 to 960 MHz range for vertical polarization in a horizontal plane of about –1 to –2 dBi, depending on the frequency point.

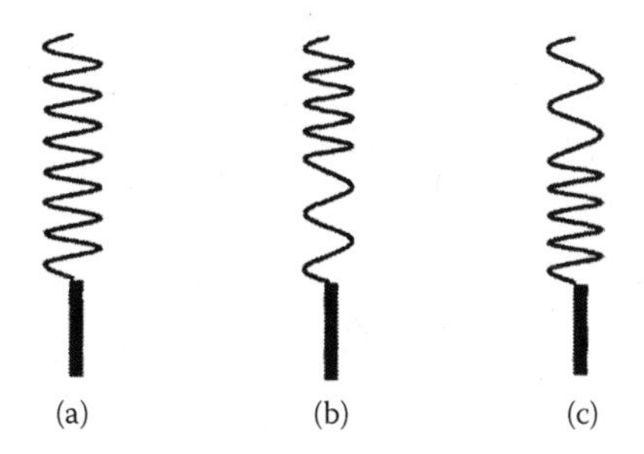

FIGURE 5.19
Helical antenna with nonuniform pitch. (From G. Zhou et al., *IEEE Antennas and Propagation Society International Symphosium*, 2000, IEEE Vol. 1. Copyright 2000 IEEE. With permission.)

5.5.2 Dual-Band Combined Design

Modification of the geometry shown in Figure 5.21 leads to the dual-band printed antenna; geometry and dimensions are shown in Figure 5.23. The modified antenna consists of two printed monopoles: The longest is responsible for radiation in the lower frequency band and the other determines radiation in the upper band. Antenna impedance and VSWR are shown in Figure 5.24. The monopole that radiates at the lower frequency band has the same geometry as that shown in Figure 5.21. The gain of this antenna in the upper frequency band is –3 to –2 dBi, depending on the frequency point; the gain in the lower band is the same as for the single-band design.

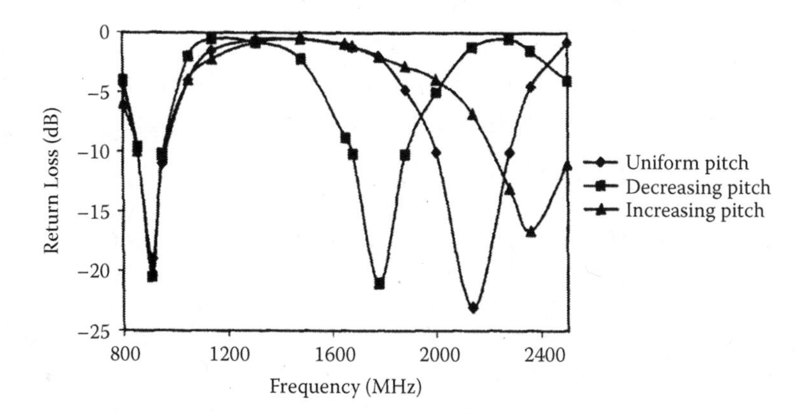

FIGURE 5.20
Return loss of helical nonuniform pitch antenna as function of pitch angle. (From G. Zhou et al., *IEEE Antennas and Propagation Society International Symphosium*, 2000, IEEE Vol. 1. Copyright 2000 IEEE. With permission.)

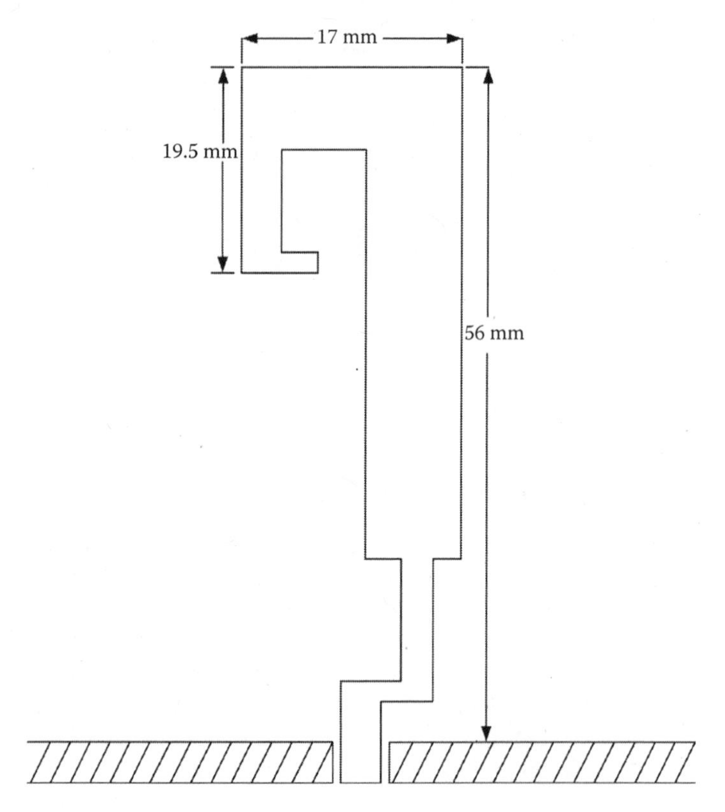

FIGURE 5.21
Single-band printed antenna design.

5.5.3 Tri-Band Design

Figure 5.25 shows antenna geometry that operates in the cellular frequency and Wi-Fi bands. It represents a modification of two previous designs: we added an additional monopole to handle radiation in the Wi-Fi band. The antenna impedance and VSWR of this antenna are presented in Figure 5.26. Because this antenna must operate with both cellular phone and Wi-Fi transceivers, the final design includes an electronic switch that connects an antenna with one of the transceivers, depending on the operating schedule. The antenna operating with VSWR less than 2:1 in bands 824 to 894 MHz and 1710 to 2700 MHz is presented in Figure 5.27 [26]. Total dimensions of the antenna are 57 mm × 41 mm.

5.5.4 CPW-Fed Multiband Design

Figure 5.28a presents a compact dual-band printed on a dielectric antenna [11]. The antenna is printed on a FR-4 substrate with a thickness of 1.6 mm

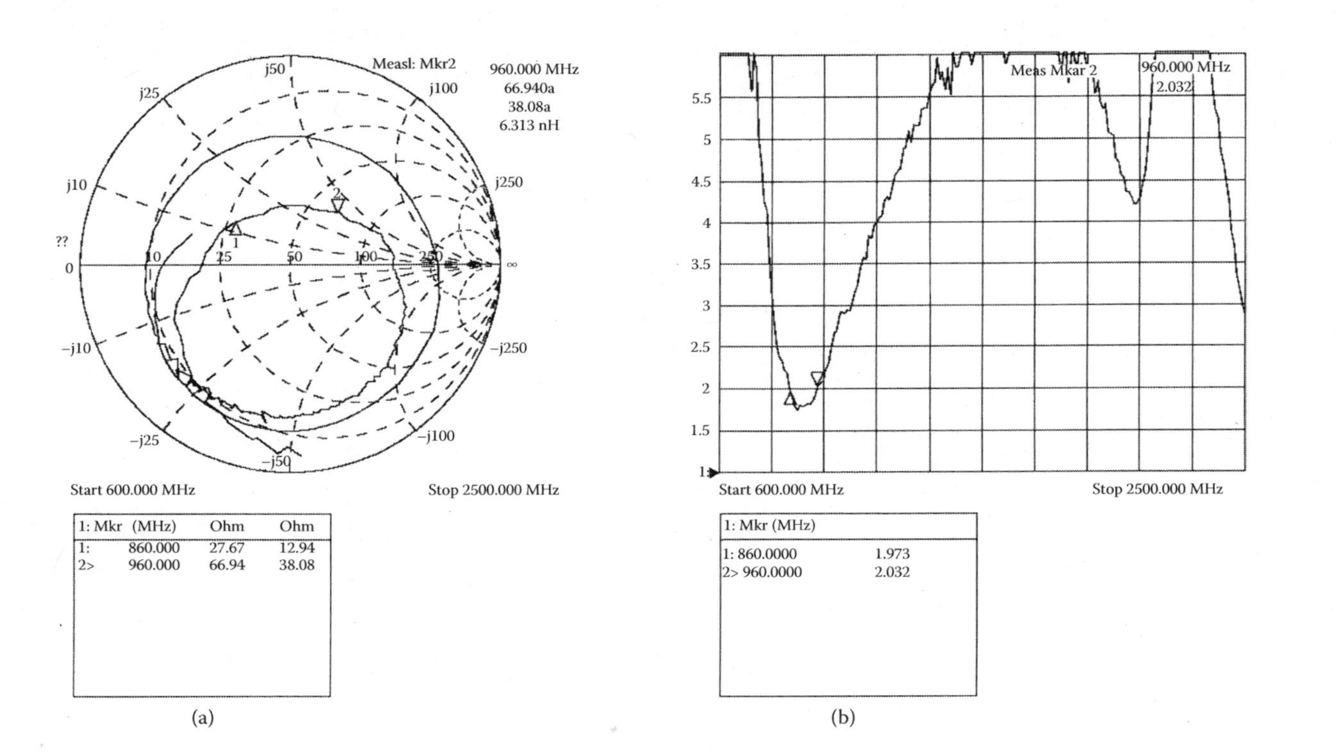

FIGURE 5.22
Output impedance and voltage standing wave ratio of single-band design: (a) Smith chart; (b) VSWR.

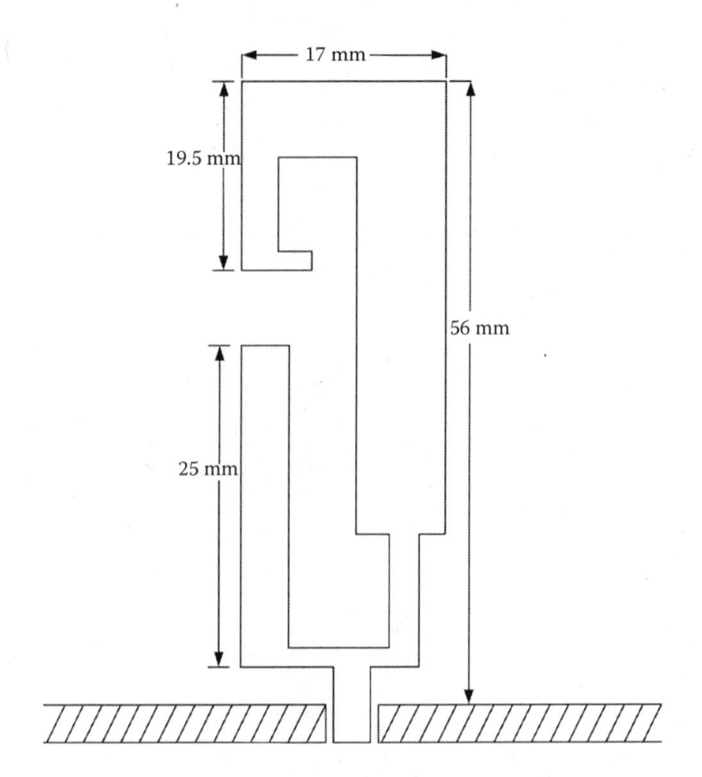

FIGURE 5.23
Geometry for dual-band application of antenna printed on dielectric board.

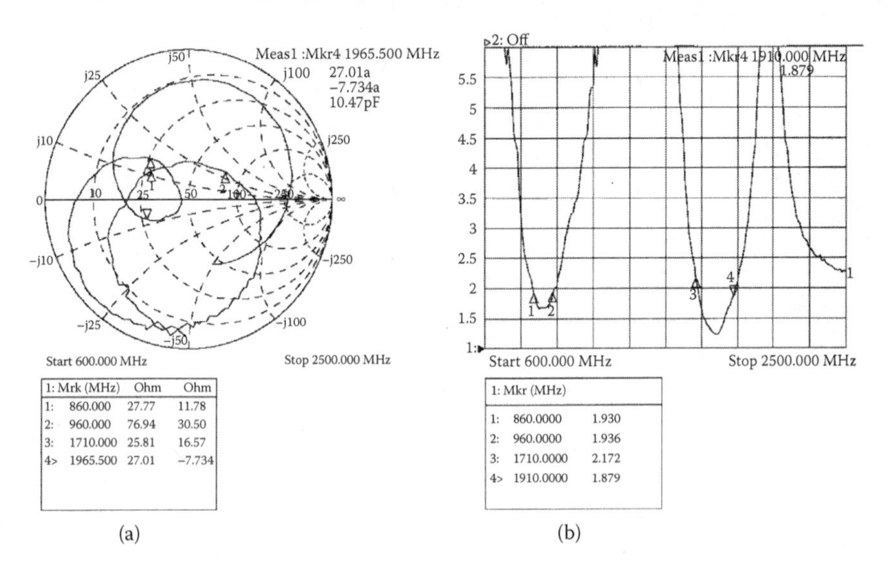

FIGURE 5.24
Impedance (a) and VSWR (b) of dual-band antenna.

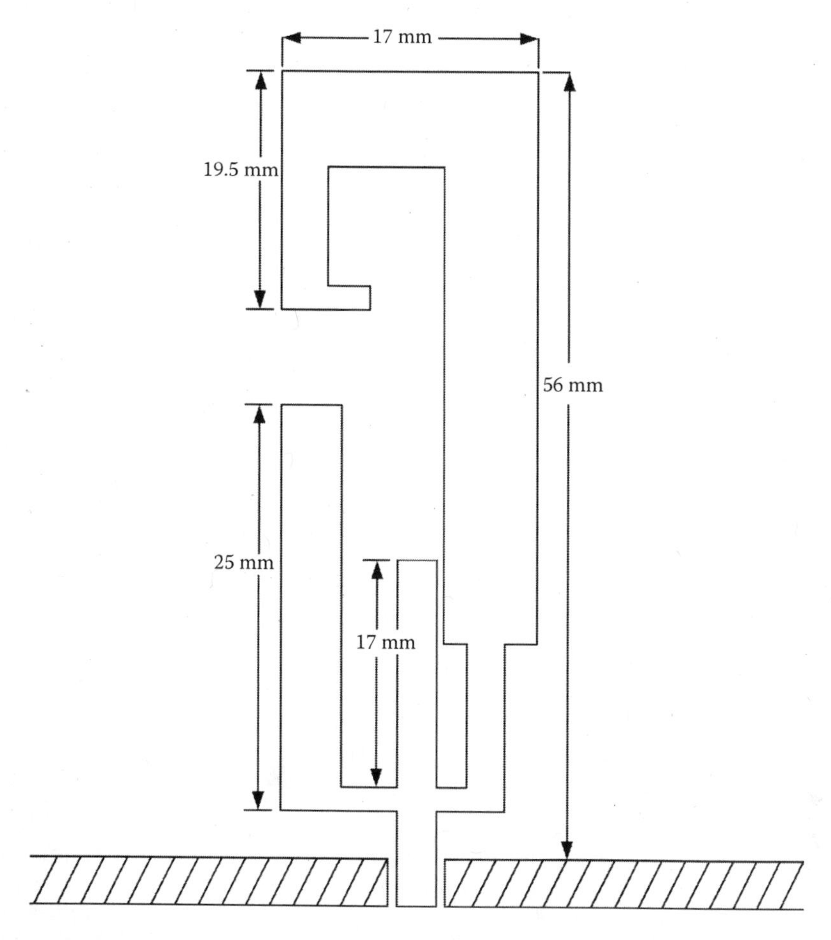

FIGURE 5.25
Topology of antenna operating in cellular and Wi-Fi bands.

and relative permittivity of 4.4. A 50 ohm CPW transmission line is used to excite the antenna. The meandered monopole antenna is 28 mm high and 11.5 mm wide. A 26 mm high conductor line extends from the end of the rectangular meander monopole with a 0.5 mm space between the line and monopole. The antenna covers two frequency bands: 875 to 965 MHz and 1706 to 1924 MHz, with return loss of more than 10 dB. The antenna can be attached to and grounded on a base that is connected to the vehicle roof. The peak gains are 1.3 dBi and 3 dBi in the low and upper bands, respectively.

The geometry of a CPW-fed slot-printed antenna for triple-band application [12] appears in Figure 5.28b. The total antenna height is ~68 mm, and the width is 60 mm. The return losses are more than 10 dB in both frequency ranges: 870 to 980 MHz and 1.74 to 1.99 GHz; the antenna gain is 1.08 to 2.37 dBi in the lower frequency range and 2.16 to 3.88 dBi in the upper band.

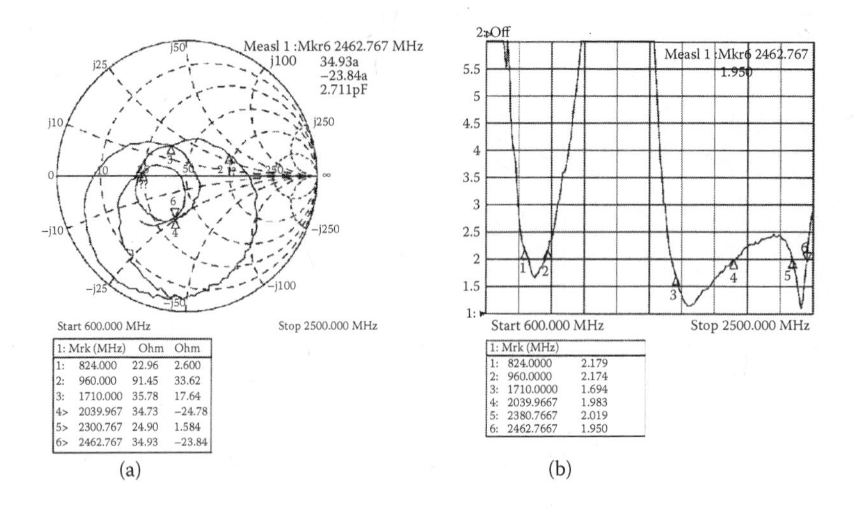

1: Mrk (MHz)	Ohm	Ohm
1: 824.000	22.96	2.600
2: 960.000	91.45	33.62
3: 1710.000	35.78	17.64
4> 2039.967	34.73	−24.78
5> 2300.767	24.90	1.584
6> 2462.767	34.93	−23.84

(a)

1: Mrk (MHz)	
1: 824.0000	2.179
2: 960.0000	2.174
3: 1710.0000	1.694
4: 2039.9667	1.983
5: 2380.7667	2.019
6: 2462.7667	1.950

(b)

FIGURE 5.26
Impedance (a) and VSWR (b) for combined cellular phone and Wi-Fi antenna printed on dielectric board.

5.5.5 Planar Inverted F Antenna

A planar inverted F antenna (PIFA) design [13] for automotive application is shown in Figure 5.29. Figure 5.29a shows installation on a roof-mount base with a GPS patch and an AM/FM mast (not shown). Figure 5.29b shows mounting with a feed module on a car windshield. The topology shown in Figure 5.29c is printed on a 0.8 mm-thick FR4 substrate. The antenna has a ground plane, a feed pin excited by a cable soldered between the ground and the end of the pin, a metal plate at the top of the pattern, and a short connection between the ground and top plate on the right-hand side. For the roof-mounted design (Figure 5.29a), the bottom ground strip can be eliminated because the metal roof mount base forms the ground. This reduces antenna

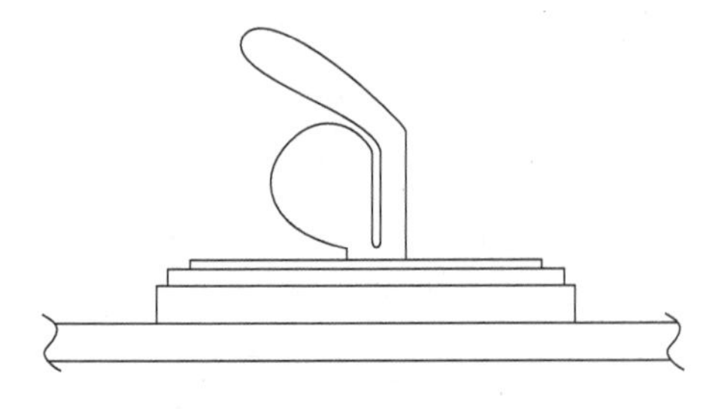

FIGURE 5.27
CPW-fed dual-band design.

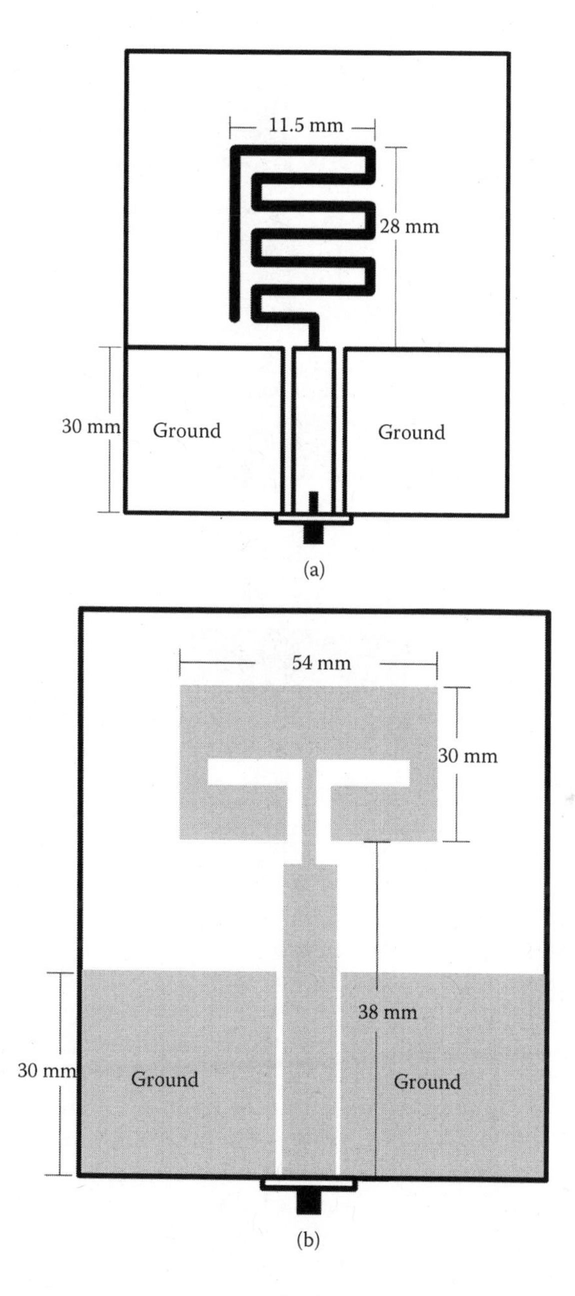

FIGURE 5.28
Geometry of multiband printed antenna for cellular application: (a) dual band design, (b) triple band geometry (From W. Chung and C. Huang, *Microwave and Optical technology Letters*, Vol. 44, No. 6, 2005. Copyright 2005. Reprinted with permission of Wiley-Blackwell, Inc.)

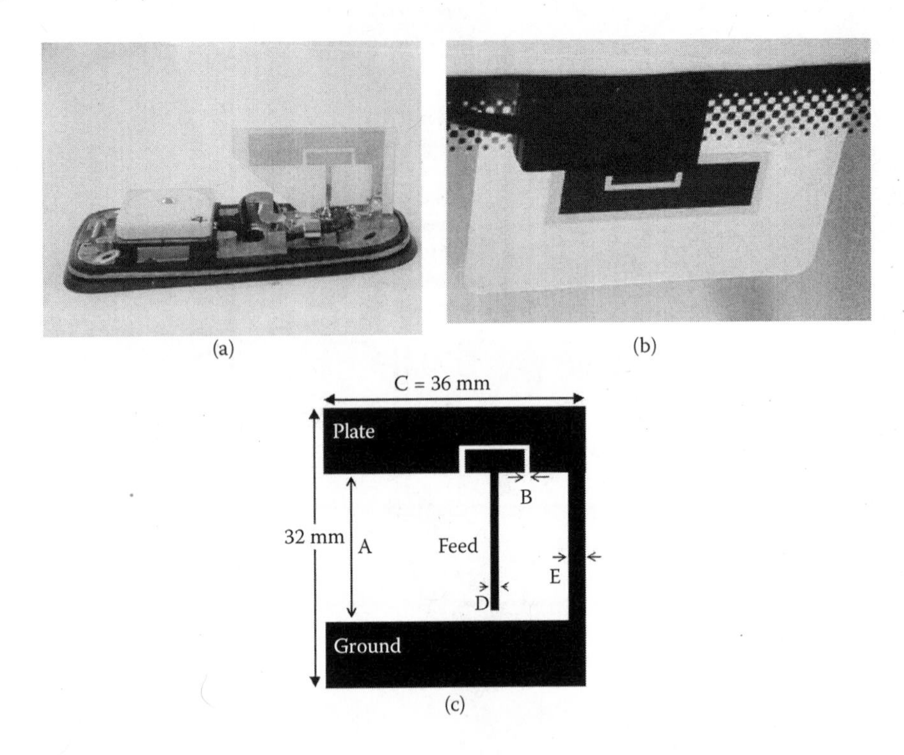

FIGURE 5.29
Planar inverted F antenna (PIFA) geometry for automotive application: (a) roof mount installation; (b) glass installation; (c) planar topology. (From R. Leelaratne and R. Langley, *IEEE Transactions on Vehicular Technology*, Vol. 54, No. 2, 2005. Copyright 2005 IEEE. With permission.)

height—an essential factor for automotive applications. A PIFA installed at the top of the front windshield (Figure 5.29b) is grounded to the car body.

Due to losses caused by glass, the antenna is mounted approximately 5 mm from the windshield to reduce induced field interaction. The distance was achieved by a spacer attached to the plastic cover around the antenna. Variations of the dimensions A, D, B, and E (Figure 5.29c) shift the resonance frequency and return loss. Generally, such antennas must be tuned to meet the requirements. The measured return losses for planar system roof-mount configuration and windshield installation are shown in Figure 5.30a and b, respectively.

The measured in horizontal plane vertically polarized radiation patterns for roof mounting are shown in Figure 5.31a. Figure 5.31b depicts vertically polarized radiation patterns in horizontal plane for a planar antenna mounted on a windshield. The screen-mounted antenna radiation pattern shows more variations in directivity caused by the interaction of the antenna with the shape of the metal car body. The gain averaged over the 360 degrees of a roof-mounted system was affected by installation position and was measured at

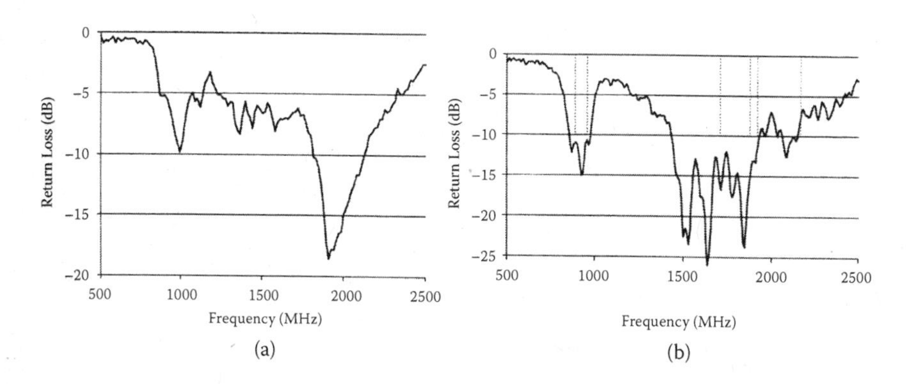

FIGURE 5.30
Return losses: (a) roof mount; (b) windshield mount. (From R. Leelaratne and R. Langley, *IEEE Transactions on Vehicular Technology*, Vol. 54, No. 2, 2005. Copyright 2005 IEEE. With permission.)

–1 dBi at 925 MHz, –2 dBi at 1800 MHz, and –3 dBi at 2100 MHz. The mean effective gain values for screen mounting were –2.6 dBi at 900 MHz, –3.1 dBi at 1800 MHz, and –2.2 dBi at the upper most band.

5.5.6 Hidden Printed Dipole

A commercial double-band dipole printed on dielectric is available from the Antenna Factor Company and can be used for hidden automotive applications. The antenna is 13 cm long and 19 cm wide. The main dipole operates at a low band and a U-shaped portion handles upper frequency band operation. The free space gain of the antenna at low frequency 850 MHz (band 860 to 960) is about 2.0 dBi and about 3 dBi at 1900 MHz (band 1770 to 1880 MHz), respectively.

5.5.7 Printed-on-Glass Design

A cellular phone antenna may be designed as a metal strip conductor on a car window [14,15]. One version printed on rear window glass is shown in Figure 5.32. The length of the vertical metal conductor is about 5.5 cm—approximately 30% less than the free space quarter wavelength. This is made possible by the dielectric effect of the glass that modifies the performance of the vertical conductor. Figure 5.33 presents the typical measured directionality of the antenna at 900 MHz frequency (horizontal plane for vertical polarization).

5.5.8 Antenova Series

Antenova [16], a leading developer and supplier of high performance antennas and RF antenna modules, presents a range of printed-on dielectric

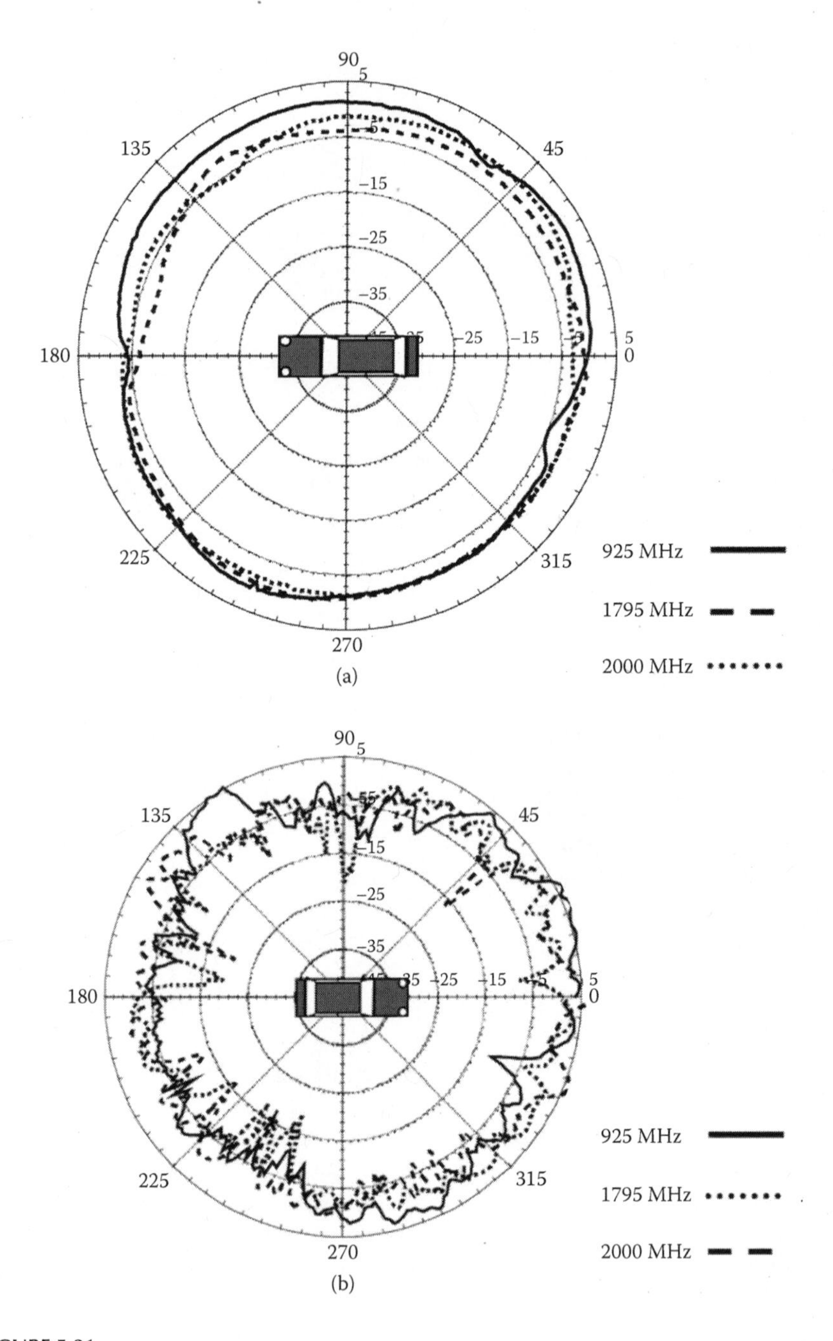

FIGURE 5.31
Radiation patterns in horizontal plane for vertical polarization: (a) roof mount; (b) glass mount. (From R. Leelaratne and R. Langley, *IEEE Transactions on Vehicular Technology*, Vol. 54, No. 2, 2005. Copyright 2005 IEEE. With permission.)

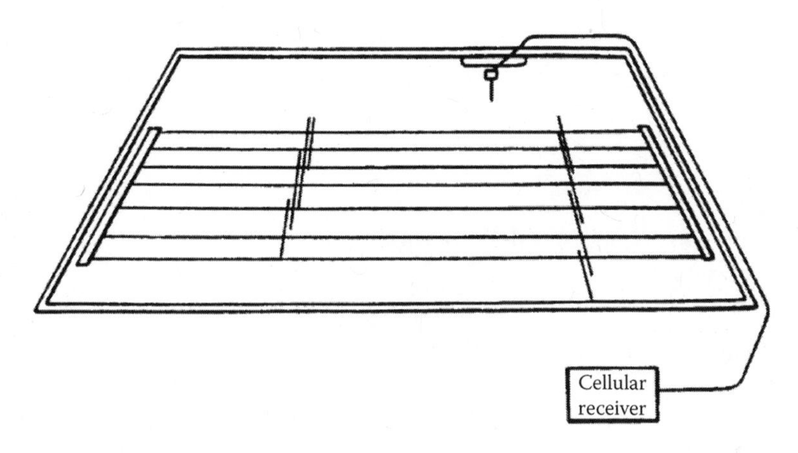

FIGURE 5.32
Cellular phone antenna printed on rear window.

substrate FR4 (0.8 or 1.6 mm thickness) antennas for cellular phone applications. Such an antenna may be hidden in the interior of a car, for example, under the front dash, in the glove compartment, and inside the backs of seats. The company offers (a) the Quadnova (quadruple band, 90 mm × 50 mm, 0.8 mm thick), (b) the 3G Trinova (triple band, 90 mm × 50 mm, 1.6-mm thick), and (c) the Pentanova (five bands, 40 mm × 15 mm × 7 mm). The antennas are

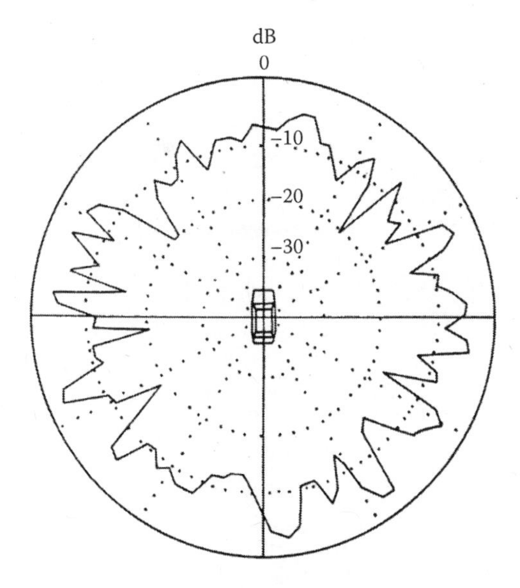

FIGURE 5.33
Radiation pattern of printed-on-glass geometry.

proposed to operate in all cellular frequency bands. Their gains, depending on the dimensions and frequency bands, are within 1 to 4 dBi.

5.6 Dual-Band Inverted F Design

Figure 5.34 presents the detailed geometry of a dual-band planar inverted F antenna (PIFA) [17]. Originally, the antenna was designed for mobile phones but can be used for vehicle roof applications. The antenna has two-dimensional planar geometry. The radiation patch is printed on a 0.8 mm-thick FR4 substrate 8 mm above the ground. The radiation patch consists of strip 1,

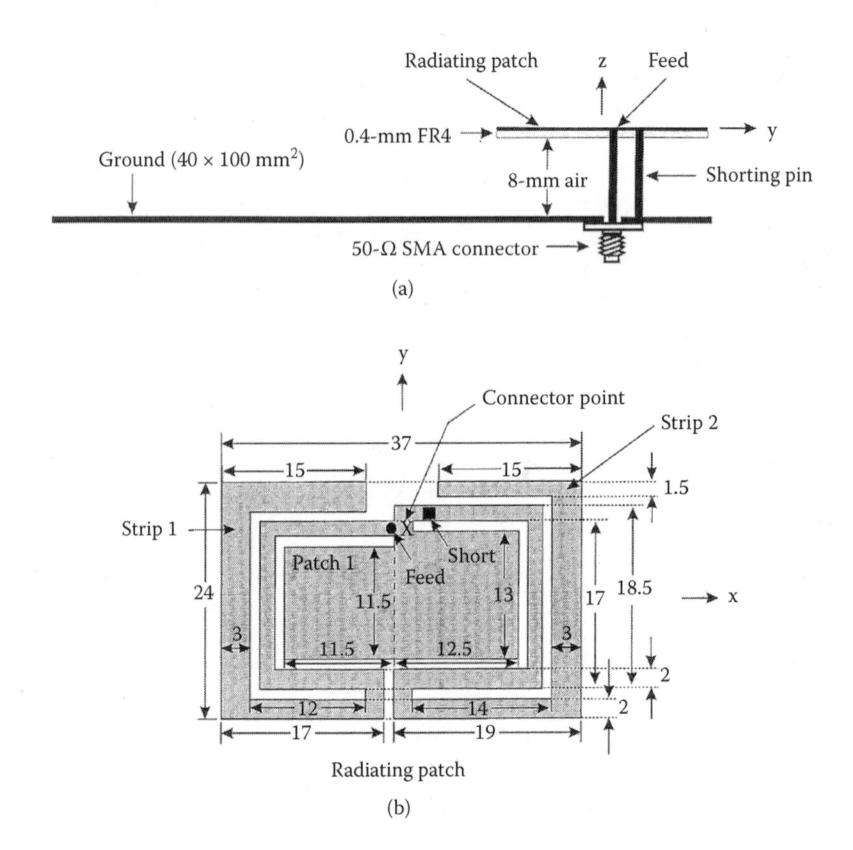

FIGURE 5.34
Geometry and dimensions of dual-band planar inverted F antenna for GSM and DCS applications: (a) side view; (b) arrangement of radiating patch. (From S. Yeh et al., *IEEE Transactions on Antennas and Propagation*, Vol. 51, No. 5, 2003. Copyright 2003 IEEE. With permission.)

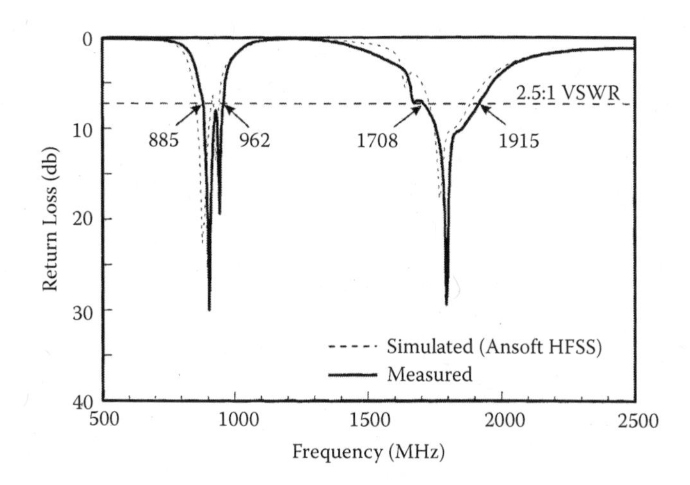

FIGURE 5.35
Measured and simulated return loss of planar inverted F antenna. (From S. Yeh et al., *IEEE Transactions on Antennas and Propagation*, Vol. 51, No. 5, 2003. Copyright 2003 IEEE. With permission.)

strip 2, and patch 1. The connection point, feed point, and short to the ground are shown.

Figure 5.35 presents the simulated and measured return loss and bandwidth of the antenna. Simulated results were obtained with Ansoft HFSS simulation software [18]. For the lower frequency band, an impedance bandwidth determined from 2.5:1 VSWR is about 77 MHz (885 to 962 MHz). In the upper frequency, the impedance bandwidth, determined from the same ratio, is about 207 MHz (708 to 1915 MHz). The measured antenna radiation patterns are presented in Figure 5.36. For the lower frequency band, the peak antenna gain reaches 2.7 dBi, and for the upper band, it is about 2.1 dBi. These measurements do not consider possible variations arising from an interior mount.

5.7 Rear View Mirror Mount

A cellular phone antenna [19] mounted in a rear view mirror is shown in Figure 5.37. The designed antenna has the following main performance parameters: 800 to 930 MHz frequency band, VSWR less than 2:1, and gain of 3 dBi.

A cellular phone antenna also can be mounted under a dashboard, in a side view mirror, or in the glove compartment. The advantages of these designs include enhancement of the aesthetic appeal of a vehicle, prevention of damage in car washes, lack of visibility to vandals, and minimized vision

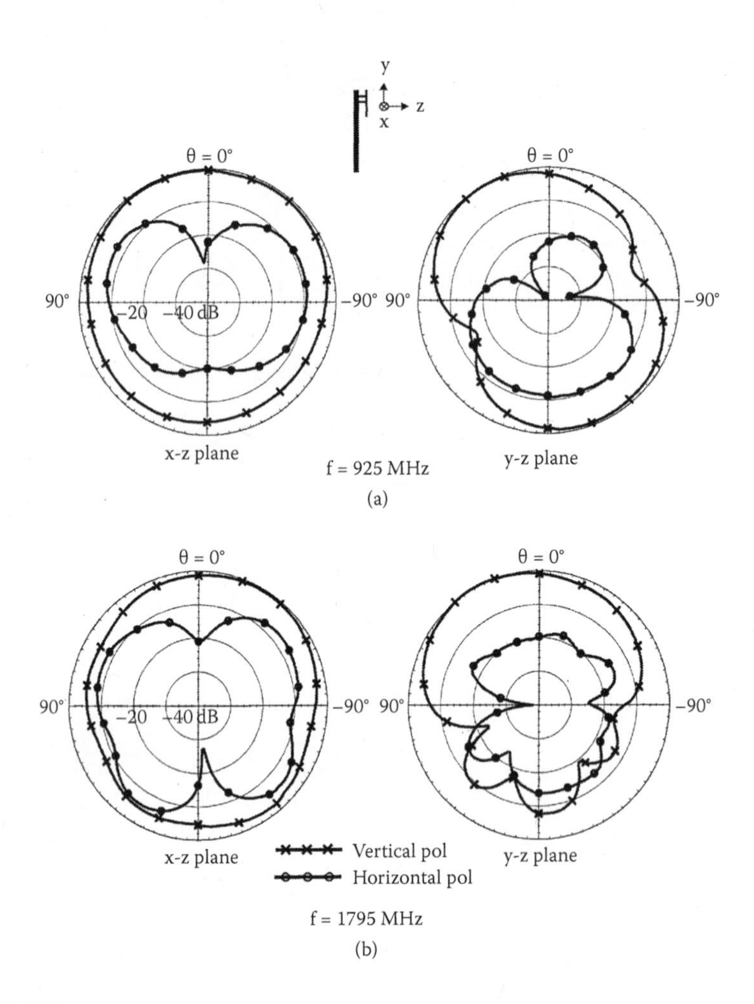

FIGURE 5.36
Measured radiation patterns (a) 925 MHz; (b) 1795 MHz. (From S. Yeh et al., *IEEE Transactions on Antennas and Propagation*, Vol. 51, No. 5, 2003. Copyright 2003 IEEE. With permission.)

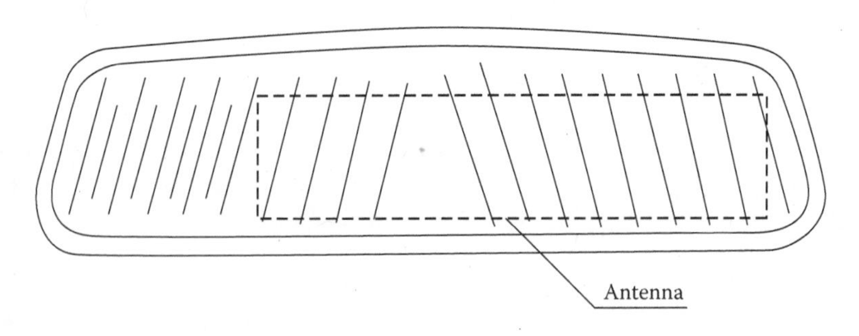

FIGURE 5.37
Antenna mounted in rear view mirror.

blockage for drivers and passengers. Of course, car body design affects antenna performance; in particular, a car body reduces antenna gain and eventually reduces the communication range (generally more than 3 dB on average around a car). A bidirectional amplifier with a reduced gain antenna can be utilized but must follow FCC regulations prohibiting cellular amplifier and antenna gains higher than the certain level, depending on the frequency band.

5.8 Bumper Installation

Installation of an antenna on a bumper is common for emergency call systems. Statistical surveys have shown that an antenna mounted on a car roof or on a bumper will survive intact in an incident. Figure 5.38 demonstrates the radiation pattern [13] of a cellular antenna mounted behind the center of a rear bumper; the car is facing in a 0 degree direction. The radiation pattern contrasts with that of a roof-mounted antenna that produces much less scattering of signals from the car body. To ensure sufficient coverage 360 degrees around a vehicle, a second antenna should be mounted on the front bumper.

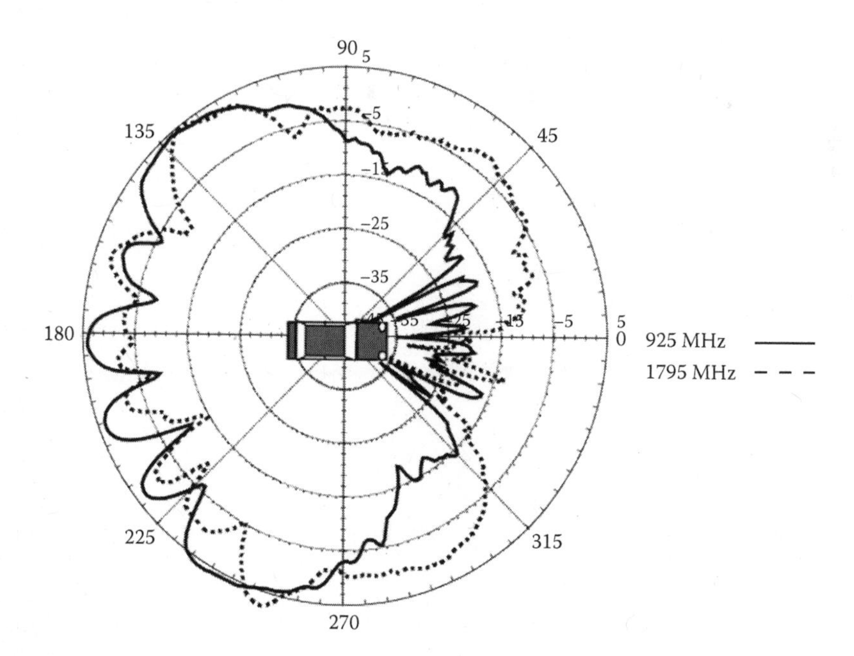

FIGURE 5.38
Measured radiation pattern for cellular antenna mounted on rear bumper. (From R. Leelaratne and R. Langley, *IEEE Transactions on Vehicular Technology*, Vol. 54, No. 2, 2005. Copyright 2005 IEEE. With permission.)

5.9 Combined FM/PCS Design

Figure 5.39 shows an integrated FM/PCS antenna [20]—a compact and attractive design for automotive use. The antenna consists of a main helical portion operating in the normal mode intended for AM/FM reception and a cellular PCS portion operating in the 1.8 GHz range. The main AM/FM portion is a nonuniform helix with a total length of 150 mm and a 5.8 mm helical diameter. The monopole for the PCS operation is designed as a meandered metal conductor to maintain the low profile. The monopole has a metal thickness (T) of 0.5 mm with an overall height (H) of 14.6 mm. The total length is 66 mm so that the monopole resonates at 1800 MHz. Technically, the design requires a splitter to provide AM/FM reception and a PCS branch to connect with the cellular receiver. Passive elements inserted in both branches provide the best match between the antenna and electronic equipment.

Figure 5.40 shows the simulated and measurement return loss of the antenna in the PCS band. The 10 dB loss bandwidth is larger than the PCS bandwidth of 120 MHz, from 1.75 to 1.87 GHz. Compared with simulated results, measurement showed a shift of approximately 50 MHz in the resonant frequency for the monopole antenna only and an additional shift of 100 MHz when the monopole was loaded with the helical antenna portion.

Figure 5.41 shows vertically polarized radiation patterns of the antenna system measured with a finite size ground plane. Figure 5.41a shows radiation patterns in the PCS frequency band with a maximum gain of 0.9 dBi. The measured radiation patterns in the FM frequency range appear in Figure 5.41b. The maximum measured gain in the FM frequency range with

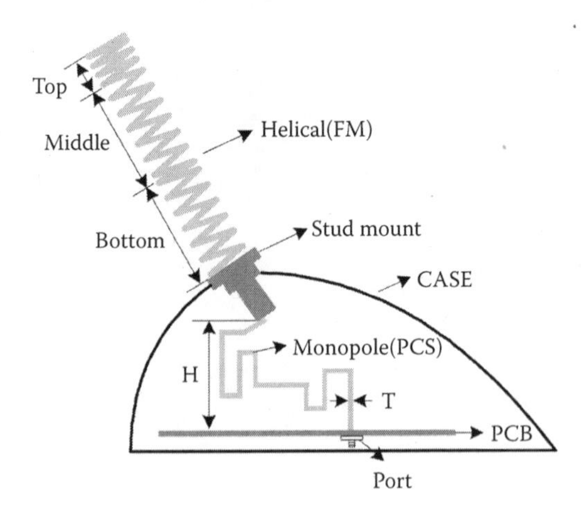

FIGURE 5.39
Integrated FM/PCS antenna design for roof application. (From Y. Hong et al., *Antennas and Propagation International Symposium*, 2007 IEEE. Copyright 2007 IEEE. With permission.)

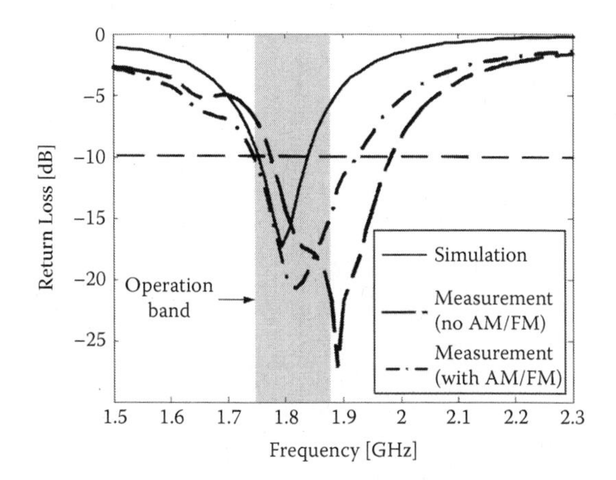

FIGURE 5.40
Measured and simulated losses of PCS antenna. (From Y. Hong et al., *Antennas and Propagation International Symposium*, 2007 IEEE. Copyright 2007 IEEE. With permission.)

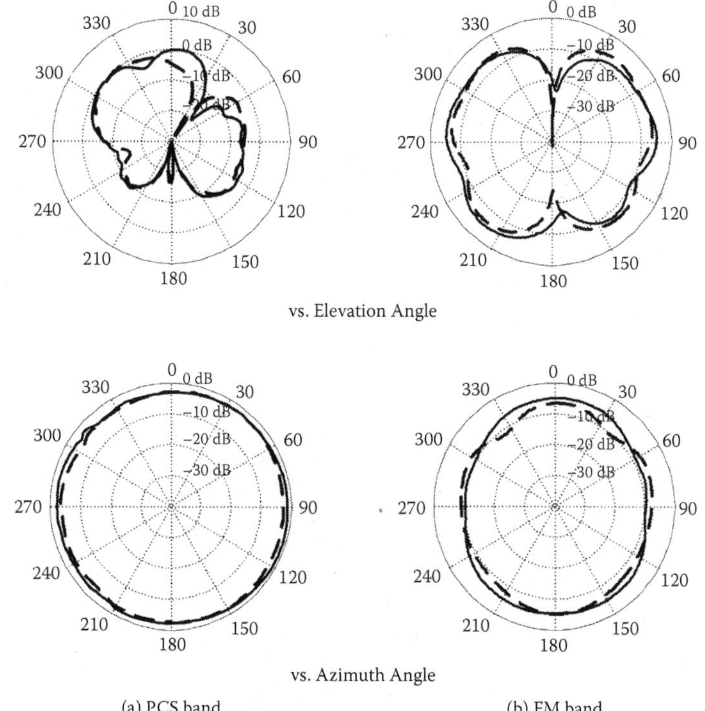

FIGURE 5.41
Measured radiation patterns of multifunctional antenna: (a) PCS band; (b) FM band. (From Y. Hong et al., *Antennas and Propagation International Symposium*, 2007 IEEE. Copyright 2007 IEEE. With permission.)

an integrated low noise amplifier (LNA) is –5.6 dBi. The LNA has a noise equal to 4.6 dB and 11 dB gain in the FM frequency range.

5.10 Integrated AM/FM/AMPS/PCS Design

Smarteq Wireless Telecom, a leader among automotive OEM antenna manufacturers [21] now markets the SportFlex™ antenna that functions in the AM/FM/AMPS/PCS frequency bands. AMPS band operation is provided by a simple 70 mm long monopole antenna mounted in a plastic tube that supports the main helical AM/FM antenna segment and is connected with the bottom end of the AM/FM antenna. The design provides more than 10 dB return loss over the entire cell frequency bands. The total height of the antenna including the AM/FM portion is about 47 cm. The AM/FM portion has an LNA with a noise level below 5 dB over the entire broadcast frequency range.

5.11 Integrated PCS/RKE Design

As shown in Reference [22], an antenna for PCS application can be integrated with a remote keyless entry (RKE) antenna. Figure 5.42 presents the geometry of a helical–monopole combined RKE/PCS antenna. The antenna, placed along the axis of a coaxial line, is fed through the ground plane. The helix is designed to operate in normal mode and is combined directly with the monopole antenna. In the helix–monopole element, L_m denotes the length of the monopole and the diameter of the wire is D_m. The helix with a total stretched length L_h, diameter D_h, and pitch angle α is connected to the bottom of the monopole.

Dual-band RKE/PCS operation can be achieved by adjusting the above geometric parameters for the helix–monopole combined antenna. However, the coupling effect of the two elements is very critical for determining antenna performance since the elements are placed in close proximity to each other. To improve impedance performance, a circular disk is loaded at the bottom of the monopole antenna. By selecting the proper disk radius b and the separation distance between the disk and the ground plane d, good impedance matching characteristics for each frequency band can be obtained. Based on measurement results, VSWR is less than 2 for the RKES band frequency (447.7375 MHz) and the PCS band (1750 to 1870 MHz). The gain is about 0.9 dBi in the RKES band and 2.65 dBi for the PCS.

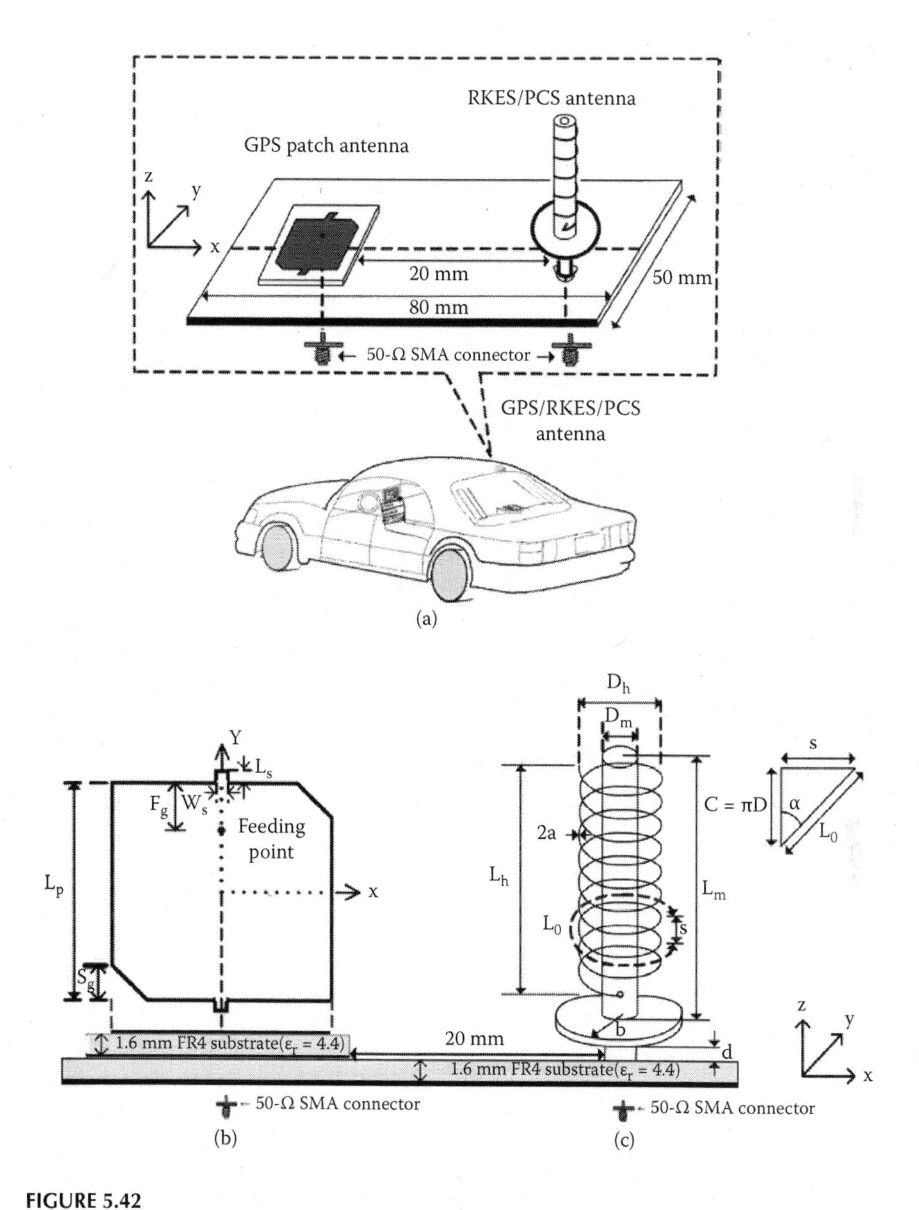

FIGURE 5.42
Integrated PCS/GPS/RKES helical monopole mounted on car roof. (From K. Oh, B. Kim, and J. Choi, *IEEE Microwave and Wireless Component Letters*, Vol. 15, No. 4, 2005. Copyright 2005 IEEE. With permission.)

5.12 Cellular Diversity Systems

5.12.1 Spatial, Polarization, and Pattern Diversity

Dietrich et al. [23] compare the efficiency of spatial, polarization, and pattern diversity for hand-held terminal applications. Their results, however, can be considered baselines for designing car cell diversity systems. The measurement results are shown for the 2050 MHz frequency band closed to cellular PCS/universal mobile telecommunications system (UMTS) applications. The investigation considered the maximum ratio combination and selection diversity.

As mentioned in Chapter 3, antenna spacing, branch envelope correlation, $\rho_e = \rho^2$ and diversity gain are the main parameters. Diversity gain [Equation (3.37 in Chapter 3] is higher when the signals received by different branches are uncorrelated (correlation coefficient = 0) and decrease as the coefficient increases. Analytical diversity gain expressions for various combining techniques in Rayleigh fading channels as a function of branch correlation and different outage are presented in Chapter 3 [Equations (3.40) and (3.41)].

Three different antenna configurations [23] were investigated. The space diversity design is shown in Figure 5.43a, the polarization diversity configuration in Figure 5.43b, and the pattern diversity configuration in Figure 5.43c. Figure 5.43a shows two identical dipoles spaced at distance *d*; spacing varied from 0.1 to 0.5 of the wavelength. A diversity polarized antenna (Figure 5.43b) consists of a vertically located dipole and a horizontally located printed "big wheel" antenna separated vertically by 0.3 of the wavelength. A big wheel has an omnidirectional radiation pattern in the horizontal plane and is

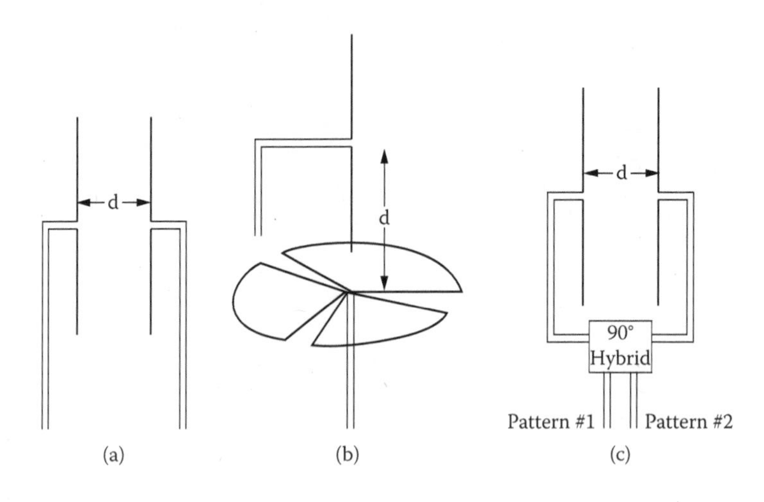

FIGURE 5.43
Diversity antenna configurations: (a) spatial; (b) polarization; (c) pattern. (From C. Dietrich et al., *IEEE Transactions on Antennas and Propagation*, Vol. 49, No. 9, 2001. Copyright 2001 IEEE. With permission.)

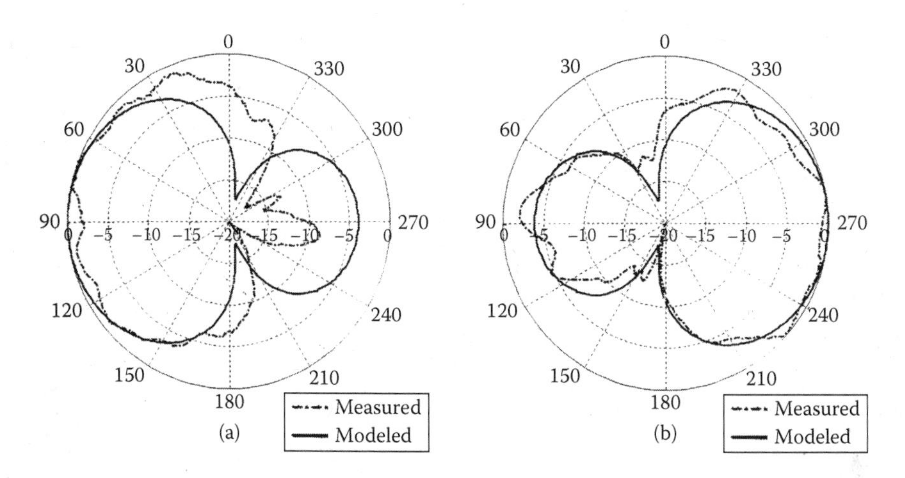

FIGURE 5.44
Measured and simulated directionality plots for antenna with pattern diversity. (From C. Dietrich et al., *IEEE Transactions on Antennas and Propagation*, Vol. 49, No. 9, 2001. Copyright 2001 IEEE. With permission.)

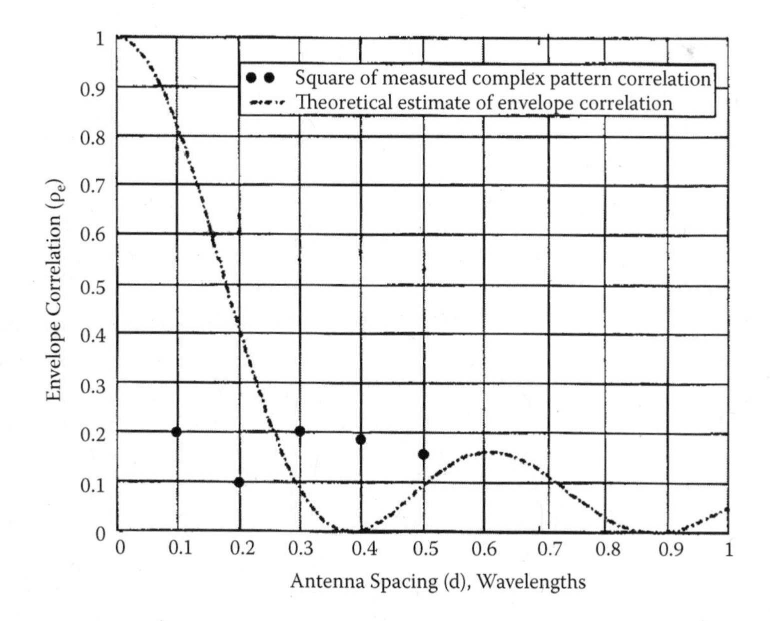

FIGURE 5.45
Correlation coefficient as function of antenna spacing. (From C. Dietrich et al., *IEEE Transactions on Antennas and Propagation*, Vol. 49, No. 9, 2001. Copyright 2001 IEEE. With permission.)

horizontally polarized in the azimuth plane. Pattern diversity involves two dipoles connected to a 90 degree hybrid as shown in Figure 5.43c, with d fixed at 0.25 of the wavelength. This yields directional patterns with opposing beam maxima (Figure 5.44a and b).

Figure 5.45 shows the envelope correlation calculated from the results of spatial diversity experiments. The measured correlations are below the theoretical curve for small antenna spacing (smaller than 0.2 λ) because the theoretical curve is calculated by assuming omnidirectional patterns. In reality, each antenna exhibits a different pattern due to distortion caused by mutual coupling. Therefore, the relative weights of the incoming multipath components received by each antenna are different, even when the antennas are closely spaced and the phases of the multipath components are similar. This reduces the probability that the signals received at both antennas will fade simultaneously.

The channels exhibited fading random signals that were approximately Rician with power ratios K ranging from 0.6 to 2. Rician distribution of the received signal is given by:

$$p(r) = \frac{r}{\sigma^2} \cdot e^{-\frac{(r^2+A^2)}{2\cdot\sigma^2}} \cdot I_0\left(\frac{A\cdot r}{\sigma^2}\right) \quad (A \geq 0, r \geq 0) \tag{5.1}$$

Rician factor K is defined as the ratio of the signal power in the dominant component over the local mean scattered power. In Rician fading channels, the fading is less, and the diversity gain is lower than in the Rayleigh fading channel.

$$K = \frac{A^2}{2\cdot\sigma^2} \quad (K(dB) = 20 \cdot \log(A) - 20 \cdot \log(\sigma) - 3) \tag{5.2}$$

Rayleigh fading is recovered for $K = 0$.

$$p(r, K=0) = \frac{r^2}{\sigma^2} \cdot e^{-\left(\frac{r^2}{2\cdot\sigma^2}\right)} \tag{5.3}$$

Measurements for the spatial, polarization, and pattern diversity techniques were generated for line of sight (LOS), non line of sight (NLOS), urban, suburban, and rural environments and other factors. Diversity gains of 7 to 10 dB (99% reliability) for spatial diversity with maximal combining technique for NLOS were measured. Comparable gains of 6 to 11 dB were obtained using polarization and pattern diversity, with small antenna spacings of 0.25 to 0.3 wavelength.

Diversity gains calculated for selection diversity were typically about 1.5 dB lower than those for a maximal ratio combined for all antenna configurations.

Polarization diversity configurations have the advantage over spatial diversity of providing gain in LOS channels with very few multipaths. The measurements indicate that polarization diversity configurations can increase the signal-to-noise ratio (SNR) by 12 dB or more in certain cases by eliminating polarization mismatch.

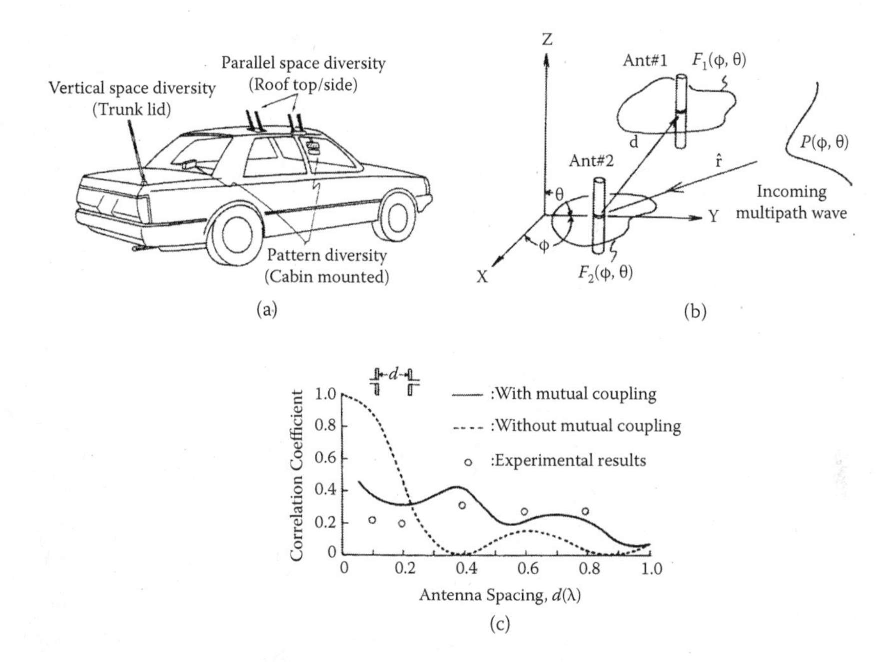

FIGURE 5.46
(a) Options for diversity antenna mounting. (b) Antenna configuration. (c) Correlation coefficient versus spacing for parallel antennas mounted on car roof. (From Y. Ebine and Y. Yamada, *IEEE Vehicular Technology Conference*, 1988 IEEE, 38th June 1988. Copyright 1988 IEEE. With permission.)

5.12.2 Monopoles for Diversity Reception

A few diversity antenna installations [24] are shown in Figure 5.46a. Three types of antenna mounted on cars are presented in the figure. Trunk mounting uses a vertical space diversity system. Two parallel diversity monopoles are mounted on the roof of the car. For the cabin-mounted type, pattern diversity is employed. One antenna is located on the back of the interior rear view mirror and another is mounted on the rear shelf. As noted earlier, the correlation coefficient is the key for estimating the improvement available from a diversity system. We know [24] that the envelope correlation coefficient between two spacing antennas is given by:

$$\rho_e = \frac{\left| \int_0^{2\pi} \int_{-\pi/2}^{\pi/2} F_1^*(\varphi,\theta) F_2(\varphi,\theta) P(\varphi,\theta) \cdot \exp(j \cdot \beta \cdot \bar{d}\,\bar{r}) d\varphi \cdot d\theta \right|^2}{\int_0^{2\pi} \int_{-\pi/2}^{\pi/2} F_1^*(\varphi,\theta) F_1(\varphi,\theta) P(\varphi,\theta) \cdot d\varphi \cdot d\theta \cdot \int_0^{2\pi} \int_{-\pi/2}^{\pi/2} F_2^*(\varphi,\theta) F_2(\varphi,\theta) P(\varphi,\theta) \cdot d\varphi \cdot d\theta} \tag{5.4}$$

where $P(\varphi,\theta)$ = distribution of incoming multipath waves, * determines the complex conjugate, $\beta = \frac{2\pi}{\lambda}$ = wave number, and $\bar{d}$ = the vector that determines the space between antennas (Figure 5.46b). $F_1(\varphi,\theta)$ and $F_2(\varphi,\theta)$ are the radiation patterns of the first and second antennas. Figure 5.46c shows the correlation coefficient for two parallel roof-mounted diversity monopole antennas. Plots are presented under the following assumptions.

Random distribution of incoming waves $P(\varphi,\theta)$ does not depend on φ and is given by:

$$P(\theta)=\frac{1}{\sqrt{2\pi S^2}}\exp\left[-\frac{(\theta-\theta_i)^2}{2S^2}\right] \tag{5.5}$$

where θ_i = direction of principal incoming waves and S = standard deviation.

The radiation patterns in the horizontal plane (vertical polarization) depend on the coupling of the antenna elements [25]:

$$F_{1,2}(\varphi)=1-\frac{z_{12}}{z_{ii}}\exp(j\cdot\beta\cdot d\cdot\cos\varphi) \tag{5.6}$$

where d = antenna spacing, z_{12} = mutual impedance, and z_{ii} (i = 1, 2) is the self-impedance.

The measurements were made in an urban area. The dotted curve shows the result when z_{12} = 0. The experimental results agree with the solid line more efficiently, particularly when the spacing between the antenna elements is less than 0.2 λ. This indicates that diversity effect can be obtained by spacing small antenna elements. The results obtained for such small spacing (less than 0.2 of wavelength) are similar to those presented in Figure 5.45 for hand-held terminals.

Details of two trunk lid diversity antennas are demonstrated in Figure 5.47. A trunk lid system consists of two sleeve antennas placed on the same axis as shown in the figure. To avoid current leakage, each antenna is equipped with a $\lambda/4$ choke. For an 800 MHz band, the total length including the mounting fixtures, is 550 mm. Experiments in residential and urban areas show that a correlation coefficient below 0.6 is obtained for antenna spacing exceeding 0.5 λ.

Diversity antennas can be installed in passenger compartments. Two antenna elements (one in the front and one in the rear) are placed in a vehicle compartment and connected with a hybrid circuit. Radiation antenna patterns are affected by significant ripples as shown in Figure 5.48a. The calculated correlation factor is less than 0.6, which meets the design objectives. The radiation pattern levels of the antennas are lower than –15 dBd. The measured correlation coefficient as a function of travel distance is shown in Figure 5.48b. An average correlation coefficient of 0.27 is achieved.

5.12.3 Switched Parasitic Elements for Diversity Design

Figure 5.49 illustrates a diversity antenna configuration with a parasitic element [25]. The pattern diversity system has one active monopole, two

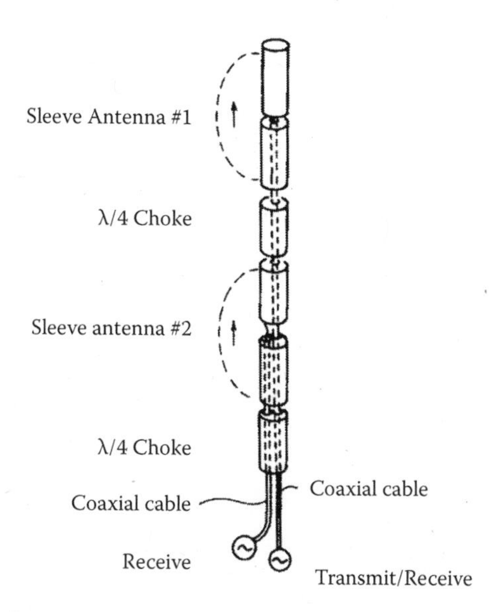

FIGURE 5.47
Configuration of trunk lid diversity antenna. (From Y. Ebine and Y. Yamada, *IEEE Vehicular Technology Conference*, 1988 IEEE, 38th June 1988. Copyright 1988 IEEE. With permission.)

parasitic switchable monopoles, and a shorted monopole is spaced nearby. The radiation patterns when the space between antenna elements is equal to 0.1 or 0.05 wavelengths are demonstrated in Figure 5.50a and Figure 5.50b. The patterns for very small distances between antenna elements are very different. A value equal to 0.05 of the wavelength corresponds to the space between the antenna elements equal to 7.5 mm.

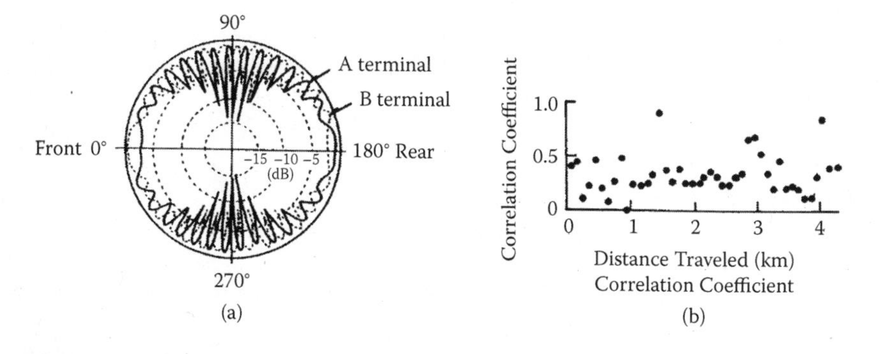

FIGURE 5.48
Cabin-mounted antenna: (a) radiation pattern; (b) correlation coefficient. (From Y. Ebine and Y. Yamada, *IEEE Vehicular Technology Conference*, 1988 IEEE, 38th June 1988. Copyright 1988 IEEE. With permission.)

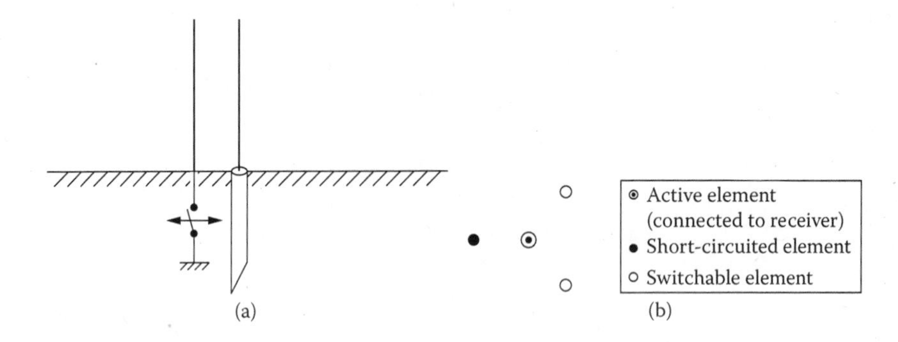

FIGURE 5.49
Geometry of parasitic element of switch for diversity application. (From R. Vaughan, *IEEE Transaction on Antenna and Propagation*, Vol. 47, No. 2, 1999. Copyright 1999 IEEE. With permission.)

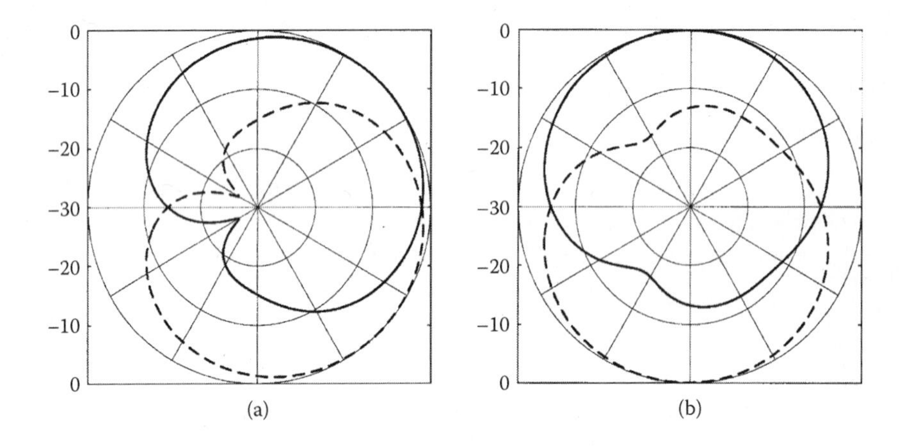

FIGURE 5.50
Radiation patterns: (a) 0.1 wavelength spacing; (b) 0.05 wavelength spacing. (From R. Vaughan, *IEEE Transaction on Antenna and Propagation*, Vol. 47, No. 2, 1999. Copyright 1999 IEEE. With permission.)

References

1. K. Fujimoto and J. James, *Mobile Antenna Systems Handbook*, Boston: Artech, 1994.
2. S. Egashira, T. Tanaka, and A. Sakitani, Design of AM/FM mobile telephone triband antenna, *IEEE Transactions on Antennas and Propagation*, 42, 538–540, 1994.
3. T. Shimazaki, Glass Mounted Antenna, U.S. Patent 4,857,939, August 1989.

4. J. Hadzoglou, Cellular Mobile Communication Antenna, U.S. Patent 4,839,660, June 1989.
5. J. Hadzoglou, M. Warner, and J. Hrabak, Dual-Band Glass Mounted Antenna, U.S. Patent 6,608,597, August 2003.
6. H. Du, Dual Band Window Mounted Antenna System for Mobile Communications, U.S. Patent 6,172,651, January 2001.
7. A. Best, Discussion on the properties of electrically small self-resonant wire antennas, *IEEE Transactions on Antennas and Propagation*, 46, 9–22, 2004.
8. P. Haapala et al., Dual frequency helical antennas for handsets, *IEEE 46th Vehicular Technology Conference*, Vol. 1, April 1996, pp. 336–338.
9. Eagleware RF and Microwave Design software, Eagleware Corporation, www.eagleware.com
10. G. Zhou et al., A non-uniform pitch dual-band helix antenna, *IEEE Antennas and Propagation International Symposium*, Vol. 1, July 2000, pp. 274–277.
11. H. Chen, Compact CPW-fed dual frequency monopole antenna, *Electronics Letters*, 38, 1622–1624, 2002.
12. W. Chung and C. Huang, CPW-fed L-shaped slot planar monopole antenna for triple band operations, *Microwave and Optical Technology Letters*, 44, 510–512, 2005.
13. R. Leelaratne and R. Langley, Multiband PIFA vehicle telematics antennas, *IEEE Transactions on Vehicular Technology*, 54, 477–485, 2005.
14. J. Kropielnicki et al., Vehicle Antenna, U.S. Patent 5,293,174, March 1994.
15. L. Nagy et al., Automotive Radio Frequency Antenna System, U.S. Patent 6,266,023, July 2001.
16. www.antenova.com
17. S. Yeh et al., Dual-band planar inverted F antenna for GSM/DCS mobile phones, *IEEE Transactions on Antennas and Propagation*, 51, 1124–1126, 2003.
18. Ansoft simulation software package, www.ansoft.com
19. P. Dennis and P. Eugene, In-Vehicle Antenna, U.S. Patent 5,649,316, July1997.
20. Y. Hong et al., Multifunctional vehicle antenna system for FM and PCS services, *IEEE Antennas and Propagation International Symposium*, 2007, pp.1092–1095.
21. www.smarteq.se
22. K. Oh, B. Kim, and J. Choi, Novel investigated GPS/RKES/PCS antenna for vehicular application, *IEEE Microwave and Wireless Component Letters*, 15, 244–246, 2005.
23. C. Dietrich et al., Spatial, polarization, and pattern diversity for wireless handheld terminals, *IEEE Transactions on Antennas and Propagation*, 49, 1271–1281, 2001.
24. Y. Ebine and Y. Yamada, Vehicular-mounted diversity antennas for land mobile radios, *38th IEEE Vehicular Technology Conference*, June 1988, pp. 326–333.
25. R. Vaughan, Switched parasitic elements for antenna diversity, *IEEE Transactions on Antennas and Propagation*, 47, 399–405, 1999.
26. A. Duzdar et al., Mobile Wideband Antennas, U.S. Patent 7,492,318, February 2009.

6

TV Antennas for Cars

6.1 Satellite TV Antennas

6.1.1 Satellite TV System Requirements

Generally, a satellite TV antenna is mounted on the roof of a vehicle. The satellite receiving system tracks the signal received by an antenna from the satellite transmitter when a car is in motion. Therefore, the system requires a special sensor to find the satellite with minimal searching. The tracking process must maintain the maximum radiation antenna pattern at a position corresponding to the satellite angle.

Direct broadcast satellite (DBS) TV antenna systems [1,2] must meet certain requirements. They must operate over 12.2 to 12.7 GHz with estimated antenna gains of 28 to 34 dBic. Beamwidth varies from 1.6 to 3.8 degrees in both elevation and azimuth directions, with the antenna gain to system noise temperature (G/T) of 10 to 12 dB/K (antenna gain is expressed in dBic; temperature in K). The antenna must provide 360 degree azimuth scanning, and the elevation scanning angle range depends on the TV satellite location. For example, complete coverage over the continental U.S. requires an elevation angle range of 20 to 65 degrees for Direct TV and 15 to 60 degrees for EchoStar [1]. The tracking accuracy must about 0.5 and 2.0 degrees in elevation and azimuth, respectively.

The system must be switchable between right-hand circular polarization (RHCP) and left-hand circular polarization (LHCP). Cross polarization isolation must be greater than 25 dB. One commercially available mechanical scanning antenna array system was described in Chapter 1. Here we present in detail an antenna array that combines mechanical scanning in the azimuth plane and electronic scanning in the elevation plane. A very important design concept is to facilitate in-motion pointing of the antenna beam toward a satellite. The simplest method uses only one beam for the tracking algorithm. This beam rotates around the main angle direction and a signal error (pointing error assessment algorithm [2]) is detected to correct the main angle direction.

Incorporation of a GPS receiver into an antenna system can provide precise tracking of a satellite. Motion tracking capability requires an elevation tracking rate of 20 degrees/sec with an acceleration of 50 degrees/sec^2 [1].

In azimuth, the tracking rate must be 45 degrees/sec with an acceleration of 60 degrees/sec^2. A horizontal drive motor, for example, can spin an antenna up to 70 degrees/sec, allowing the unit to stay locked onto the satellite even when the vehicle turns rapidly. The antenna should be concealed within the roof of the vehicle, with the goal of occupying no more than 5 cm in height and 60 cm in diameter.

6.1.2 Ridged Waveguide Antenna Array

Figure 6.1 depicts an antenna design using a ridged waveguide [1,3,4] with an electronic scan capability in the elevation plane. An antenna array consists of 32 ridge waveguides, with right-hand circular polarization (RHCP) and left-hand circular polarization (LHCP) outputs located at opposite ends of the antenna. A single waveguide with radiating elements is shown in Figure 6.2. It couples the energy from all radiating elements and combines it. The radiating surface of the waveguide antenna uses multiple radiating elements spaced uniformly along the waveguide axis. The radiating elements are X-shaped cross slots, as shown in the figure.

The phase centers of the cross slots are positioned in a straight line along the waveguide axis and between the centerline of the waveguide ridge section and one of the walls. The radiation elements are placed about half a waveguide wavelength apart. The waveguide is about 2 cm wide and 1 cm high; ridge height is around 0.7 cm, and the length of the radiating element is about 1 cm. The total length of the waveguide is about 60 cm with about 30 cross slot elements spaced 2 cm apart.

The angle beamwidth of one waveguide at the plane of the waveguide location is 2.3 to 2.4 degrees. RHCP and LHCP polarization outputs are shown in Figure 6.1. Both ends of the 32 waveguides have antenna output

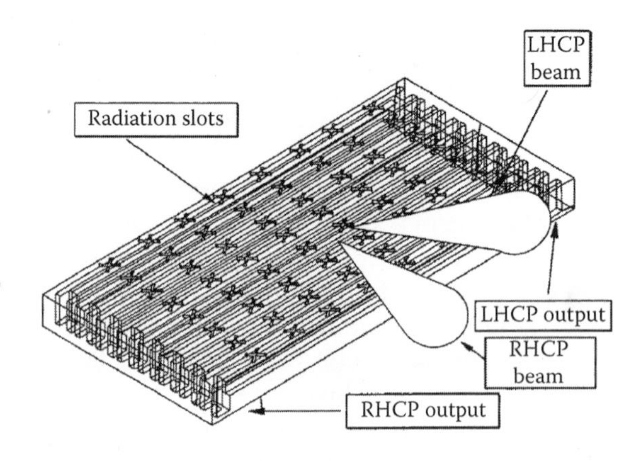

FIGURE 6.1
Antenna array topology. (From J. Wang and J. Winters, *Vehicular Technology Conference*, 2004, IEEE, 60th. Copyright 2004 IEEE. With permission.)

FIGURE 6.2
Radiating element of waveguide. (From J. Wang and J. Winters, *Vehicular Technology Conference*, 2004, IEEE, 60th. Copyright 2004 IEEE. With permission.)

probes that couple the output signal from the waveguide to the beam-forming combiner that joins signals from all 32 waveguides, providing the maximum radiating energy toward the TV satellite. Each antenna probe has a low noise amplifier (LNA), and (depending on the structure of the beam forming network) a passive combiner with the output connected to a single receiver or active combiner with a receiver. The antenna must have a small profile (less than 7.5 cm) and be embedded in a vehicle roof or between the roof and the head liner and meet all requirements for DBS antennas.

6.1.3 Electronic Beam-Controlled Phase Array

Thirty-two equally spaced waveguide antenna probes constitute a linear receiving antenna array. Each antenna probe has its own receiver, LNA, and electronically controlled phase shifter to shift the signal received by each single waveguide. All receiver outputs are combined in a network constituting an electronically controlled beam in the elevation plane. The output of the combining network is connected with a satellite receiver.

Figure 6.3 illustrates a linear passive phase array with electronic beam control. The system consists of N linear equally spaced identical antenna receiving elements. Each element has a phase shifter controlled electronically by a special beam control block with computer. Each phase shifter has a special electrical control circuit (for example, a pin diode) that can change the phase of the received signal. The computer calculates the phases for each element in such a way to direct the beam (maximum receiving) at a predetermined angle. The control block converts the calculated phases into the electrical signals, and these control signals change the phase shifter states, collimating maximum energy received by the array at the given direction.

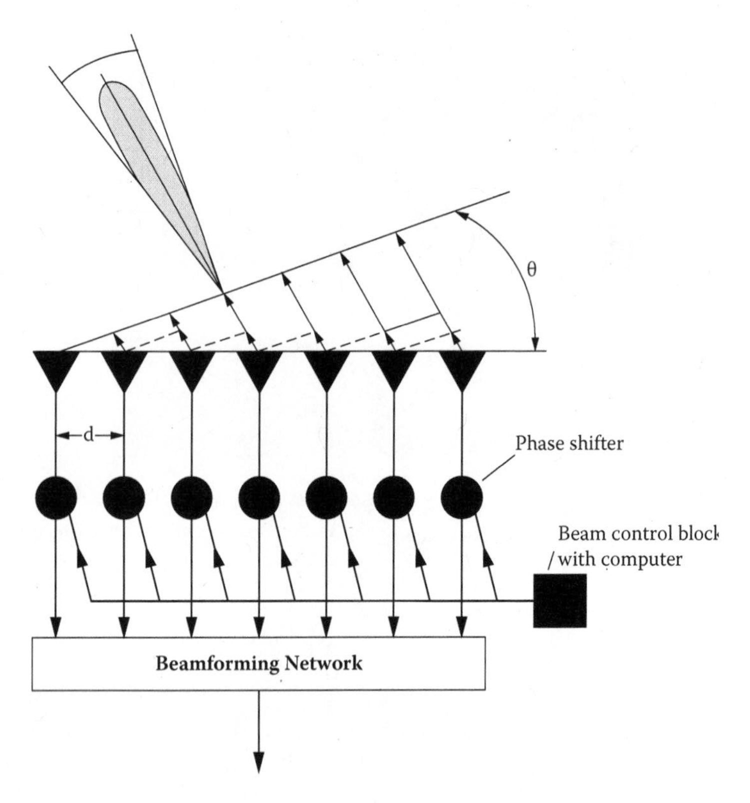

FIGURE 6.3
Linear passive phased antenna array.

Equation (6.1) is the simplest algorithm that estimates antenna element phase values, maximizing the receiving energy at a certain angle direction. Let us assume that the space between the elements of the linear antenna array equals d and the TV signal is emitted by a satellite with angle direction θ (Figure 6.3). The progressive phase shift among the equally spaced adjacent antenna elements is given by:

$$\Delta\phi = k \cdot d \cdot \sin\theta \tag{6.1}$$

where $k = 2 \cdot \pi/\lambda$. The antenna array factor [5] can be calculated as:

$$AF = \frac{1}{N} \cdot \frac{\sin\left(\frac{N}{2} \cdot (\Delta\phi - \Delta\phi_0)\right)}{\sin\left(\frac{\Delta\phi - \Delta\phi_0}{2}\right)} \tag{6.2}$$

where $\Delta\phi_0 = k \cdot d \cdot \sin\theta_0$ represents a progressive phase shift provided by the electronic shifters. Equation (6.2) is equal to 1 at the maximum, when $\Delta\phi = \Delta\phi_0$. Thus, if a computer has information about the elevation angle of the satellite θ, the car computer block calculates phase shift $\Delta\phi_0 = k \cdot d \cdot \sin\theta_0 = \Delta\phi$ between the adjacent elements and phase values for an n-th phase shifter equal to $\Delta\phi_{0n} = n \cdot \Delta\phi_0$ (n = 1, 2, 3…N). The calculated phases are converted to electrical signals that change the state of the phase shifters to direct the antenna beam at angle θ.

Let us assume that each of the 32 antenna array elements has a microwave pin diode phase shifter. Such shifters are generally digital. The digital phase shifter has 1, 2, 3, or more bits of phase values. For example, a 1 bit digital phase shifter realizes only two phases: 0 and 180 degrees; a 2 bit digital shifter can realize four phases: 0, $\pi/2$, π, $3\pi/2$; and a 3 bit digital phase shifter can realize 0, $\pi/4$, $\pi/2$, $3\pi/4$, π, $5\pi/4$, $3\pi/2$, and $7\pi/4$ phases. According to the requirements, the tracking minimal shift angle must be 0.5 degrees. The minimal scanning angle of the electronically controlled beam antenna array with digital phase shifters can be estimated. Let us assume the phase shifter provides a minimal digital phase shift equal to Δ. The minimal scan angle $\delta\theta_0$ may be estimated from the following expression [6]:

$$\delta\theta_0 \approx \Delta/(k \cdot d \cdot N) \tag{6.3}$$

In our case, Δ must be equal or less than $\pi/2$ because antenna linear dimensions $d \cdot N \approx 60$ cm, wavelength about 2.5 cm, and $\delta\theta_0$ must be less than or equal to 0.5 degrees. For this application, it is possible to use a 2 bit digital phase shifter with a minimal discrete phase equal to 90 degrees.

6.1.4 G/T Estimation

The γ= G/T ratio (or figure of merit) for an active satellite TV antenna system is required to be more than 10 dB/K. G is the antenna gain and T is the antenna system noise temperature. Actually, G/T determines the SNR at the output of a receiver system. The higher the antenna G/T, the less power is required from the satellite, leading to a reduction in receiving antenna size. The G/T value for an active linear antenna array with N identical elements (each element has a LNA) can be estimated from the following formula [7]:

$$\left(\frac{G}{T}\right) = \frac{N \cdot G_e \cdot \eta}{T_a + T_0 \cdot \left(L_f \cdot F - 1 - L_f/g + N \cdot L_f/\left(g \cdot \sum_{k=1}^{N} \frac{1}{L_k}\right)\right)} \tag{6.4}$$

where G_e = antenna element gain including mismatches and scan factor but excluding the effects of active antenna components, T_a = antenna input temperature, T_0 = room temperature, L_f = ohmic loss of the passive element between the antenna element and the LNA (e.g., insertion loss of the electronically control phase shifter or cable between element and LNA), F = noise

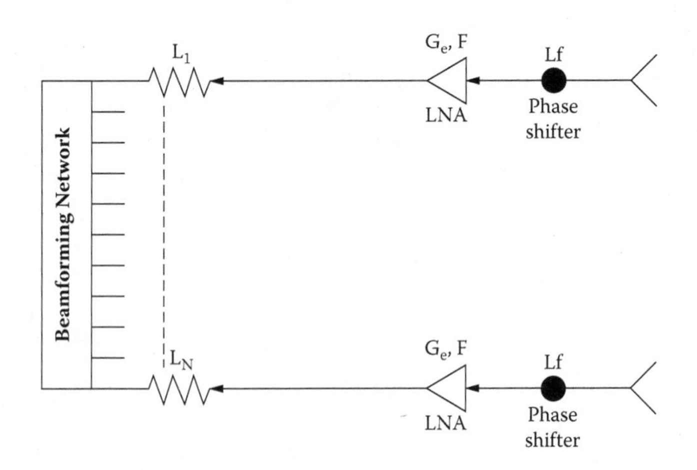

FIGURE 6.4
Equivalent circuit of active antenna array. (From J. Lee, *IEEE Transactions on Antennas and Propagation*, Vol. 41, No. 2, 1993. Copyright 1993 IEEE. With permission.)

figure of single amplifier element, g = amplifier gain, L_k (k = 1, 2, 3…N) = loss of feeder between amplifier and beam-forming network, and η = efficiency of antenna aperture. Figure 6.4 shows the equivalent circuit of an active linear antenna array. If we assume that all antenna channels are identical ($L_k = L_N = L$, n = 1, 2, 3…N, and $\eta = 1$), the formula can be simplified:

$$\left(\frac{G}{T}\right) = \frac{N \cdot G_e}{T_a + T_0 \cdot (L_f \cdot F - 1 - L_f/g + L_f \cdot L/g)} \tag{6.5}$$

According to the graph shown in a paper by Pritchard and Ogata [8], the value of the input antenna temperature for the 12 GHz frequency range is about 45 K. When an amplifier gain is sufficiently large, the ratio (6.5) can be expressed as:

$$\left(\frac{G}{T}\right) \cong \frac{N \cdot G_e}{T_0 \cdot (L_f \cdot F - 1)} \tag{6.6}$$

The estimated gain for an antenna array consisting of 32 waveguides with 30 radiating elements in each waveguide is 28 to 32 dBic. Assume that the noise of each LNA is about 0.6 dB (typical value for a high quality device), the phase shifter losses L_f are about 2 dB, and the room temperature is 290 K. In this case, the figure of merit ratio G/T is about 4.2 to 8.2 dB/K. If losses L_f are close to 1 (meaning a lossless connection between antenna element and amplifier, then the G/T value is around 11.6 to 15.6 dB/K and meets the antenna system requirement of 10 to 12 dB/K.

6.2 Antennas for Digital Terrestrial TV

6.2.1 Introduction

Digital video broadcasting (DVB) is a new combination of digital technologies that provides better quality pictures in comparison with analog TV. More countries are using DVB to transmit television and broadcast programs. As noted earlier, DVB covers the VHF band from 110 to 270 MHz and the UHF band from 350 to 870 MHz (depending on country). A broadband, low cost, compact antenna system is key for high quality TV reception in a vehicle.

Generally, mobile TV antennas exhibit more problems receiving TV signals than fixed household antenna systems. A TV mobile antenna is installed lower than a typical household antenna and thus the received signal is lower. The correction factor for received electric field strength in the UHF band for reception at 10 m and 1.5 m above the ground is 16 dB [9]. This means that the received electric strength for a low-mounted antenna could be 16 dB less than that of a high-mounted antenna. Also, the signal received by a moving antenna suffers from significant fluctuation. Figure 6.5 shows variations of a 509 MHz frequency signal (bandwidth about 430 KHz) received in a vehicle moving at 36 km/hour. The figure shows that the electrical field strength drops many times over a short traveling distance of 10 m. Therefore, typical antenna systems for TV terrestrial applications use the diversity technique, and each antenna element includes an LNA to increase the dynamic range of the receiving signals.

A broadband amplifier must provide signal amplification in a wide frequency band with minimal inserted noise. A high input impedance amplifier [10] with a gain of 15 to 20 dB in the entire frequency band may be a good candidate for broadband TV.

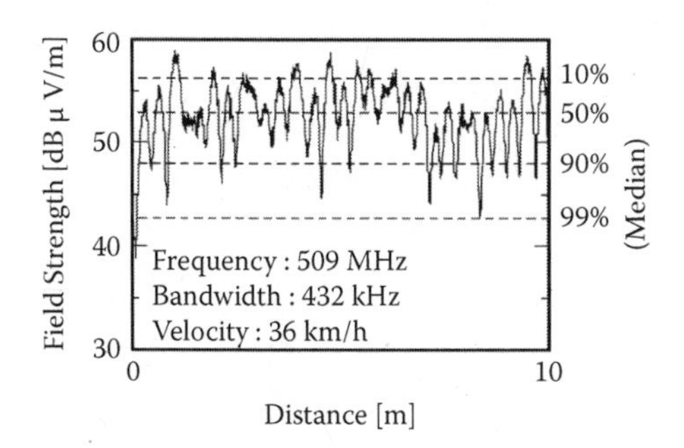

FIGURE 6.5
Field strength variation in mobile reception area. (From N. Itoh and K. Tsuchida, *Proceedings of the IEEE*, Vol. 94, No. 1, 2006. Copyright 2006 IEEE. With permission.)

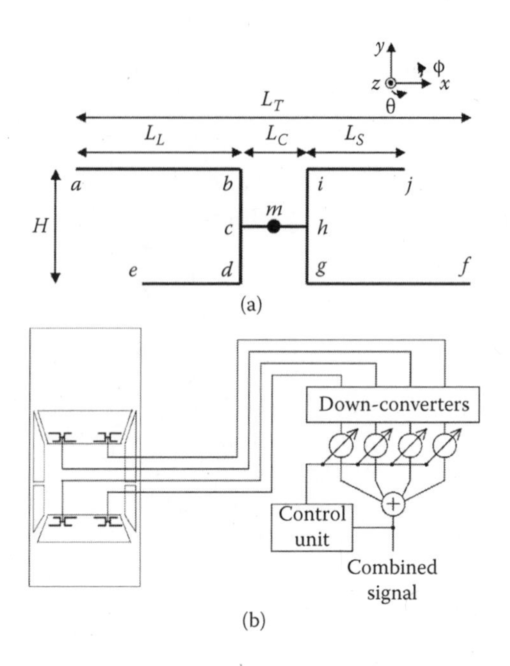

FIGURE 6.6
H-shaped antenna and control system for combining received signals: (a) antenna element topology; (b) control system. (From H. Iizuka et al., *Institute of High Frequency Technology and Mobile Communication University of the Bundeswehr Munich*, Germany. Copyright 2006 IEEE. With permission.)

6.2.2 H-Shaped Glass Configuration

An H-shaped antenna array element [11] for digital terrestrial automotive reception is shown in Figure 6.6a. Four symmetrically mounted elements of the antenna array (Figure 6.6b) exhibit gain for the sides, front, and rear of the car. Two Π-shaped wires (a–b–c–d–e and f–g–h–i–j) of the array are connected by a wire c–m–h. The parts a–b–c and f–g–h are longer than c–d–e and h–i–j. Voltage of 1V is excited at the feed point m. A control system for combining the signal received by four antenna elements and antenna elements mounted at the tops of the front and rear windows are shown in Figure 6.6b. The topology of the four antennas covering 360 degrees in the X–Y plane across the frequency band is demonstrated in Figure 6.7. The antenna array, built for diversity applications, has a feature in which a figure 8 radiation pattern is rotated with increasing frequency. Such a design in a diversity system can avoid a weak receiving signal around 360 degrees.

Each resonance mode of the single antenna element is excited with its own half wavelength length. The wire a–b–c–m–h–g–f is resonated at the lower frequency (series resonance). Resonance at the middle frequency points

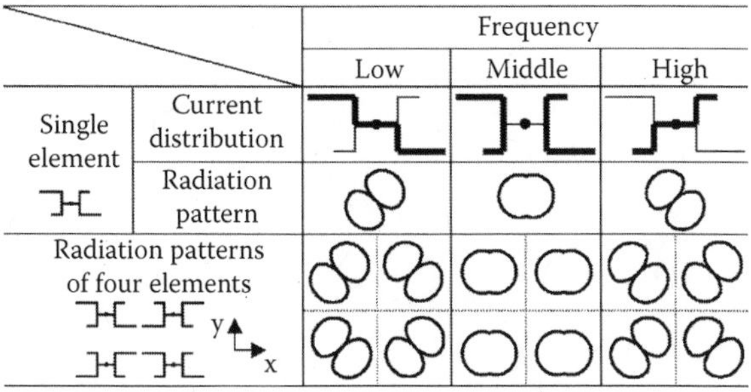

		Frequency		
		Low	Middle	High
Single element	Current distribution			
	Radiation pattern			
Radiation patterns of four elements				

FIGURE 6.7
Concept covering 360 degrees in horizontal plane with four antennas. (From H. Iizuka et al., *Institute of High Frequency Technology and Mobile Communication University of the Bundeswehr Munich*, Germany. Copyright 2005 IEEE. With permission.)

is realized by parallel connection of two Π-shaped wires. At the high-frequency range, resonance is realized by wires e–d–c–m–h–i–j. The radiation pattern of a single element rotates with increasing frequency. The design covers 360 degrees. The antenna elements are mounted at the tops of the front and rear windows; element dimensions are shown in Table 6.1.

Radiation patterns of one element calculated in free space using a FEKO commercial simulator are shown in Figure 6.8. Radiation patterns in the X–Y plane (horizontal polarization) are presented at 484, 576, and 638 MHz frequencies. The solid line corresponds to 484 MHz, the point curve presents the directionality for 576 MHz, and the dotted line corresponds to 638 MHz. Reflection losses are included when the antenna is terminated with characteristic impedance of 110 ohm. The figure 8 radiation pattern is rotated clockwise as the frequency increases. The combined directivity of the antenna array consisting of four antenna elements according to the

TABLE 6.1

Dimensions of Antenna Wire Elements

Total length L_T of antenna	257 mm
Length L_L of wires a–b and f–g	114 mm
Length L_S of wires d–e and i–j	79.5 mm
Length L_C of wires c–m–h	29 mm
Height H of antenna element	60 mm
Wire radius a	0.5 mm

Source: H. Iizuka et al., *Institute of High Frequency Technology and Mobile Communication University of the Bundeswehr Munich*, Germany. Copyright 2005 IEEE. With permission.

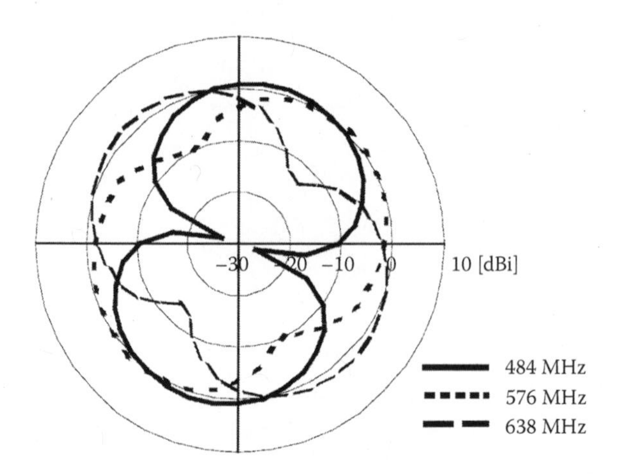

FIGURE 6.8
Calculated radiation patterns of H-shaped antenna in horizontal plane. (From H. Iizuka et al., *Institute of High Frequency Technology and Mobile Communication University of the Bundeswehr Munich*, Germany. Copyright 2005 IEEE. With permission.)

MRC method [12] is:

$$D_{MRC}(\theta,\varphi)=\frac{1}{\sqrt{\sum_{n=1}^{4}A_n^2(\theta_0,\varphi_0)}}*\sum_{n=1}^{4}A_n(\theta_0,\varphi_0)A_n(\theta,\varphi)e^{j\psi_n} \tag{6.7}$$

$$\psi_n = kx_n(\sin\theta\cos\varphi-\sin\theta_0\cos\varphi_0)+ky_n(\sin\theta\sin\varphi-\sin\theta_0\sin\varphi_0) \tag{6.8}$$

where (θ_0, φ_0) = angle direction of the arrival signal, $A_n(\theta, \varphi)$ = amplitude of the signal received from direction (θ, φ), x_n and y_n are antenna positions with index n, $K = 2\pi/\lambda$, λ is the wavelength. The peak plot $D_{pp}(\theta, \varphi)$ is defined as the plotted pattern of the peak $D_{MRC}(\theta_0, \varphi_0)$ of the combined pattern (6.7):

$$D_{pp}(\theta,\varphi)=D_{MRC}(\theta_0,\varphi_0)_{\theta=\theta_0,\varphi=\varphi_0}=\sqrt{\sum_{n=1}^{4}A_n^2(\theta,\varphi)} \tag{6.9}$$

An omnidirectional pattern of the peak plot in the φ plane indicates that the combined pattern steering beam has the same gain for any direction in the plane. The average gain of the peak plot G_{AVE} is given by:

$$G_{AVE}=\frac{1}{2\pi}\int_{0}^{2\pi}|D_{pp}(\theta,\varphi)|^2 d\varphi \tag{6.10}$$

Figure 6.9a shows calculated minimum and averaged gains of the peak plots of the four antenna elements in free space. The minimum gain represented

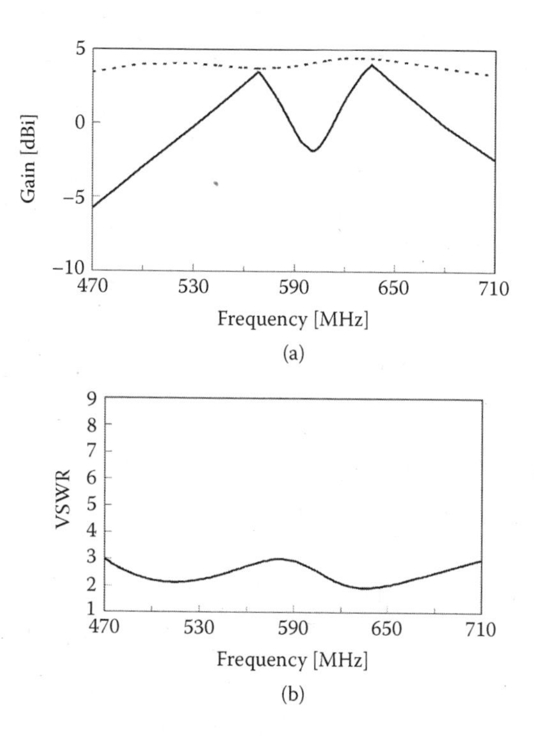

FIGURE 6.9
(a) Calculated averaged and minimum gains of peak plots for four antennas in free space; (b) VSWR of single antenna with respect to 110 ohm. (From H. Iizuka et al., *Institute of High Frequency Technology and Mobile Communication University of the Bundeswehr Munich*, Germany. Copyright 2005 IEEE. With permission.)

by a solid line is higher than −5.8 dBi across the 470 to 710 MHz band. The average gain of the peak plot (pointed curve) varies from 3.3 to 4.5 dBi. Figure 6.9b shows the calculated VSWR of a single antenna with respect to 110 ohm. The 110 ohm was chosen to provide the widest bandwidth. The VSWR is less than 3 between 470 and 710 MHz.

In a car, the plane with antenna element is inclined from the horizontal plane. Figure 6.10 shows the relative amplitude of the radiation pattern (horizontal polarization) of the antenna with variation of the rotation angle of the antenna plane from the horizontal X–Y plane around the X axis. Data are presented for two orientations relatively to X–Y coordinates: pointed curve corresponds to the antenna plane placed in X–Y plane with the length L_T oriented along axis X; the solid curve corresponds to the antenna turned by 90 degrees (length L_T is oriented along axis Y). Figure 6.11 shows the measured radiation patterns of the single antenna at 484, 576, and 638 MHz) in the horizontal plane for horizontal polarization. The measured VSWR was less than 3 between 470 and 710 MHz.

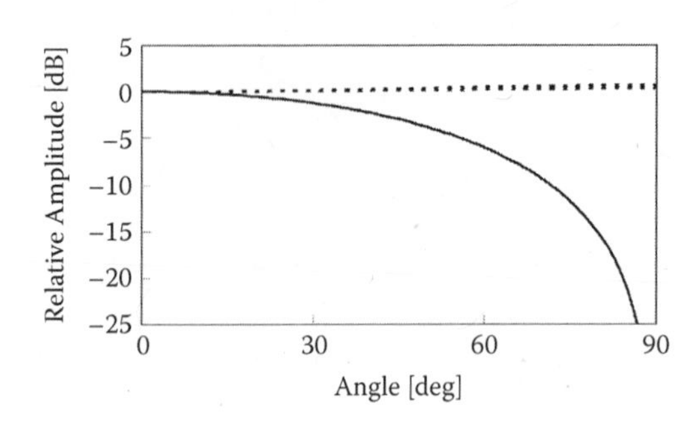

FIGURE 6.10
Relative amplitude of radiation pattern versus angle between mounting plane and horizontal plane. (From H. Iizuka et al., *Institute of High Frequency Technology and Mobile Communication University of the Bundeswehr Munich*, Germany. Copyright 2005 IEEE. With permission.)

The experimental car has an angle of approximately 30 degrees from the horizontal plane for both the front and rear windows. The degradation of the radiation pattern in comparison with the value shown in Figure 6.10 (solid curve) is estimated to be –1.3 dB. However, the pointed line shows a little (0.6 dB) variation.

The single antenna presented in Figure 6.6a was etched on FR4 board substrate with a dielectric constant of 4.6 and thickness of 0.8 mm. The width of the lines was 1 mm. The lengths L_L, L_S, and L_C were experimentally adjusted to 102, 67, and 30 mm, respectively, with $H = 60$ mm. The prototype antenna

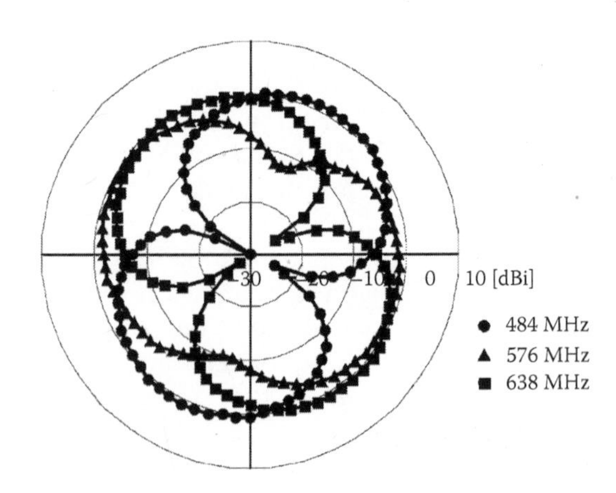

FIGURE 6.11
Measured radiation patterns of prototype antenna. (From H. Iizuka et al., *Institute of High Frequency Technology and Mobile Communication University of the Bundeswehr Munich*, Germany. Copyright 2005 IEEE. With permission.)

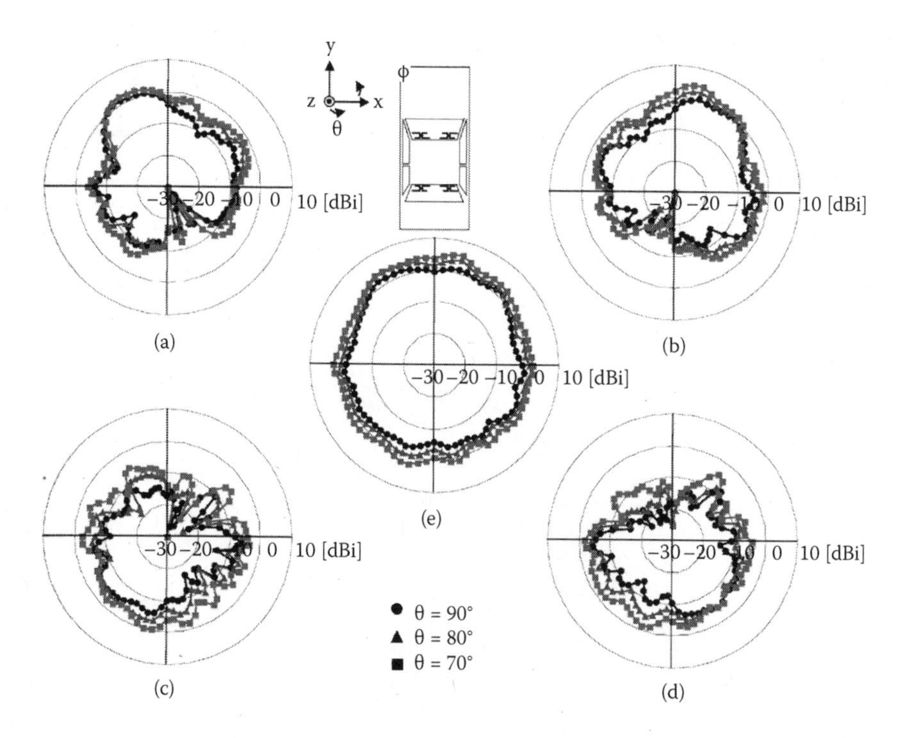

FIGURE 6.12
Measured radiation patterns in inclined planes at 530 MHz. Four prototype antennas are mounted at tops of front and rear car windows: (a) front left; (b) front right; (c) rear left; (d) rear right. (From H. Iizuka et al., *Institute of High Frequency Technology and Mobile Communication University of the Bundeswehr Munich*, Germany. Copyright 2005 IEEE. With permission.)

has LC balun transforming 110 to 50 ohm. The insertion loss of the LC balun was 0.3 dB across the frequency band.

Four prototype antennas were mounted at the tops of the front and rear windows of a car as shown in Figure 6.6b. The distance between the edge of the metal roof and the nearest sides of the antennas is about 20 mm. The antenna elements are spaced 600 mm along the X axis and 1600 mm along the Y axis. Figure 6.12 presents the typical measured radiation pattern of the four antennas in the horizontal plane for different θ angles equal to 70, 80, and 90 degrees at 530 MHz. The peak plot combined with these four radiation patterns is shown in Figure 6.12e. We see an almost omnidirectional radiation pattern. The minimum gains of the peak plot were –2.3 dBi for $\theta = 80$ degrees, –3.7 dBi for $\theta = 80$ degrees, and –6 dBi for $\theta = 90$ degrees.

The module shown in Figure 6.13a consists of an antenna element, balun, filter, and LNA. The LNA improves the noise of the system by reducing the loss cable effect. Impedance matching of the LNA for optimum noise is

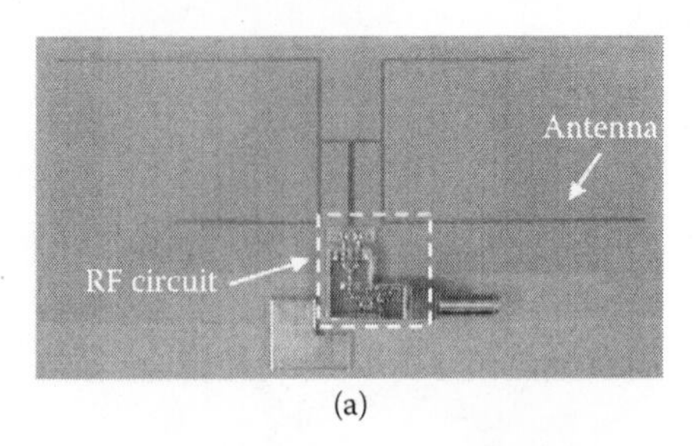

(a)

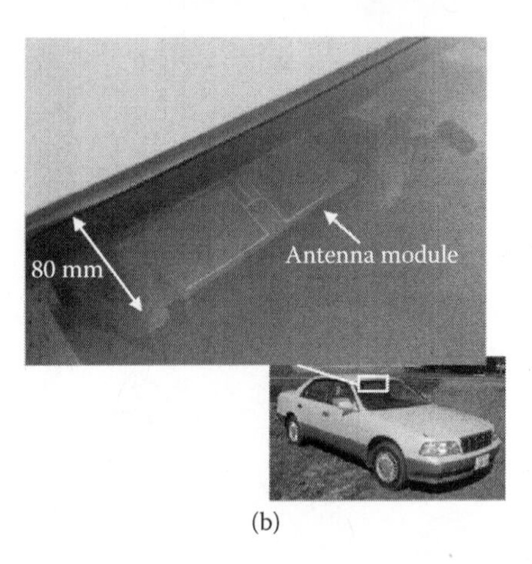

(b)

FIGURE 6.13
(a) Prototype of antenna module; (b) module mounted on front window. (From H. Iizuka et al., *Institute of High Frequency Technology and Mobile Communication University of the Bundeswehr Munich*, Germany. Copyright 2005 IEEE. With permission.)

controlled by passive reactance elements. Figure 6.13b shows the mounted module. In conclusion, note that:

1. The combined plots of four antennas are almost omnidirectional.
2. The effect of the angle of the antenna plane from the horizontal plane yields a gain reduction about 1.3 dB.
3. The minimum gain and averaged gain of the peak plot in the horizontal plane are –6 and –1.5 dBi, respectively—acceptable values for a digital automotive terrestrial system.

6.2.3 Simple Meander Window Glass Design

Figure 6.14 shows two options of geometry covering a frequency between 470 and 770 MHz [14] for a printed-on-glass (rear side) antenna. The antenna A has a coaxial cable-fed multipath wire loop shorted to the ground; B's height

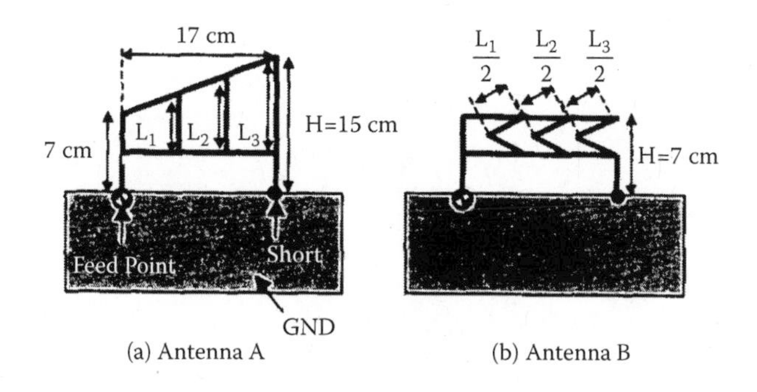

FIGURE 6.14
Geometry of printed-on-glass antenna element. (From S. Matsuzava et al., *Antennas and Propagation Society Symposium*, 2004 IEEE. Copyright 2004 IEEE. With permission.)

(*H*) is reduced by bending the wires to approximately half of the antenna A dimension. The input impedance of antennas A and B is stable around 100 ohm from 370 to 870 MHz, and both antennas have return loss below –10 dB between 470 and 770 MHz.

Two prototype antennas were printed on the inside of the rear side window of a passenger van. The antennas are grounded to the van body. Figure 6.15 shows the directionality in horizontal plane of antenna B with mounting positions shown by arrows. The 485 MHz, horizontally polarized, horizontal radiation patterns show high directivity away from the body of the van. A peak gain of 4 dBi was achieved by each antenna element. The diversity technique can provide an omnidirectional radiation pattern.

The adaptive array radiation pattern in horizontal plane for horizontal polarization (Figure 6.16) was estimated by combining the measured radiation patterns of four antennas mounted on the side glass. An average gain

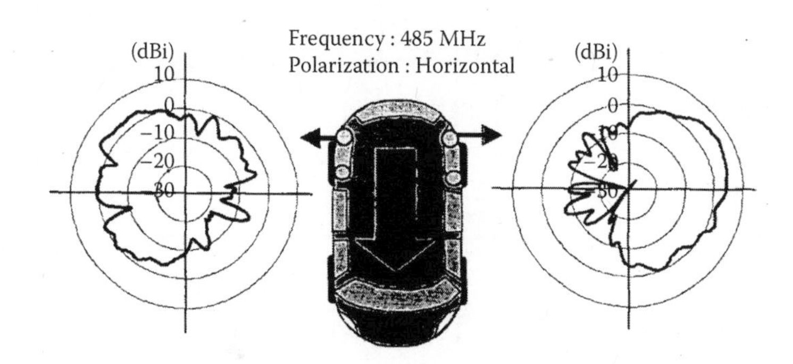

FIGURE 6.15
Radiation patterns in horizontal plane for antennas installed on rear side glass of passenger van (485 MHz). (From S. Matsuzava et al., *Antennas and Propagation Society Symposium*, 2004 IEEE. Copyright 2004 IEEE. With permission.)

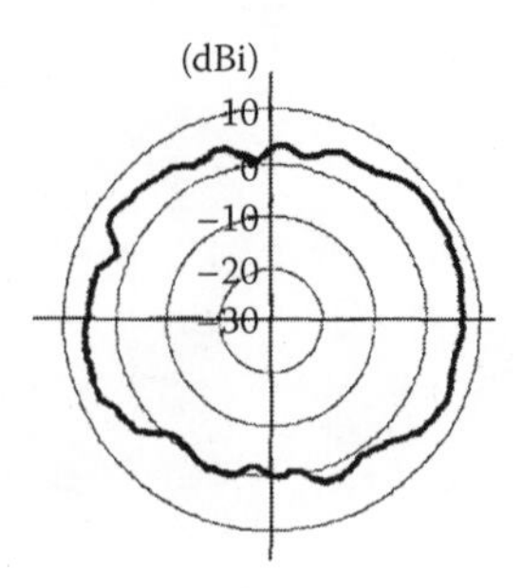

FIGURE 6.16
Overlap radiation estimated by combining patterns of four antennas installed on passenger van.

of 4.6 dBi was obtained in the horizontal plane. Because the antennas were installed on the side glass, the forward and rear gains were lower than the side gain, about 0 dBi. As stated in [14], final field tests conducted on public roads around Nagoya (Japan) confirmed that the designed adaptive array of four antenna elements was capable of mobile digital television (DTV) reception.

6.3 Printed-on-Glass Patent Examples

Several printed-on-glass TV antennas for automotive use are described in patents [15–18]. Figure 6.17 presents a wideband printed-on-glass antenna [15] that operates in a frequency range from 65 MHz to 2 GHz. The antenna

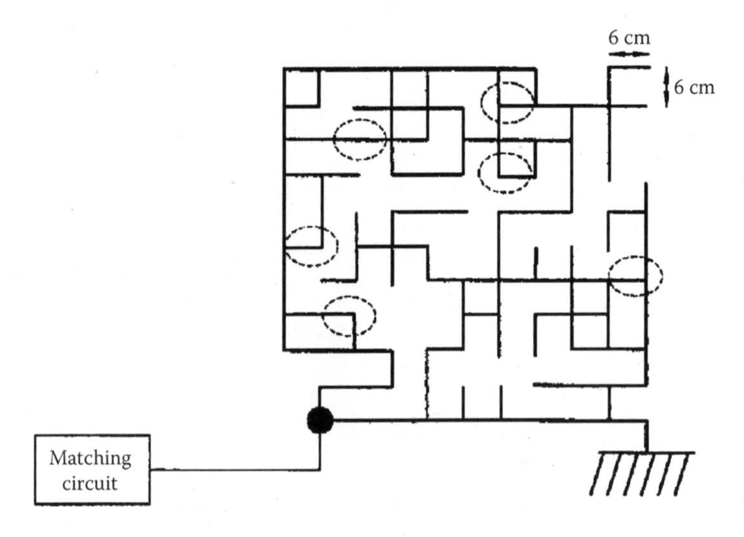

FIGURE 6.17
Wide band printed-on-glass TV antenna.

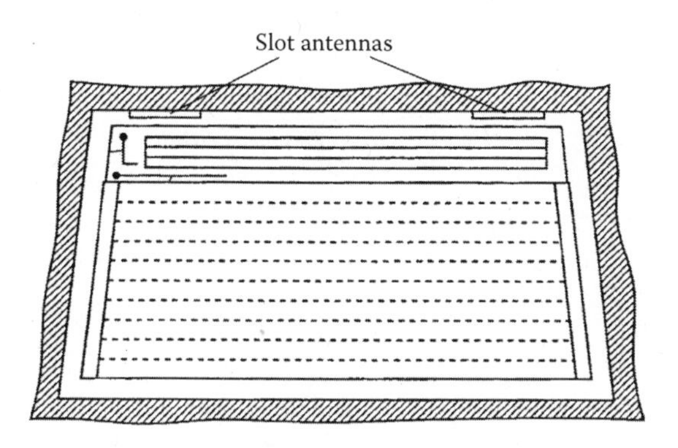

FIGURE 6.18
Window glass slot antennas.

dimensions are 600 mm × 600 mm. The antenna includes a loop portion with one side connected to a feeding point and the other connected to the ground. Tuning arms within the loop are connected with the loop and provide wideband reception. The antenna output has a tuning circuit connected to a feeding point and configured to select a frequency radio signal.

An interesting solution [16] is a printed-on-glass TV antenna shown in Figure 6.18. The glass antenna for receiving a DTV band consists of a rectangular slot antenna; one side is formed by an edge of the vehicle body surrounding the window glass and the residual three sides are formed by the printed metal line connected to the vehicle body as shown in Figure 6.19. The width of slot *W* is equal to 10 mm, the metal line width is 1 mm, and the feeding point *x* is L/20.

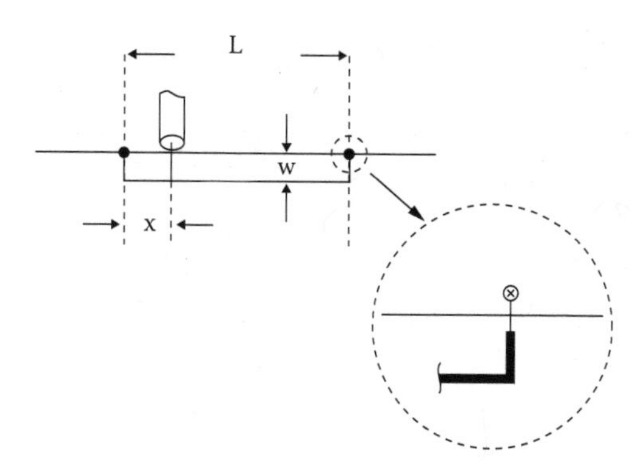

FIGURE 6.19
Design of connection to car body.

TABLE 6.2

Sensitivity of Window Glass Antenna

L	Average Sensitivity (dB)
0.3 λk	–13.9
0.5 λk	–14.5
0.8 λk	–11.7
1.0 λk	–15.4

Table 6.2 shows the average sensitivity for a single-slot antenna for 470 to 686 MHz using a horizontally polarized wave. Sensitivity is indicated in terms of the ratio of receiver input signal level with a dipole as a reference antenna to the receiver input signal level of the slot antenna. An average sensitivity is calculated by averaging the sensitivity values of all frequencies in the bandwidth.

The shortening factor k value depends on the dielectric substrate properties; λ represents wavelength. For a glass plate, k is about 0.6 [18]. The frequencies that vary from 470 to 686 MHz correspond to the wavelengths from 63.8 to 43.7 cm. Radiation pattern measurements of the single-slot antenna show that the ratio of maximum and minimum values over 360 degrees in azimuth is 30 to 40 dB, depending on the frequency band point.

Figure 6.20 presents four slot antennas constituting a diversity system: two slots are printed on the upper edge of the front glass and two slots on the upper edge of the rear glass of the vehicle.

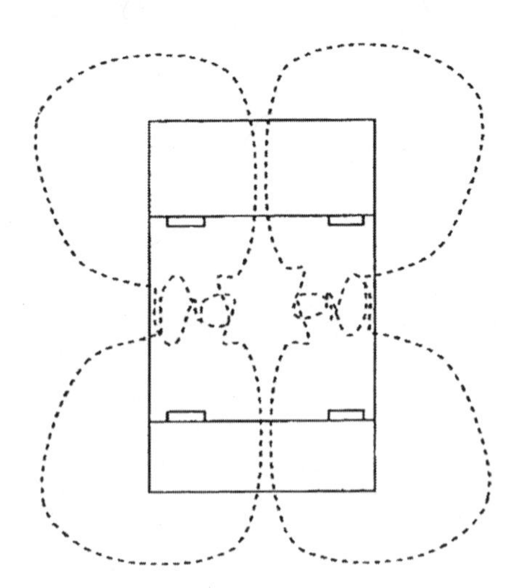

FIGURE 6.20
Four slot antennas comprising diversity system.

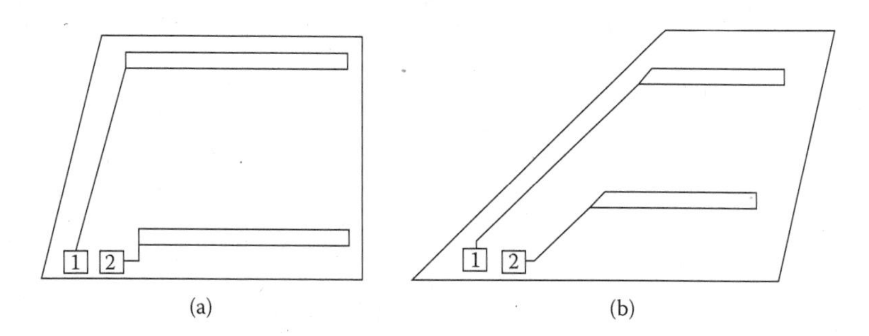

FIGURE 6.21
Two options for rear side printed-on-glass diversity TV antenna elements.

Patent [19] describes a diversity system for TV based on printed antennas mounted on the rear quarter window glass. Figure 6.21 is one of the drawings included in the patent. Two antenna elements are mounted in the rear quarter window and can be combined within the diversity reception system. The system selects the received signal with the higher reception level from two antenna elements. Sometimes, the antenna elements are provided on two side windows. The antenna has sensitivity in the TV band (90 to 230 MHz) of about 25 dBu, and 32 dBu in the 470 to 770 MHz band (for a horizontally polarized wave). Sensitivity is defined as the induced voltage at the antenna's open terminals under field strength of 60 dBu (0 dBu = 1 uV/m). A conventional monopole has an antenna sensitivity of 43 to 45 dBu in the VHF bands and 25 to 30 dBu in the UHF bands.

6.4 Bumper Diversity System

Figure 6.22 illustrates an antenna element pattern printed on a flexible foil [20]. Four identical antennas are mounted on the front and rear bumpers as schematically demonstrated in Figure 6.23. The antenna element dimensions are 470 mm × 100 mm. The typical folded dipole intended for TV reception in the VHF band (90 to 222 MHz in Japan) is about 1300 mm long. The suggested antenna is 830 mm shorter than the folded dipole.

The designed antenna has a balun that transforms the input impedance of 300 ohm to output impedance equal to 50 ohm. The measured VSWR in the 90 to 108 MHz band is less than 4:1; less than 6:1 in the 170 to 220 MHz band; and less than 4:1 in the 470 to 770 MHz band. Based on experimental results, the minimal antenna sensitivity for a horizontally polarized

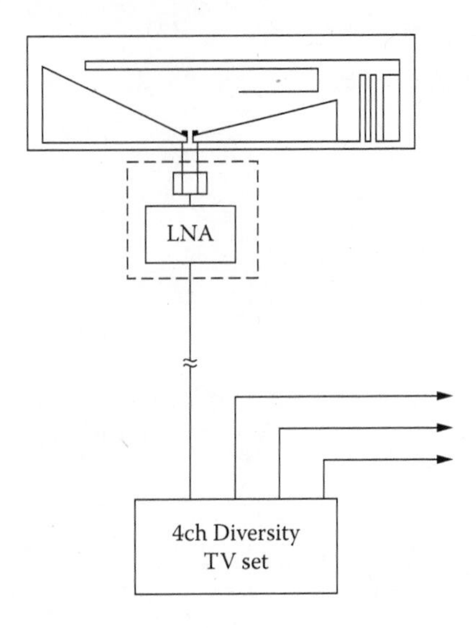

FIGURE 6.22
Automotive antenna element mounted on flexible foil. Arrows indicate connections with other antennas.

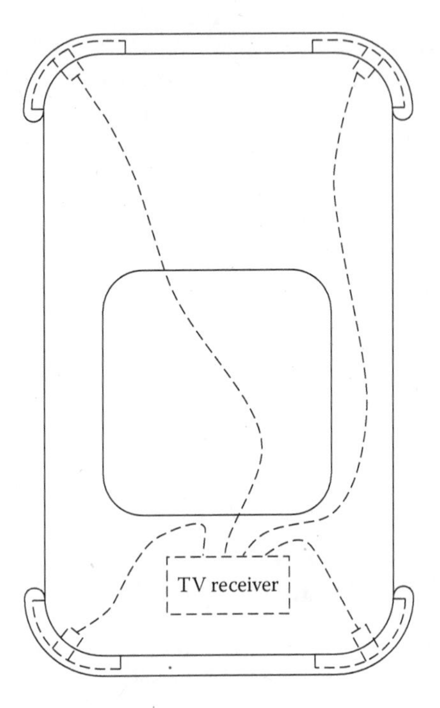

FIGURE 6.23
Four identical foil antennas mounted on car bumpers.

wave for any of the four bumper mounted antenna elements is more than 40 dBu. Each antenna element includes an LNA as part of a high quality diversity system.

6.5 Compact Mast Monopole

As mentioned in previous chapters, one disadvantage of a hidden antenna is a non-omnidirectional radiation pattern. Diversity systems improve reception quality but require a few antenna elements and additional electronic circuits that increase the cost of such systems. A small mast antenna [21] is a cost-effective solution. A base-mounted antenna is presented in Figure 6.24. The system has a monopole antenna with a height of 120 mm (a quarter of the wavelength at the 600 MHz middle frequency point, a grounded base with a matching circuit, and an amplifier.

The operating frequency band is around 400 to 800 MHz. Figures 6.25a and 6.25b show the antenna impedance and VSWR in this frequency range when using the matching circuit shown in Figure 6.25c. Simple topology of the matching circuit provides VSWR in the entire frequency range less than 3:1. The measured gain of the antenna mounted on a 1 m diameter circular metal plate for a vertically polarized wave in the horizontal plane is more than 0 dBi in the operating band.

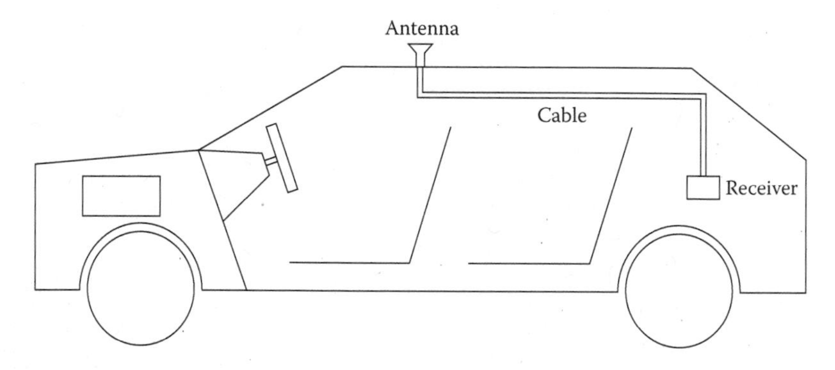

FIGURE 6.24
Monopole antenna for TV application mounted on car roof.

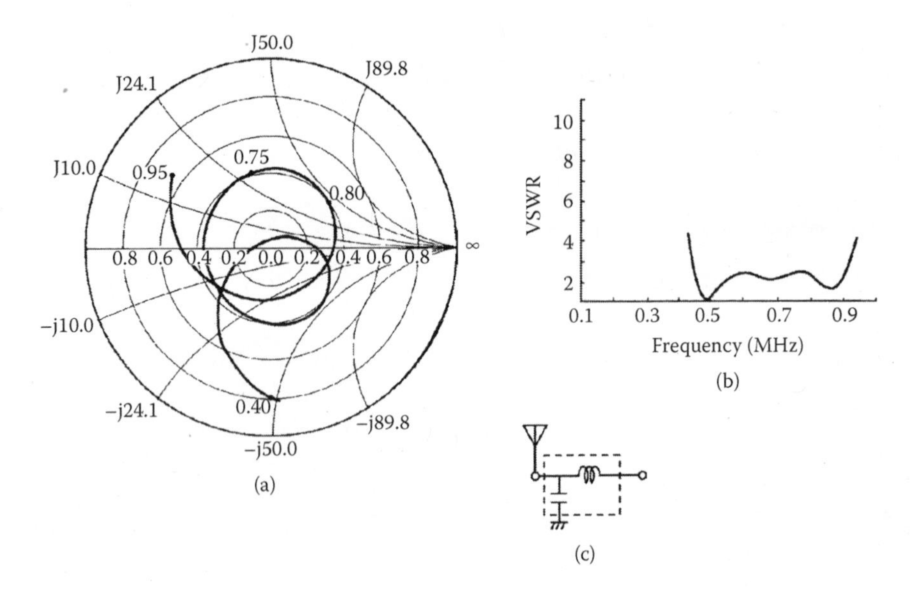

FIGURE 6.25
Monopole antenna parameters: (a) Impedance; (b) VSWR; (c) Matching circuit.

References

1. J. Wang and J. Winters, An embedded antenna for mobile DBS, *60th IEEE Vehicular Technology Conference*, 2004, pp. 4092–4095.
2. F. Hules, G. Streelman, and H. Yen, Direct broadcast reception system for automotive OEMs, *IEEE Antennas and Propagation Symposium*, 2005, pp. 80–83.
3. J. Wang et al., Vehicle Mounted Satellite Antenna Embedded within Moon Roof or Sun Roof, U.S. Patent 7,227,508, June 2007.
4. J. Wang et al., Vehicle Mounted Satellite Antenna Embedded within Moon Roof or Sun Roof, U.S. Patent 7,391,381, June 2008.
5. C. Balanis, *Antenna Theory Analysis and Design*, New York: John Wiley & Sons, 1997, pp. 257–271.
6. B. Hatcher, Granularity of beam positions in digital phased arrays, *Proceedings of IEEE*, 56, 1795–1800, 1968.
7. J. Lee, G/T and noise figure of active Array antennas, *IEEE Transactions on Antennas and Propagation*, 41, 241–244, 1993.
8. W. Pritchard and M. Ogata, Satellite direct broadcast, *Proceedings of IEEE*, 78, 1116–1130, 1990.
9. N. Itoh and K. Tsuchida, HDTV Mobile reception in automobiles, *Proceedings of IEEE*, 94, 274–280, 2006.
10. L. Reiter et al., Compact antenna with novel high impedance amplifier diversity module for common integration into narrow dielectric parts of a car skin, Institute of High Frequency Technology and Mobile Communication, University of Bundeswëhr Munich.

11. H. Iizuka et al., Modified H-shaped antenna for automotive digital terrestrial reception, *IEEE Transactions on Antennas and Propagation*, 53, Part 1, 2542–2548, 2005.
12. W. Stutzman and G. Thiele, *Antenna Theory and Design*, New York: John Wiley & Sons, 1981.
13. J. Imai et al., Experimental results of diversity reception for terrestrial digital broadcasting, *IEICE Transactions and Communications*, E85-B, 2527–2530, 2002.
14. S. Matsuzava et al., Radiation characteristics of on-glass mobile antennas for digital terrestrial television, *IEEE Antenna and Propagation Symposium*, 2, 1975–1978, 2004.
15. Y. Noh, Wide Band Glass Antenna for Vehicle, U.S. Patent 7,242,358, July 2007.
16. R. Doi et al., Glass Antenna and Glass Antenna System for Vehicles, U.S. Patent 7,071,886, July 2006.
17. H. Yamashita et al., Heating Line Pattern Structure of Defogger, U.S. Patent 7,211,768, May 2007.
18. H. Toriyama et al., Development of printed-on-glass TV antenna system for car, *IEEE Vehicular Technology Conference Digest*, 334–342, 1987.
19. H. Yotsuya et al., Window Antenna for a Vehicle with Dual Feed Points, U.S. Patent 4,727,377, February 1988.
20. K. Kudo et al., TV Antenna Apparatus for Vehicle, U.S. Patent 5,977,919, 1999.
21. S. Fukushima, Antenna Device, U.S. Patent Application 20080238802, October 2008.

7

Satellite Radio Antennas

7.1 Basic Passive Antenna Requirements

Satellite radio antennas include passive antenna elements and low noise amplifiers [1].

7.1.1 Parameters for Sirius Service (2320 to 2332.5 MHz)

For left-hand circular polarization (LHCP) antenna gain ($\theta = 0$ degrees corresponds to the zenith angle direction) must be more or equal to (1) 2 dBic for θ angles between 0 and 20 degrees; (2) 4.5 dBic for θ angles between 20 and 40 degrees; (3) 0 dBic for θ angles between 40 and 70 degrees. Antenna gain for vertical polarization (VP) must be at least –5 dBi at θ angles between 90 and 80 degrees.

7.1.2 Parameters for XM Service (2332.5 to 2345 MHz)

For LHCP, antenna gain is not less than 2dBic for angles θ between 30 and 65 degrees. Terrestrial antenna gain for linear polarization ($\theta = 90$ degrees) is equal to or more than –2dBi. All antenna parameters must be measured on a metal ground plane with a diameter of 1 m.

7.2 System With Two Antenna Elements

As mentioned in Chapter 1, a satellite radio antenna must receive LHCP signals from a satellite and vertically polarized (VP) signals transmitted from terrestrial power stations. Early satellite radio antennas contained one element for reception of the signal transmitted by satellite and another for reception of the signal transmitted terrestrially. Antennas with two elements that meet OEM requirements are shown in Figure 7.1 and Figure 7.2.

A combination of quadrifilar and monopole structure is shown in Figure 7.1a to Figure 7.1d. A quadriflar antenna consists of four helixes

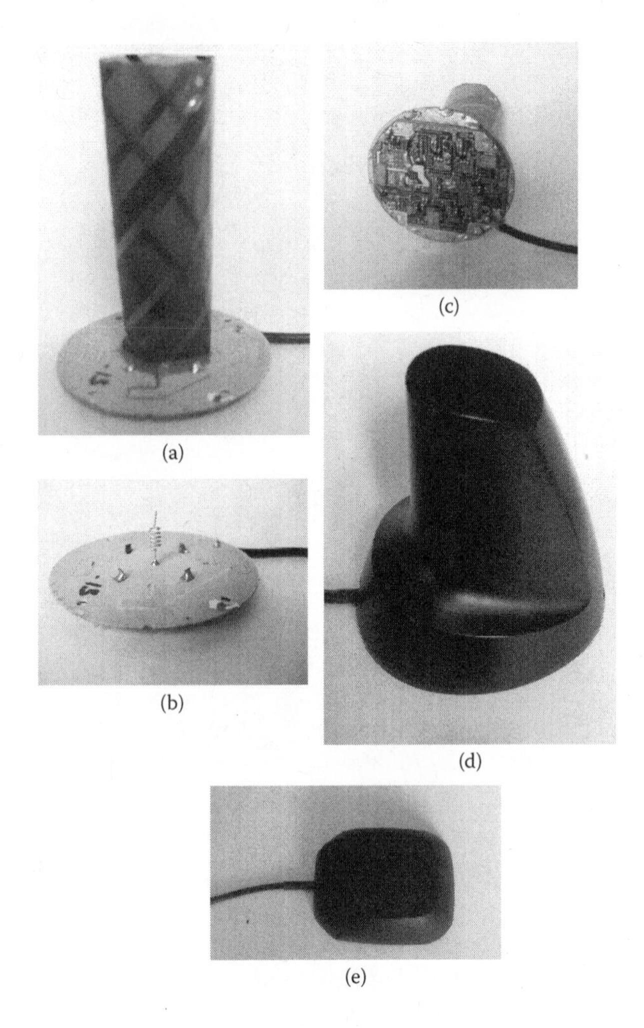

FIGURE 7.1
Combination of quadrifilar antenna and monopole structure: (a) satellite antenna portion; (b) terrestrial antenna portion; (c) amplifier location; (d) antenna in case; (e) antenna based on patch design.

spaced equally on a cylinder. This geometry, combined with an output feed network, creates LHCP. The helix monopole portion (Figure 7.1b) mounted inside the quadriflar antenna receives signals from the VP antennas mounted on terrestrial towers. Figure 7.1c shows an amplifier circuit. The quadrifilar design with a case is shown in Figure 7.1d. Physical system dimensions (with case) are length ≈90 mm, width ≈90 mm, and height ≈100 mm.

Figure 7.2 shows the configuration of a satellite radio antenna with crossed dipoles and separately mounted monopoles. LHCP is provided by the cross

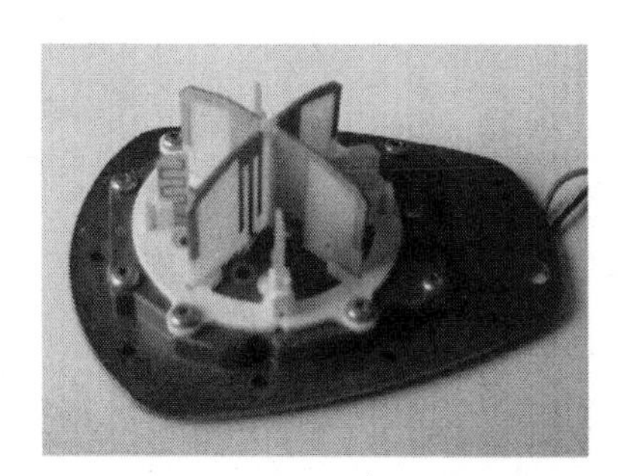

FIGURE 7.2
Antenna with cross printed dipole elements for satellite reception and monopoles for terrestrial reception.

dipoles mounted on a printed dielectric board and VP is realized by four meandered monopoles mounted on a dielectric substrate. One monopole is positioned within each quadrant of the cross dipole and has length approximately equal to 0.25 wavelength. Antennas shown in Figure 7.1 and Figure 7.2 include two separate amplifiers and therefore two outputs: one connected with the first receiver input and the other connected with the second.

7.3 System With Single Patch Antenna Element

Today, antenna engineers use design solutions based on a single antenna for receiving both satellite and terrestrial signals. An example is a compact microstrip patch antenna printed on a low loss ceramic substrate [3,4]. Figure 7.1e shows an example of this type antenna in its case.

The single patch antenna element receives circular polarized signals from a satellite located at the high elevation angle direction and receives VP signals transmitted from a terrestrial station positioned at the low elevation angle direction ($\theta \approx 90$ degrees). The proper choice of ceramic can ensure acceptable satellite gain performance and satisfactory terrestrial gain value.

A microstrip patch antenna consists of a very thin metallic patch placed above the conducting ground plane. The patch and ground plane are separated by a dielectric with constant ε_r. The simplest microstrip geometry of the half wave linear polarized rectangular patch is shown in Figure 7.3a. The length L, width W, and thickness h are the major parameters. A coaxial

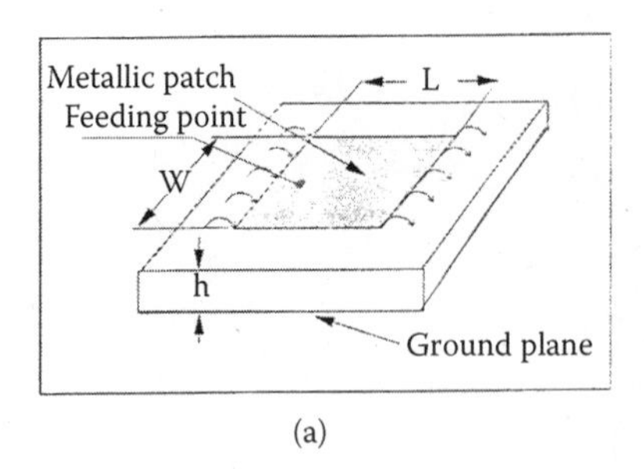

(a)

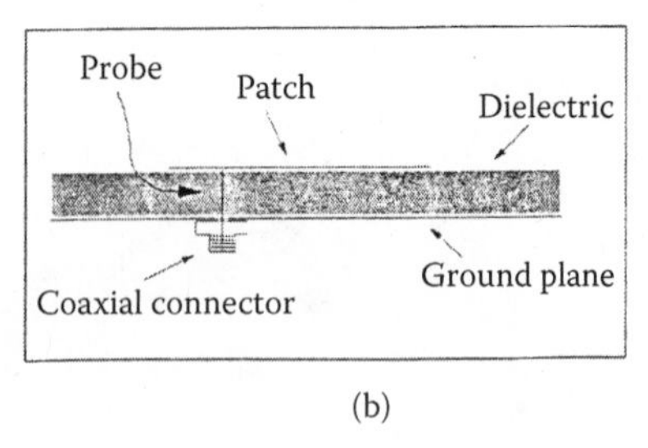

(b)

FIGURE 7.3
Rectangular linear polarized patch antenna geometry: (a) Patch showing fringe fields accounting for radiation, (b) Patch excited by coaxial connector.

connector and RF coaxial cable provide the antenna feed (Figure 7.3b). The patch is printed on the top side of the dielectric substrate and the bottom of the board is the ground. The amplifier of the active patch antenna is mounted under the ground of the patch. The patch provides CP wave reception. The feeding probe operates as a monopole and together with the patch receives the VP wave from the low elevation angles (Reference [1], p. 60]. The SDARS antenna design includes only one low noise amplifier (LNA) and therefore only one output is connected to a receiver. The current typical SDARS antenna for automotive use is a rectangular patch.

7.4 Simplified Engineering Formulas for Rectangular Passive Patch Parameters

7.4.1 Geometry Dimensions for Linear Polarization

The patch acts as a resonant $\lambda/2$ parallel-plate transmission line. Radiation from the patch is emitted from two slots, at the left and right edges as shown in Figure 7.3a. The resonant length L of the patch is a critical value and typically is slightly smaller than $\lambda/2$, where λ is the wavelength in the dielectric ($\lambda = \lambda_0/\sqrt{\varepsilon_r}$, λ_0= wavelength in free space). The length of the patch [5] is given by:

$$L = \frac{300}{2 \cdot f_r \cdot \sqrt{e_{eff}}} - 2\Delta_l \tag{7.1}$$

where

$$\Delta_l = 0.412 \cdot h \cdot \frac{(\varepsilon_{eff} + 0.3) \cdot (W/h + 0.264)}{(\varepsilon_{eff} - 0.258) \cdot (W/h + 0.813)} \tag{7.2}$$

TABLE 7.1

Comparisons of Measurement and Simulation Results

No.	Equation L (mm)	Experiment L (mm)	Equation W (mm)	Experiment W (mm)	Frequency f_r (GHz)	Diel. Const. ε_r	Thickness h (mm)
1	12	12	11.23	11.2	7.05	2.55	2.42
2	15.77	15.80	13.85	13.75	5.1	2.55	4.76
3	10.1	10	9.05	9.1	4.6	10.2	1.27
4	25.08	25	20.01	20	3.97	2.22	0.79

Source: M. Kara, *Microwave and Optical Technology Letters*, 12, 1996. Reprinted with permission of Wiley-Blackwell, Inc.)

$$\varepsilon_{eff} = \frac{\varepsilon_r + 1}{2} + \frac{\varepsilon_r - 1}{2} \cdot \left(1 + 12\frac{h}{W}\right)^{-\frac{1}{2}} \tag{7.3}$$

$$W = \sqrt{h \cdot \frac{300}{f_r \cdot \sqrt{\varepsilon_r}}} \cdot \left[Ln\left(\frac{300}{f_r \cdot \sqrt{\varepsilon_r} \cdot h}\right) - 1\right] \tag{7.4}$$

where f_r = resonance frequency in GHz, ε_r = dielectric constant of substrate, h = thickness in millimeters, and L and W are expressed in millimeters.

The feed point for a linear polarized antenna is at the axis of symmetry along the L length side as demonstrated in Figure 7.3a. A good initial approximation for a 50 Ω feed point position is about one third of the distance from the center of the antenna to a radiating edge. The best way to define a more accurate feed point location is to use computer software for simulating microstrip antennas, for example, the IE3D package [6] or the more simple computer program included with Sainati's book [7]. Table 7.1 presents a few examples [8] and compares results of calculations based on Equations (7.1) to (7.4) and measurements for different patch geometries. The experimental results are in good agreement with the approximate engineering calculations.

7.4.2 Circular Polarization Geometry

Two practical methods are used to create a single feed patch for receiving a circularly polarized wave. These methods are shown in Figure 7.4a, and b (see References [3], pp. 505 and 506, and [9], pp. 39 and 40). The first method uses a rectangular patch, and the ratio L_{cp}/W_{cp} is chosen so that two orthogonal linear modes are excited with equal amplitude and a 90 degree phase shift. Orthogonal modes are combined in the far zone and produce circular polarization. The feed point is placed on diagonal 1 (LHCP) or diagonal 2 (RHCP) as shown in Figure 7.4a. The ratio L_{cp}/W_{cp} in a rectangular CP

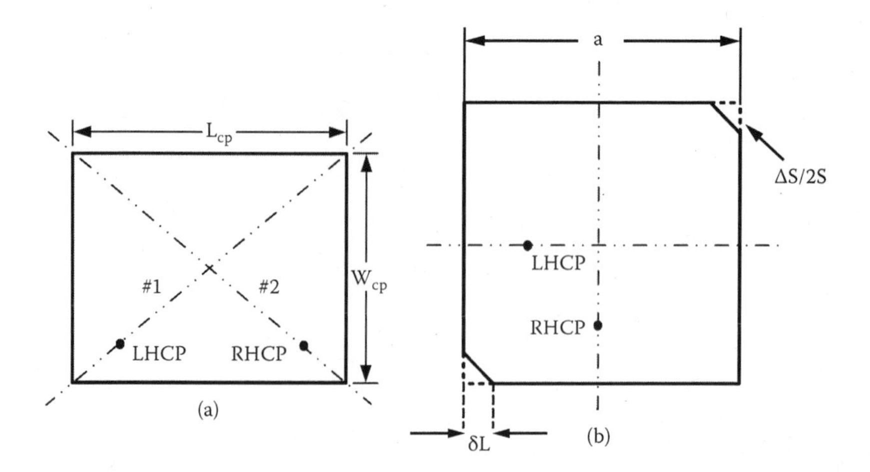

FIGURE 7.4
Configurations of single feed circular polarization patch antennas: (a) rectangular patch; (b) patch with cut-off corners.

microstrip antenna (Figure 7.4a) is given in Reference [5], p. 770:

$$W_{cp} = \frac{L_{cp}}{1 + 1/Q_T} \tag{7.5}$$

Total quality factor Q_T will be defined in the next paragraph. The second technique consists of removing a pair of corners from a conductor patch to create a truncated corner square patch antenna as shown in Figure 7.4b. This method creates two orthogonal linear modes that must be adjusted to provide identical amplitudes and a 90 degree phase difference between them. The ratio between ΔS and S (Figure 7.4b) equals:

$$\frac{\Delta S}{S} = \frac{1}{2 \cdot Q_T} \tag{7.6}$$

where S = area of the square patch antenna prior the corner removal (Reference [3], p. 39). The zero order approximation for a side length equal to a can be given by:

$$a = \sqrt{L \cdot W} \tag{7.7}$$

where L and W values are calculated according to Equations (7.1) to (7.4). Cutting edges (Figure 7.4B), with δL are calculated in accordance with the ratio given by:

$$\delta L \approx \frac{\sqrt{L \cdot W}}{\sqrt{Q_T}} \tag{7.8}$$

7.4.3 Performances of Passive Antenna Elements

Asymptotically, the pattern directivity D for a microstrip antenna ([5], p. 751) is given by:

$$D \approx 6.6 \text{ (8.2 dBi)}; \; W << \lambda_0 \tag{7.9}$$

The half power beamwidth of an antenna is equal to the angular width between directions, where the radiated fields are reduced to $1/\sqrt{2}$ of the maximum value ([5], p. 751):

$$\theta_H \approx 2 \cdot \cos^{-1}\left(\frac{1}{2 + k_0 \cdot W}\right)^{\frac{1}{2}} \tag{7.10}$$

$$\theta_E \approx 2 \cdot \cos^{-1}\left(\frac{7.03}{3 \cdot k_0^2 \cdot L^2 + k_0^2 \cdot h^2}\right)^{\frac{1}{2}} \tag{7.11}$$

where $k_0 = \frac{2\pi}{\lambda_0}$ and θ_H and θ_E are the half power beamwidth in the H and E planes, respectively. E and H radiation pattern planes are named for linear polarized waves. The E plane contains the electric field vector and direction of maximum radiation ([5], p. 29). The H plane contains the magnetic field vector and direction of maximum radiation; it is perpendicular to the E plane. The beamwidth of the antenna can be increased by choosing a patch with smaller L and W values if it uses a substrate with a higher dielectric constant. As beamwidth increases, gain and directivity decrease. Antenna gain for matching with a load microstrip antenna is equal to $G = \eta_{23} \cdot D$, (Equation 2.8, Chapter 2) in which efficiency for a microstrip antenna η_{23} can be expressed ([5], p. 762) as:

$$\eta_{23} = \frac{Q_T}{Q_r} \tag{7.12}$$

Total quality factor Q_T is determined (p. 761) by:

$$\frac{1}{Q_T} = \frac{1}{Q_r} + \frac{1}{Q_d} + \frac{1}{Q_c} + \frac{1}{Q_{sw}} \tag{7.13}$$

where Q_r = quality factor due to radiation losses; Q_c = quality factor due to conduction losses; Q_d = quality factor due to dielectric losses; and Q_{sw} = quality factor due to surface waves. For very thin substrates, ($h \ll \lambda_0$) Q_r is usually the dominant factor (p. 761). Simulation results that estimate different quality factors are shown in Reference [9], pp. 54 and 55. For example, for a microstrip rectangular antenna operating at 2.45 GHz with $L \approx W = 56.46$ mm, $\varepsilon_r = 2.6$, $h = 0.76$ mm and dielectric loss $\tan\delta = 0.0025$, the ratio $Q_T/Q_r \approx 0.76$, $Q_T/Q_{sw} \approx 0.02$, $Q_T/Q_c \approx 0.09$, and $Q_T/Q_d \approx 0.12$. The patch antenna bandwidth for linear polarization can be estimated by the following [5, p. 762]:

$$BW_{LP} = \frac{\Delta f}{f_0} = \frac{VSWR - 1}{Q_T \cdot \sqrt{VSWR}} \tag{7.14}$$

For circular polarization, bandwidth is given by:

$$BW_{CP} = \frac{\sqrt{2(VSWR - 1)}}{Q_T} \tag{7.15}$$

The bandwidth of a CP patch antenna is always greater than that of a linear polarized (LP) microstrip [10]:

$$\frac{BW_{CP}}{BW_{LP}} = \sqrt{\frac{2VSWR_{\max}}{VSWR_{\max} - 1}} \tag{7.16}$$

7.4.4 Circular Polarization Design Guidelines

The procedure for circular polarization (CP) design may be summarized as follows:

1. According to Equations (7.1) through (7.4), determine L and W for an antenna with linear polarization.
2. Determine the location of the feed point that provides the best matching of a linear polarized antenna to 50 ohm. Feed point location typically is calculated using IE3D electromagnetic software that significantly expedites the design process.
3. Calculate directivity and gain. Determine efficiency and quality factor Q_T.
4. Based on Equations (7.5) and (7.6) in Section 7.4.2, make an initial estimation of the CP antenna geometry.
5. Determine the location of the feed point using a simulation to determine the CP of the patch antenna.

It is necessary to emphasize that the suggested steps serve only as baselines for determining patch antenna topology. Experiments are required to complete the design of a CP antenna. Therefore, after mathematical

FIGURE 7.5
Milling Machine Quick Circuit 5000 prototyping system.

simulation, it is preferable to fabricate an antenna prototype and test it. A prototype can be built, for example, using Ti-Tech's Milling Machine Quick Circuit 5000 (Figure 7.5). This system is a very convenient tool for "cut and try" fabrication of passive and active antennas printed on dielectric circuit boards.

The Quick Circuit is a desktop prototyping system for use in the design phase. It can drill, mill, and route traces and spaces as fine as 0.1 mm in width. The machine is controlled by a computer that uses design geometry data converted to Gerber or DXF files for the board prototype built. Preparing the board prototype with Milling Machine is rapid (a few minutes for a board size of a few centimeters) and the cutting process can be repeated to achieve the best experimental tuning of board parameters.

7.5 Simulated Example of SDARS Patch

This numerical example is based on IE3D simulation software. Figure 7.6a presents a patch antenna mounted on an infinite grounded dielectric plane. Geometry parameters are length $L = 13.4$ mm, width $W = 12.8$ mm, height $h =$ 4 mm, and dielectric $\varepsilon_r = 18.1$. The ground plane is an infinite perfect electrical conductor. The coordinates of the feed point are $dx = 4.6$ mm and $dy = 4.3$ mm (the point with coordinates $x = 0$ and $y = 0$ is located at the corner of the patch).

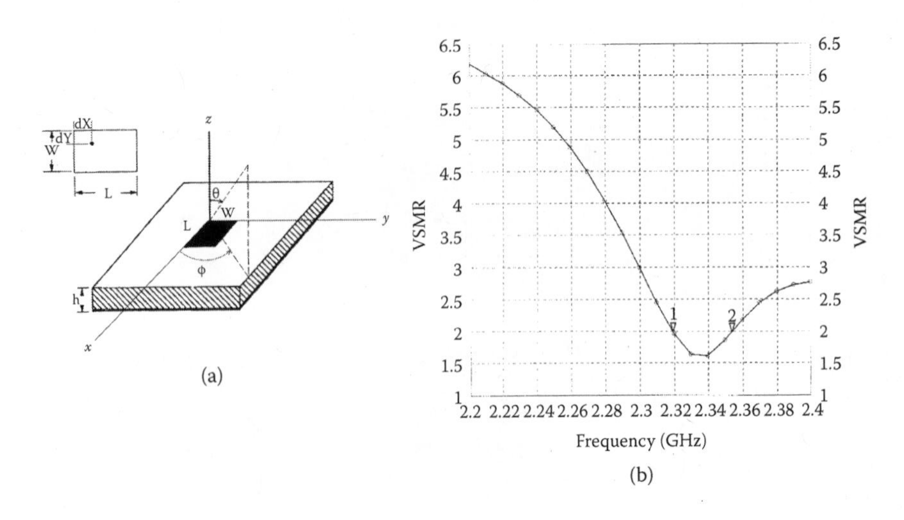

FIGURE 7.6
(a) Simulated antenna geometry; (b) Simulated VSWR of patch antenna.

The calculated VSWR is shown in Figure 7.6b. LHCP and RHCP radiation patterns in two orthogonal planes ($\varphi = 0$ degrees and $\varphi = 90$ degrees) are presented in Figure 7.7 (2.33 GHz) and Figure 7.8 (2.34 GHz). The estimated maximum directionality values are equal to 6 dBic for LHCP

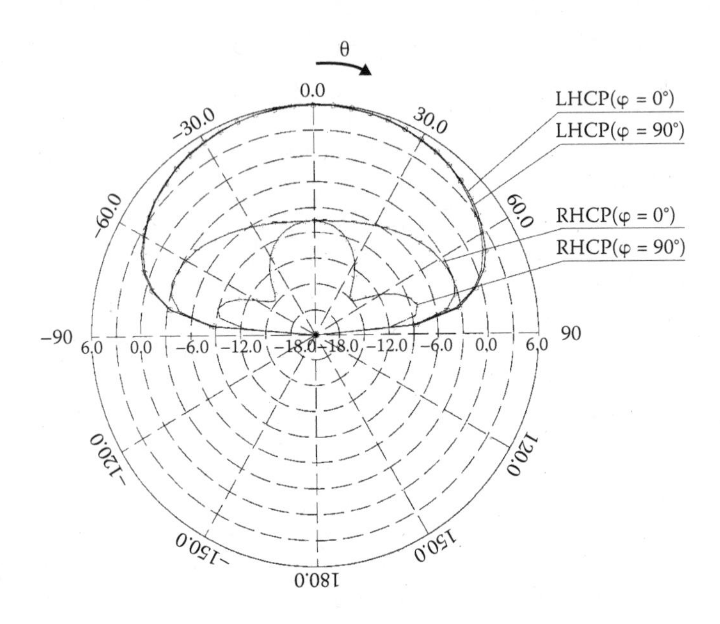

FIGURE 7.7
Simulated antenna radiation pattern in two orthogonal planes, 2.33 GHz frequency point.

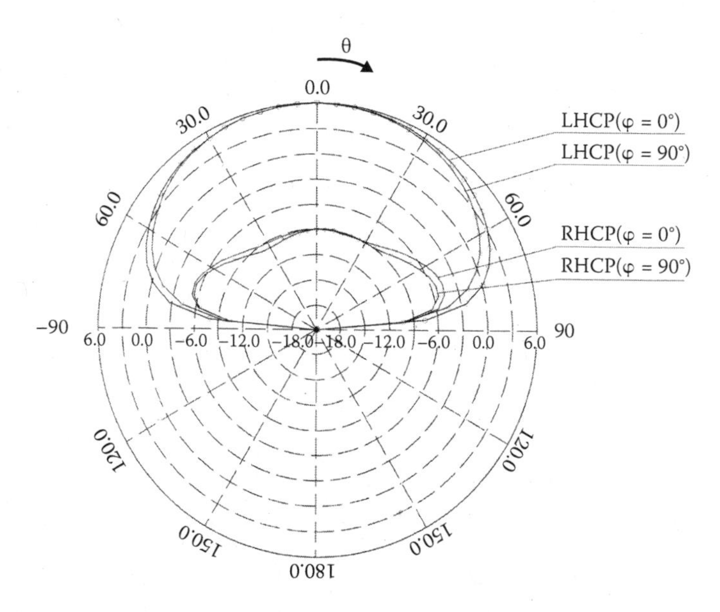

FIGURE 7.8
Simulated antenna radiation pattern in two orthogonal planes, 2.34 GHz frequency point.

(both frequency points). The cross component (RHCP) equals –7 dBic for 2.33 GHz and –6 dBic for 2.34GHz. Figure 7.7 and Figure 7.8 show that the radiation pattern value vanishes at a θ value of ±90 degrees. This is justified by image theory. Of course, simulation results serve only as baseline values for designing SDARS patch antennas. The final geometry that meets the performance requirements is determined by experimental "cut and try" tuning.

7.6 Commercially Available Passive Elements

A variety of passive patch SDARS antenna elements can be purchased on the electronic market, for example:

TOKO — The XM passive patch (Part DAV2338CL10T) has the following parameters:

- Frequency range: 2338 ± –10 MHz
- Center frequency without case: 2348 MHz
- Center frequency with case: 2338 MHz
- Bandwidth: 20 MHz

- Impedance: 50 ohms
- Return loss 10 dB
- Gain for LHCP: ~6.7 dBic at zenith; 6.2 dBic at 30 degrees from zenith; 6 dBic at 40 degrees; 2.5 dBic at 60 degrees; and –4 dBic at horizon
- Dimensions: 25 mm × 35 mm × 4 mm
- Measuring condition: 70 mm^2 ground plane

INPAQ—This company produces patch antennas that operate in the 2320 to 2332.5 Sirius frequency band. The gain at zenith ($\theta = 0$ degrees) is about 5 dBic and gain at $\theta = 80$ degrees elevation is –3 dBic. The axial ratio is less than 3 dB. Measuring condition is 50 mm^2 ground plane.

AMOTECH — The frequency band is 2326 ± –4 MHz; zenith gain is ~ 6.1 dBi. The axial ratio is less than 3 dB. The dielectric constant is 9.8 and dimensions are 25 mm × 25 mm × 4 mm. Measuring condition is 70 mm^2 ground plane.

7.7 Vertical Polarization Gain of Patch Antenna at Horizon Angle Direction

Duzdar et al. [11] examined vertical polarization (VP) gain as a function of patch thickness. Antennas with different heights are etched on identical dielectric material substrates. Results show that VP gains of the 4, 5, 6, and 7 mm height antennas were measured as –2.88, –2.58, –2.48, and –2,38 dBi, respectively. The VP gain of the $\lambda/4$ monopole at the same elevation angle was +0.7 dBi. These results were obtained in an anechoic chamber with a ground plane of 1 m diameter and show that increasing the dielectric thickness causes an increase in VP gain at low elevation angles.

Figure 7.9 shows a second antenna design option that allows increases of VP gain [12]. The common patch antenna used for satellite radio reception has a parasitically enhanced perimeter that extends from the circuit board and encompasses the antenna. This perimeter is called a parasitic fence and the fence is soldered to the ground plane. Results for a patch antenna with a fence were compared with performance of an antenna element without a fence. The antenna with a fence increased the VP gain values at low elevation angles ($\theta = 80$ to 90 degrees) by 0.5 to 1.8 dB. It also increased the LHCP gain for the θ angle sector within 80 and 90 degrees by 0.3 dB and reduced antenna gain for θ between 0 and 70 degrees by an average value of 0.4 dB.

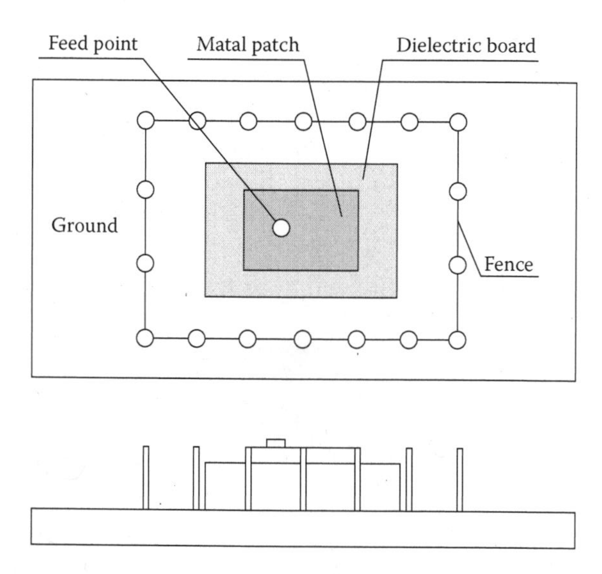

FIGURE 7.9
Patch antenna module with parasitic fence.

Petros et al. [2] presented measurement results for a microstrip patch antenna etched on a low loss ceramic substrate. The VP gain averaged –3 dBi for 2332 and 2339 MHz over 360 degrees—very close to the –2 dBi specification.

7.8 Ground Size Effects

The parameters of an antenna mounted on a car roof vary based on ground plane size [13–15]. For example, the radiation pattern measured in the elevation plane shown in Figure 7.10 on a small 1.5 $\lambda_0 \times 1.5\ \lambda_0$ grounded substrate (2340 MHz) produced scattered radiation in the backward direction (in comparison with an infinite grounded dielectric, Figure 7.7 and 7.8). Kuboyama et al. investigated resonance frequency, radiation pattern, and gain as a function of ground plane size [14]. They reported that, for a patch antenna with a ground plane equal to the patch metallization, the resonance frequency was higher than that of an infinitely sized ground plane antenna.

Detailed investigation of the finite size ground plane effect on XM SDARS antenna performance was studied [15]. An antenna gain of 2 dBic for θ angles between 30 and 65 degrees is close to the optimum 3 dBic that theoretically may be achieved for a small antenna mounted on an ideal infinite metal plate. Figure 7.11 shows results of a simulation comparison of LHCP

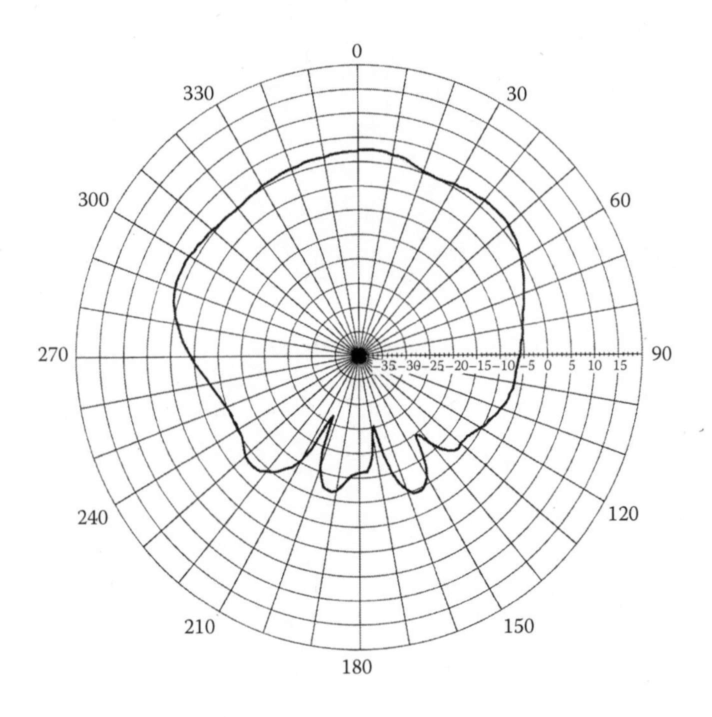

FIGURE 7.10
Measured antenna radiation pattern of patch antenna mounted on 1.5 × 1.5 grounded substrate.

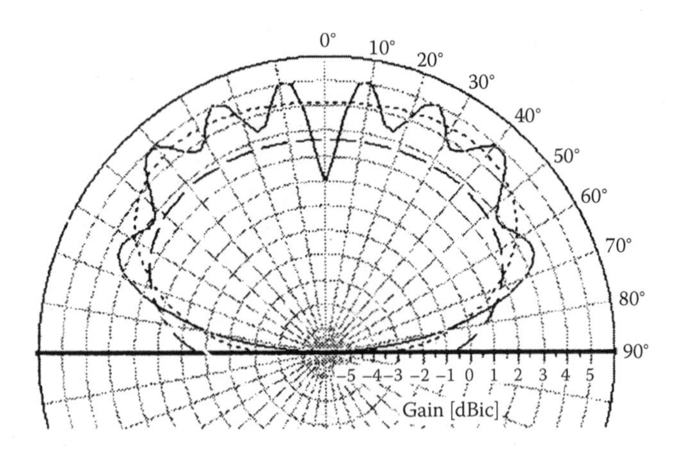

FIGURE 7.11
Simulated elevation antenna radiation patterns: (a) infinite perfect electrical conducting ground plane (dotted line); (b) without ground plane in free space (broken line); (c) 1 m in diameter ground plane (solid line). (From M. Daginnus, R. Kronberger, and A. Stephan, *IEEE Antennas and Propagation Society International Symposium*, 2002. Copyright 2002 IEEE. With permission.)

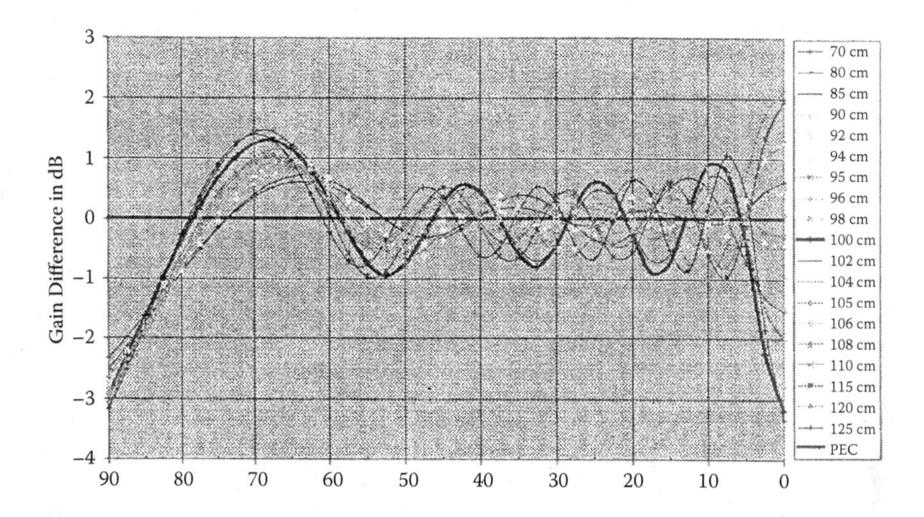

FIGURE 7.12
Satellite antenna gain differences of different circular ground plane sizes. (From M. Daginnus, R. Kronberger, and A. Stephan, *IEEE Antennas and Propagation Society International Symposium*, 2002. Copyright 2002 IEEE. With permission.)

radiation patterns of an antenna mounted on an infinite perfect electrical conductive ground plane (pointed line), a 1 m diameter ground plane (solid line), and an antenna without a ground plane (dotted line). For the finite 1 m diameter ground plane, ripples were about ± 1 dB within the entire range of elevation angles. Figure 7.12 shows a set of simulations with varying ground plane diameters. The location of the ripples appears to depend on the ground plane size. Ripples ±1 dB are typical for different sized metal planes. The simulation presented in [15] shows that SDARS antennas must be placed at least 15.2 cm from the sheet metal edge to provide satisfactory performance.

7.9 On-Vehicle Antenna Location

Of course, vehicle size and shape affect antenna parameters. Reflections and shadows of a metal vehicle body can significantly change a radiation pattern. A variety of measurements of an XM SDARS antenna at 2335 MHz [16] are presented in Figure 7.14 to Figure 7.17.

Figure 7.14 shows the measured radiation pattern of the antenna mounted at position 1 (Figure 7.13). Directionality is not as smooth as the curve in Figures 7.7 and 7.8. Based on measurement results, the XM gain requirement (2 dBi with θ at 30 to 65 degrees) cannot be maintained for all elevation angles.

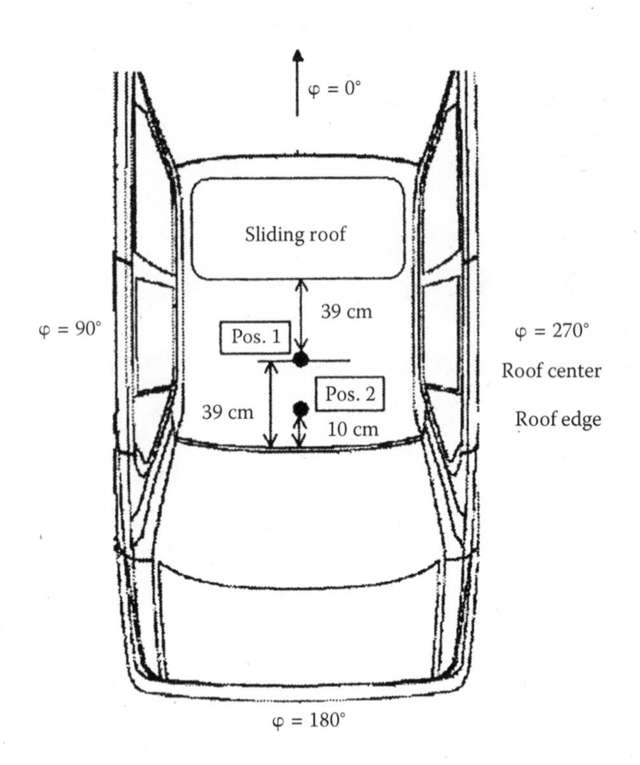

FIGURE 7.13
Vehicle antenna mounting locations. (From R. Kronberger, G. Hassmann, and S. Schulz, *IEEE Antennas and Propagation Society International Symposium*, 2002. Copyright 2002 IEEE. With permission.)

The radiation pattern for the antenna at position 2 is presented in Figure 7.15. Comparison of both radiation patterns shows that the curve for position 2 near the roof edge has a slightly increased ripple. In addition, directionality for the pattern in Figure 7.15 shows reduced gain value of 2 to 5 dB toward the front.

Figure 7.16 demonstrates the directionality of an antenna mounted on a trunk lid. The radiation pattern is blocked by the car body and shows significant reduction in the gain at the front. VP radiation of the terrestrial antenna portion along the horizontal direction is shown in Figure 7.17 (dotted line indicates directivity of the quarter wavelength monopole). The deviation from the omnidirectional pattern (maximum ripple ~12 dB for both the terrestrial antenna portion and the monopole) is caused by the shape of the car. Measurements also show a reduced average over 360 degrees gain of the antenna mounted on the vehicle (from 1 to 2.2 dB, depending on vehicle location). Therefore, only field tests can determine final antenna location.

Table 7.2 presents gain comparison data (dB scale) in the elevation plane (LHCP, $\varphi = 0$ degrees) for different roof-mounted antennas: a quadrifilar helical design shown in Figure 7.1 (antenna 1 in the table), a cross dipole presented

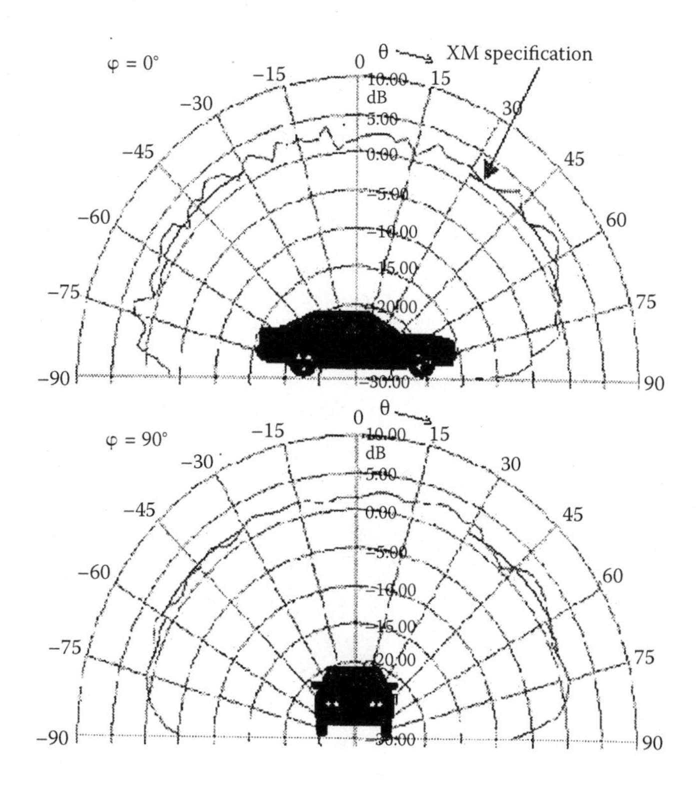

FIGURE 7.14
LHCP Radiation pattern of antenna at position 1. (From R. Kronberger, G. Hassmann, and S. Schulz, *IEEE Antennas and Propagation Society International Symposium*, 2002. Copyright 2002 IEEE. With permission.)

in Figure 7.2 (antenna 2), and a simple patch printed on FR4 dielectric material with thickness of 1.6 mm (antenna 3). The quadrifilar antenna design showed the highest gain at θ angle range within 20 to 60 degrees; the worst performer was the thin patch antenna mounted on the dielectric substrate. However, the differences among the three designs were not very significant. The patch design is the simplest solution from the view of a manufacturer.

7.10 Compact Dual-Polarized Antenna

Figure 7.18 shows a combined compact dual-polarized antenna with increased VP gain [17] that operates in both XM and Sirius frequencies. The system has two low noise amplifiers: one for the satellite antenna portion and the other for the terrestrial antenna. Amplifiers are mounted under the ground.

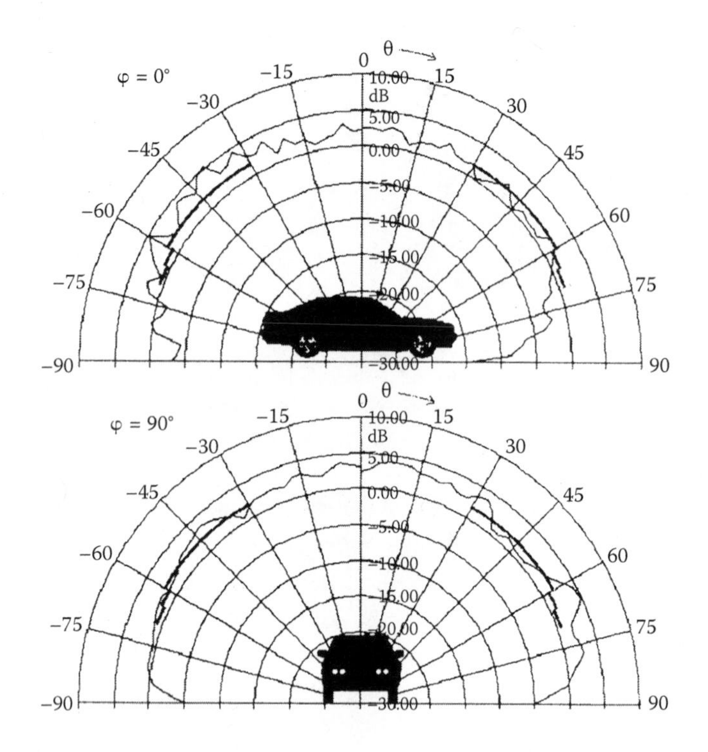

FIGURE 7.15
LHCP radiation pattern of antenna at position 2. (From R. Kronberger, G. Hassmann, and S. Schulz, *IEEE Antennas and Propagation Society International Symposium*, 2002. Copyright 2002 IEEE. With permission.)

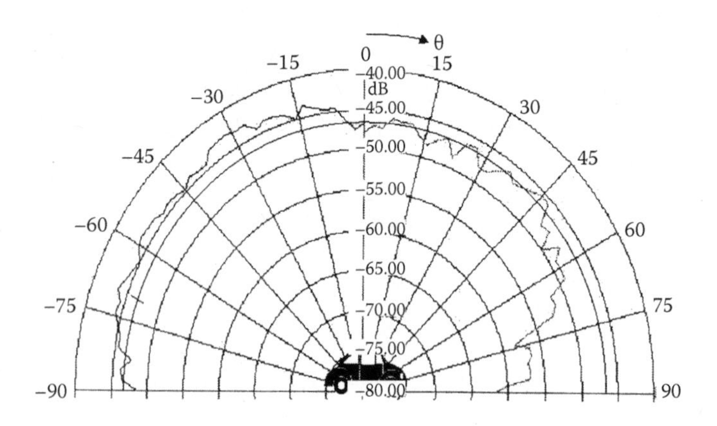

FIGURE 7.16
LHCP radiation pattern of antenna mounted on trunk lid. (From R. Kronberger, G. Hassmann, and S. Schulz, *IEEE Antennas and Propagation Society International Symposium*, 2002. Copyright 2002 IEEE. With permission.)

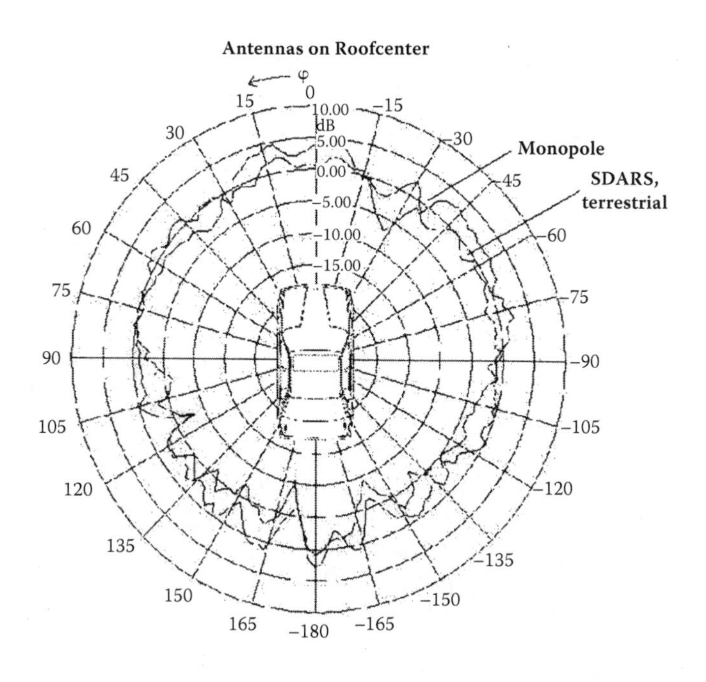

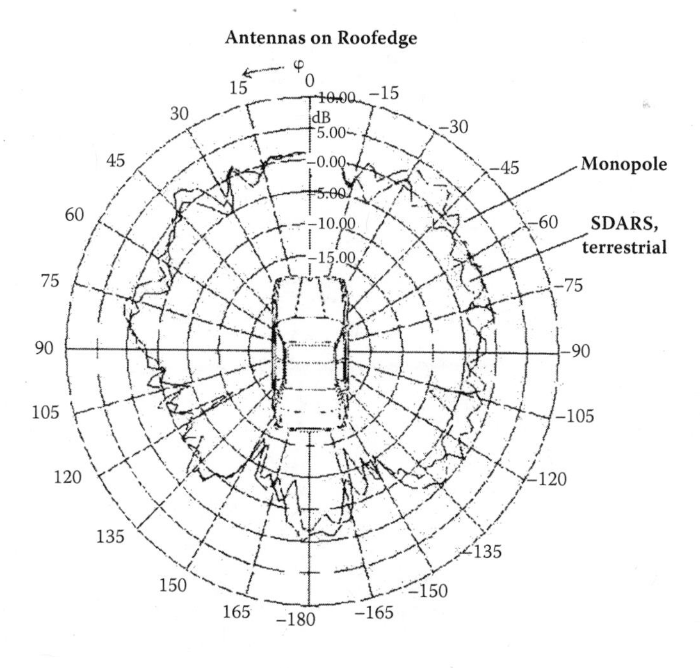

FIGURE 7.17
Vertical polarization terrestrial radiation pattern at horizon and directionality of reference monopole. (From R. Kronberger, G. Hassmann, and S. Schulz, *IEEE Antennas and Propagation Society International Symposium*, 2002. Copyright 2002 IEEE. With permission.)

TABLE 7.2

Comparison of Three Roof-Mounted SDARS Antenna Designs

Frequency (MHz)	2332	2340	2345
Gain (20 degrees) – gain (60 degrees) Antenna 1	2.3	2.45	2.42
Gain (20 degrees) – gain (60 degrees) Antenna 2	3.55	3.48	3.6
Gain (20 degrees) – gain 60 degrees) Antenna 3	3.91	4.04	4.16

As seen in the figure, the VP monopole antenna is placed in the center of the satellite annular ring microstrip antenna.

Both antenna portions have a common ground plane and different feed ports. A $\lambda/4$ delay line provides circular polarization of the satellite portion. The inner and outer diameters of the annular ring patch are 8 and 30.6 mm, respectively, and the antenna is embedded inside 6 mm thickness FR-4. The distance h_2 between the ring and ground is 3 mm and the distance h_3 between the $\lambda/4$-strip line and the ground is 0.5 mm. The 6 mm high monopole is a cylindrical element with a 1 mm diameter. The monopole has a circular patch with a 13.4 mm diameter at the top. The monopole patch is shorted with two pins to the

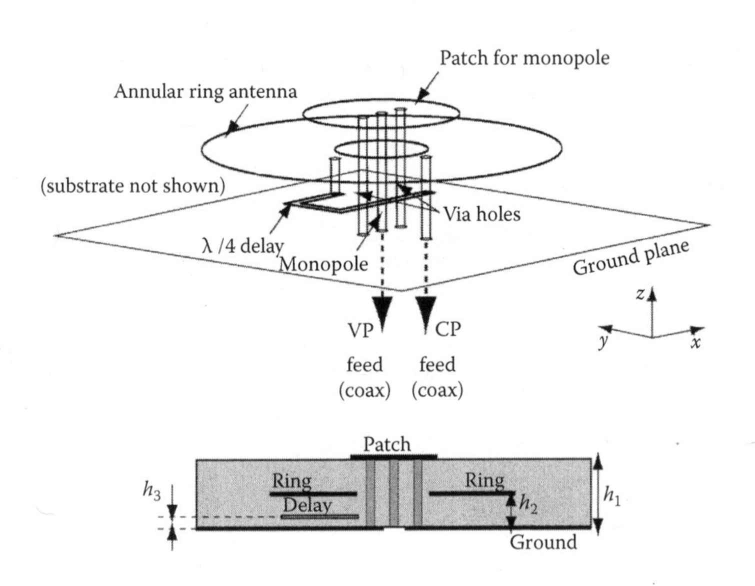

FIGURE 7.18
Combined dual polarized antenna pair. (From Y. Hong, *IEEE Transactions on Microwave Theory and Techniques*, 54, Part 1, 2006. Copyright 2006. IEEE. With permission.)

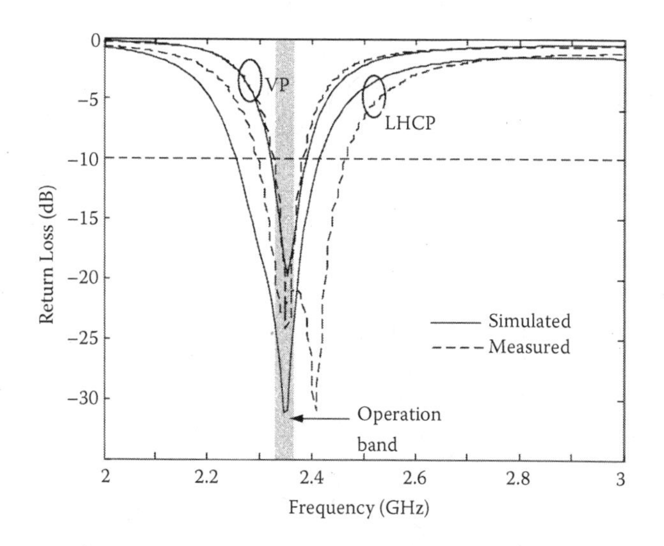

FIGURE 7.19
Simulated and measured return losses. (From Y. Hong, *IEEE Transactions on Microwave Theory and Techniques*, 54, Part 1, 2006. Copyright 2006. IEEE. With permission.)

ground. The spacing between the centers of these pins is 4.9 mm for optimum performance at 2330 MHz. The size of the ground plane is 70 × 70 mm.

Figure 7.19 shows return losses for both antenna portions. The graph shows good agreement between the simulated and measured results. A shift of approximately 80 MHz is seen in the peak S11 frequency for LHCP. Figure 7.20 demonstrates the design of a separate antenna portion system

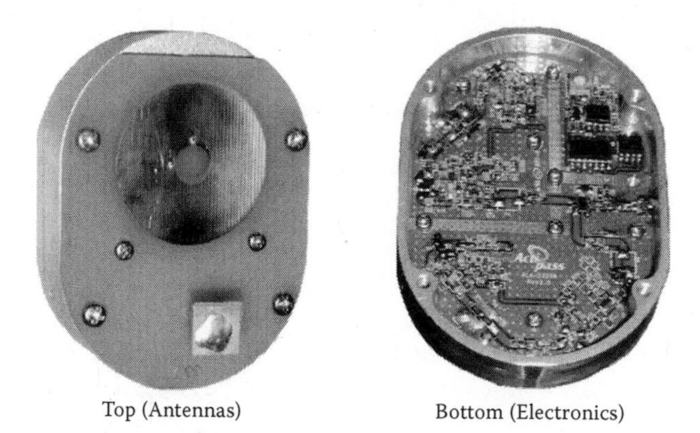

FIGURE 7.20
Top and bottom of fabricated dual-path dual-polarized antenna. (From Y. Hong, *IEEE Transactions on Microwave Theory and Techniques*, 54, Part 1, 2006. Copyright 2006. IEEE. With permission.)

with integrated amplifiers. The overall dimensions of the package are 7 cm × 5 cm × 3.6 cm with the monopole protruding 1.6 cm above the top surface. The electronics are located under the antenna ground plane. The maximum gains (antenna plus amplifier) of the system were measured as 31.3 and 21.9 dBi for the LHCP and VP modes, respectively.

7.11 Low Profile Cross Slot Antenna for SDARS Application

Figure 7.21a shows a cross-slot antenna that operates in satellite frequency range [18]. It presents a cavity-backed crossed-slot aperture with a single coaxial cable. The design described in the paper is built on a substrate of Duroid 5880. The cavity is 63 mm square and 3 mm thick; the two slots are 51 and 54 mm in length and 1 mm wide. The edges of the cavity are electroplated with copper. The feed is offset from the center of the cavity by 23 mm, along the diagonal of the cavity. The antenna is fed from the coaxial cable by attaching the outer conductor to the back surface of the cavity and the inner conductor to the front surface through the metal plated via.

The circular polarization is provided by two slots so that radiation from one is 90 degrees out of phase with the other. Radiations from two orthogonal slots with slightly different lengths are combined in the far zone, producing circular polarization. Toward the horizon, the antenna system (as shown in [18]) produces linear polarization.

Figure 7.22 presents the measured return losses of an antenna design based on the dimensions shown in Figure 7.21a. The experimental prototype is tuned for 2340 MHz and is well matched for this frequency point. As noted in Reference [18], the input impedance can be tuned by moving the feed point along the cavity diagonal. This antenna was designed for LHCP,

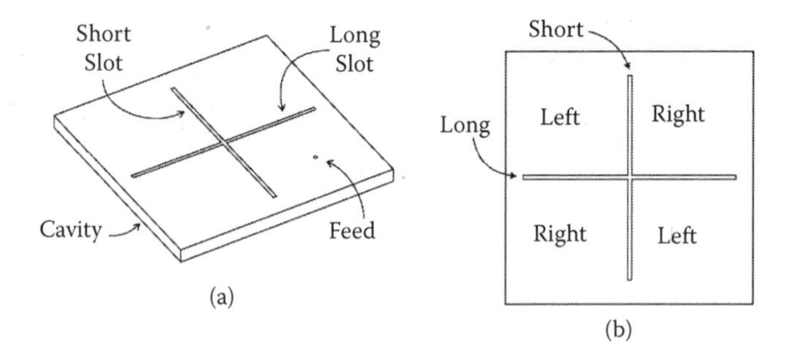

FIGURE 7.21
Cross-slot design for SDARS radio: (a) antenna geometry; (b) feed point locations for LHCP and RHCP. (From D. Sievenpiper, H. Hsu, and R. Riley, *IEEE Transactions on Antennas and Propagation*, 52, 2004. Copyright 2004. IEEE. With permission.)

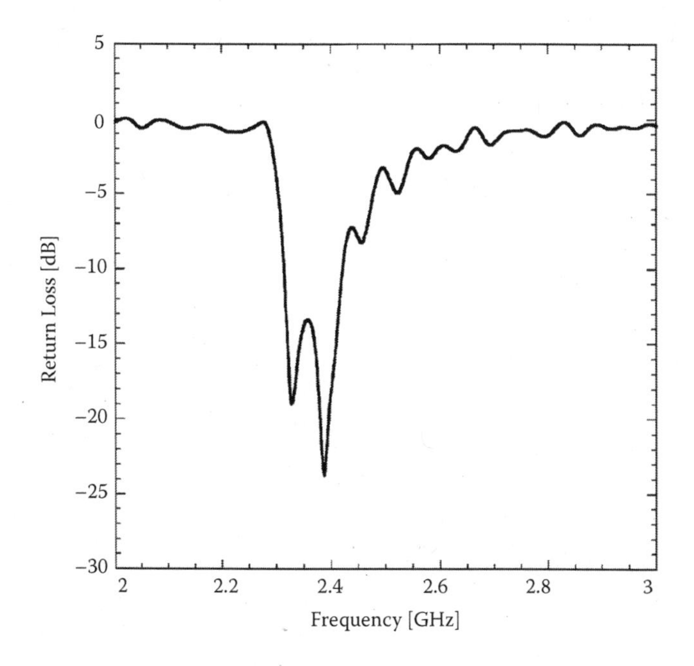

FIGURE 7.22
Measured return losses. (From D. Sievenpiper, H. Hsu, and R. Riley, *IEEE Transactions on Antennas and Propagation*, 52, 2004. Copyright 2004. IEEE. With permission.)

but RHCP can be produced by changing the location of the feed point shown in Figure 7.21a. For the radiation pattern measurements, the antenna was placed in the center of a 1 m square ground plane. The LHCP radiation pattern is shown in Figure 7.23. Measurements indicate that in the range ±70 degrees from zenith, the antenna has an average gain of +2.2 dBic; the cross component (RHCP) shown in Figure 7.23 by dotted line is lower by 11.3 dB. The radiation VP pattern near the horizon (within 20 degrees) has a gain of about +0.5 dBi. At 2340 MHz, the efficiency was estimated at 94%.

7.12 Glass Mount System

Figure 7.24 presents a glass-mounted satellite radio antenna [19]. The radiation element of the antenna includes a top metallization portion (e.g., square truncated corner square patch) and a bottom (excitation) portion that is electromagnetically coupled with the top metallization through the rear windshield. As shown in Figure 7.24, the bottom metallization portion has a slot and, in operation, the combination of the slot and microstrip line located adjacent to the amplifier circuit (shown in phantom in Figure 7.24) excites electromagnetic waves received by the top portion.

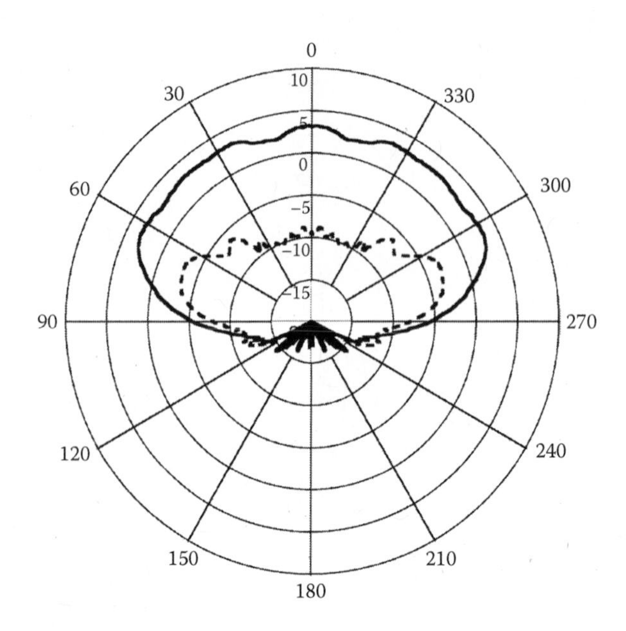

FIGURE 7.23
Measured LHCP radiation pattern. (From D. Sievenpiper, H. Hsu, and R. Riley, *IEEE Transactions on Antennas and Propagation*, 52, 2004. Copyright 2004. IEEE. With permission.)

Glass-mounted antennas present some advantages including eliminating the need for a proper seal around an installation. However, this design shows increased losses associated with RF coupling between the top and bottom antenna portions that reduces the effective passive antenna gain.

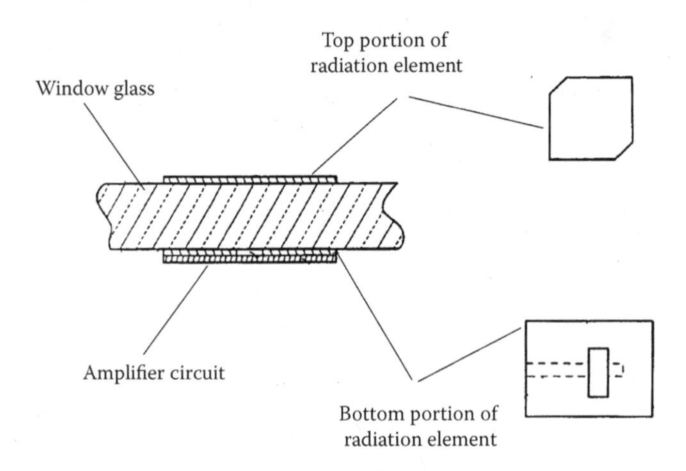

FIGURE 7.24
Block diagram of glass-mounted satellite radio antenna.

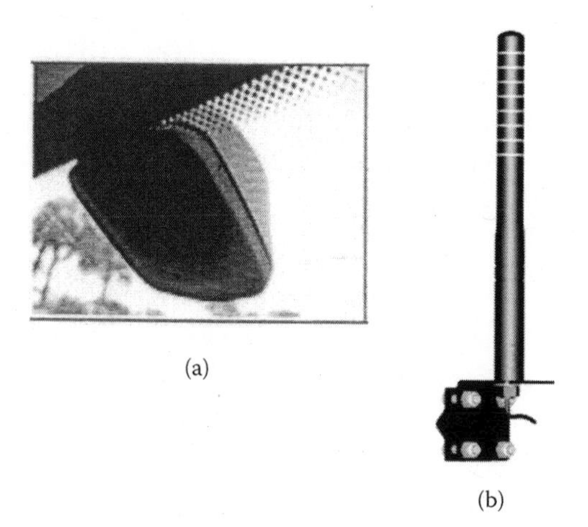

FIGURE 7.25
SDARS antennas from Think Wireless (TW): (a) antenna mounted on glass; (b) antenna mounted on truck.

Figure 7.25a shows a commercially available glass-mounted SDARS antenna from TW Think Wireless Inc.

7.13 XM Truck Antenna

Figure 7.25b shows an XM truck antenna from TW Think Wireless Inc. The minimal gains are 2 dBic at 20 to 60 degrees elevation and –2 dBi at the horizon. Amplifier gain is about 30 dB; height is 24.1 cm; radome diameter is 2.1 cm.

7.14 Active Design

7.14.1 Antenna Amplifier Requirements

Figure 7.26a demonstrates an electrical block circuit of an amplifier for satellite radio reception. A typical low noise amplifier for SDARS application has three stages for satellite reception. Table 7.3 shows detailed typical specifications for a Sirius amplifier used with a single antenna for satellite and terrestrial reception. As a rule, the first stage along with the input matching circuit provides a low noise figure for the entire amplifier circuit. The second and third stages are matched to provide maximum output power. A SAW band pass filter with very narrow bandwidth around 50 MHz between the second and third stages handles out-of-band rejection. Figure 7.27 shows

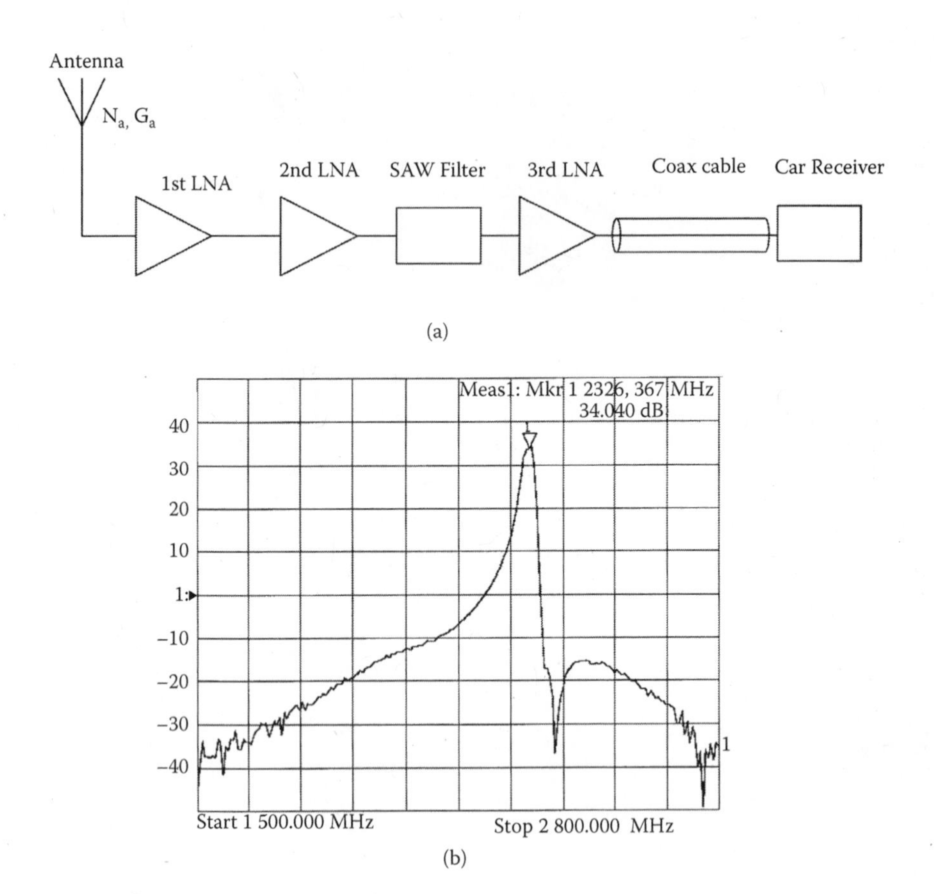

FIGURE 7.26
SDARS antenna amplifier and gain versus frequency response: (a) block diagram of electrical circuit; (b) amplifier gain.

an example of SAW filter response for a SDARS XM application. Frequency response parameters for a SAW filter for XM SDARS application include central frequency response of 2338.755 MHz; maximum insertion attenuation of ~2.5 dB at 2332.5 to 2345 MHz; and maximum amplitude ripple of 0.8 dB. Attenuation levels at various frequencies are:

- 88 to 108 MHz → 65 dB
- 880 to 960 MHz → 60 dB
- 1710 to 1990 MHz → 50 dB
- 2305 MHz → 11 dB; 2315 MHz → 10 dB; 2320 MHz → 4.4 MHz
- 2450 MHz → 25 dB; 3060 MHz → 49 dB

TABLE 7.3

Active Antenna Amplifier Parameters for Sirius Satellite Radio (2320 to 2332.5 MHz)

Gain without Cable and Connectors	Gain with Cable and Connectors	LNA Noise Figure	Gain Ripples across Entire F Band	LNA 1 dB Compression Point and Third Order Intercept
~36 dB	23 dB minimum	1 dB maximum	1 dB maximum	P1 = −17 dB IP3 = −10 dB
Active Antenna System Noise Temperature	**Out-of-Band Rejection (dB) Central F (MHz)**	**VSWR at Amplifier Output**	**Power Supply Current**	**Power Supply Voltage**
100 *K*	F < 2175 → −30 F = 2227 → −15 F = 2340 → −8 F = 2400 → −15 F = 2500 → −27 F > 2700 → −40	<2:1	150 mA maximum	5 V

LNA = Low noise amplifier. F = frequency. P1 = compression point. IP3 = third-order intercept.

Figure 7.26b shows a typical gain of a commercially available antenna amplifier for a Sirius application.

A design with two separate antennas for satellite and terrestrial reception utilizes two amplifiers: one for satellite reception and a terrestrial amplifier

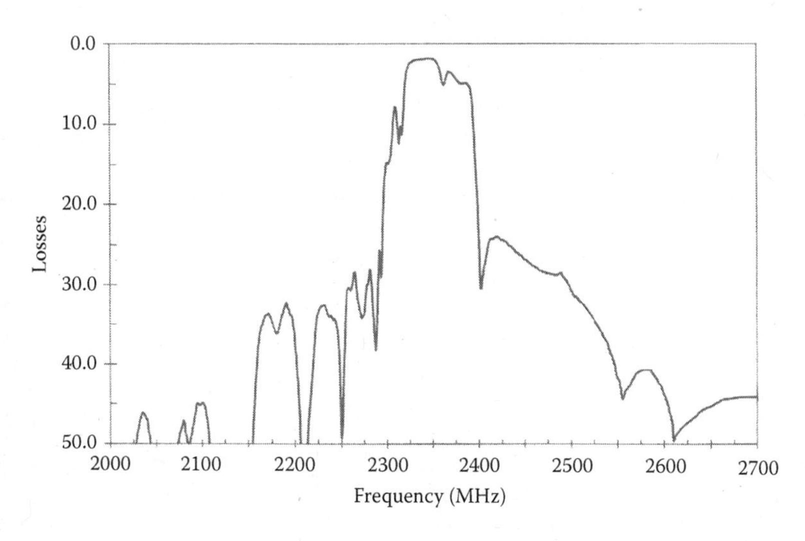

FIGURE 7.27
Typical SAW filter response.

(with only two amplifier stages) exhibits a gain ~28 dB and noise figure ~2 dB. Usually Fakra connectors are used between the receiver and active antenna.

Different companies recommend using their products in SDARS amplifier systems. For example, ultralow noise GaAs HJ FET NE3509 M4 and NE3508M04 discrete elements from the California Eastern Lab (CEL) can be good candidates for the first stage. Based on specifications, these discrete elements exhibit a gain of ~15 dB and a minimum noise figure of 0.35 dB. The CEL three-stage satellite amplifier is built on the base of three identical discrete NE3509MO4S elements. The Tyco MAALSS0013 integrated amplifier operating in the 2.2 to 2.4 GHz band produces a gain of 17 dB, noise figure equal to 0.9 dB (minimum), a 1 dB compression point of 10 dBm, and a typical IP3 output equal to 26 dBm. Infineon Technologies suggests a low noise amplifier based on a SiGe BFP640F transistor at the first and second stages and BFP650 at third stage.

Reference [17] discusses use of an ATF 54143 FET element for the first stage of a SDARS active antenna. The authors use only two stages for satellite and terrestrial amplifiers. Their results show that the gains for satellite and terrestrial amplifiers are equal, at 30 dB and 19 dB, respectively, with noise values of 0.9 and 1.7 dB. The second stage utilizes an MGA-86563MMIC integrated element from Avago Technology with gain equal to 20 dB and a noise of 2 dB. The first and second stage terrestrial amplifiers are the same; they are built on ATF 54143 transistor bases.

When designing an amplifier with a high gain, stability becomes very important. As stated earlier, a high gain circuit has loss resistive elements to provide good stability. Stability factor K can be calculated from the measured S parameters in a wide frequency band (each frequency to a particular S parameter).

All S parameter values can be measured using a standard network analyzer. Some of the modern analyzers include special built-in software that calculates stability factor K on the bases of measured S parameters. Preliminary estimation of stability can be predicted on the basis of circuit simulation with the help of software packages. For example, the Genesys package can accomplish simulation on the basis of S parameters of discrete elements or integrated amplifier elements (typically known from the data sheet).

7.14.2 Amplifier Simulation Results

Today, a few powerful software tools can help in designing LNA SDARS amplifiers and expedite the design process. One of these, the Genesys software, provides linear and nonlinear simulation, realizes low pass, high pass, and band pass filter synthesis, helps to match amplifier with antenna, and estimates amplifier noise.

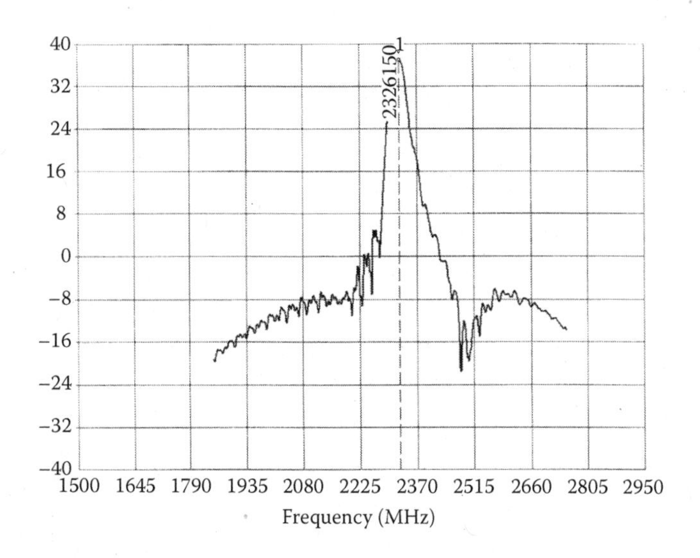

FIGURE 7.28
Simulated antenna amplifier gain.

Figure 7.28 shows calculated gain of the three stages of a SDARS amplifier in the Sirius frequency range of 2320 to 2332.5 MHz using the Genesys tool. The first and second stages are based on low noise PHEMT transistor ATF34143, the third stage is accomplished with a GaAs MMIC amplifier element MGA-86563. The MGA-86563 is an integrated element matched to 50 ohms that can be used without impedance matching as a high performance 2 dB NF gain block. The SAW narrow band filter with a frequency response, as shown in Figure 7.27, is placed between the second and third stages. *S* parameter noise data for both amplifier elements are available in the literature. Figure 7.28 shows the amplifier gain of a 50 ohm impedance system (source and load). Figure 7.29 presents the input and output VSWRs of the amplifier. The simulation considers the dimensions of strip lines that connect the elements of the circuit on the 0.8 mm thick FR-4 dielectric board. Simulation results revealed amplifier gain around 37 dB; input and output VSWRs are less than 2 over the entire frequency range; noise figure was about 0.9 dB in band.

7.15 Commercially Available Modules

Companies designing and producing commercially available XM and Sirius automotive antenna modules include Delphi, Belkin, Metra, Directed Electronics, Receptec LLC (now Larid Technologies), Micro-Ant, and others.

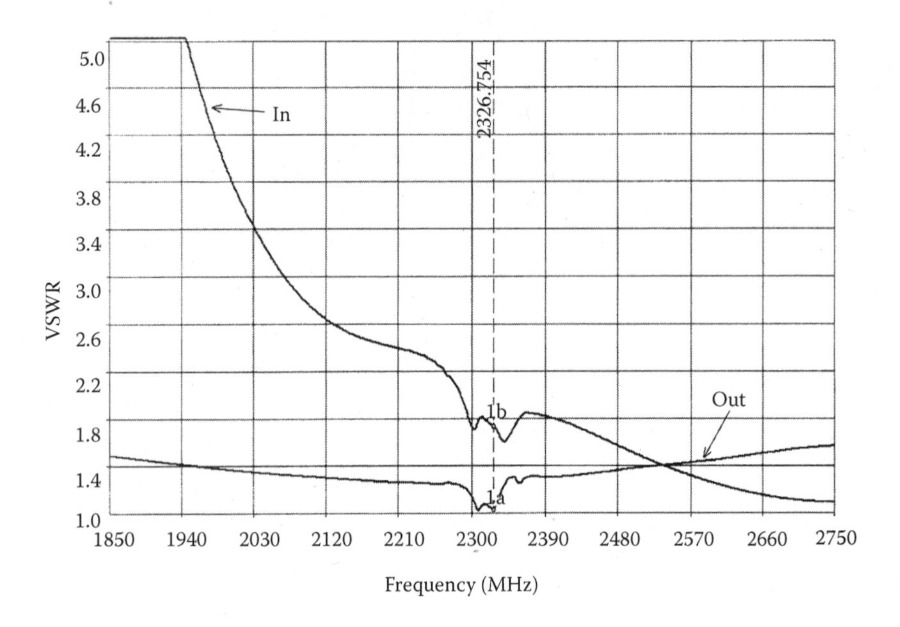

FIGURE 7.29
Input and output VSWRs of simulated amplifier.

A typical antenna module includes a passive patch, low noise amplifier, and RF cable connected with a satellite radio.

7.16 SDARS Parameter Measurements

As stated previously, certain antenna parameters must be measured to meet requirements: passive antenna gain for different elevation angles; radiation pattern; output VSWR of the antenna system; antenna temperature; amplifier gain, intermodulation distortion P_{IP3}; and 1 dB compensation power point P_{1dB}.

7.16.1 Gain and Radiation Pattern Measurements in Anechoic Chamber

Usually passive and active antenna parameters are tested (without a vehicle) in an anechoic chamber and mounted on a 1 m diameter metal screen for testing. The antenna and screen are placed on a turntable that can be rotated in both azimuth and elevation planes. A signal generator is connected with a transmitting antenna; the receiving antenna is connected with a spectrum analyzer. During the design process, radiation pattern and directionality are measured for E and H plane linear polarized waves, LHCP, and

RHCP radiation patterns. The distance between the satellite antenna with the 1 m metal screen and the transmitting antenna must exceed 4 m, which determines the far zone for the antenna on the metal screen. The gain of the antenna without a car is determined by the traditional relative method (Chapter 2).

7.16.2 Gain and Radiation Pattern Measurements of Antenna Mounted on Car

The elevation radiation pattern measurements of a satellite antenna mounted on or in a car are performed with a special radio transparent arch (Figure 7.30 and Figure 2.2). A car with an antenna is placed on a turntable that does not rotate and is maintained at a predetermined azimuth angle φ. A spectrum analyzer connected with an antenna output records the elevation radiation pattern values as the transmitting antenna with the generator moves along the arch, changing the elevation angle. Measurements of directionality in horizontal plane are achieved by rotating the antenna mounted on a car on a turntable over 360 degrees in azimuth. The transmitting antenna stays at the given elevation angle.

FIGURE 7.30
Measurements of antenna on vehicle. (From R. Kronberger, G. Hassmann, and S. Schulz, *IEEE Antennas and Propagation Society International Symposium*, 2002. Copyright 2002. IEEE. With permission.)

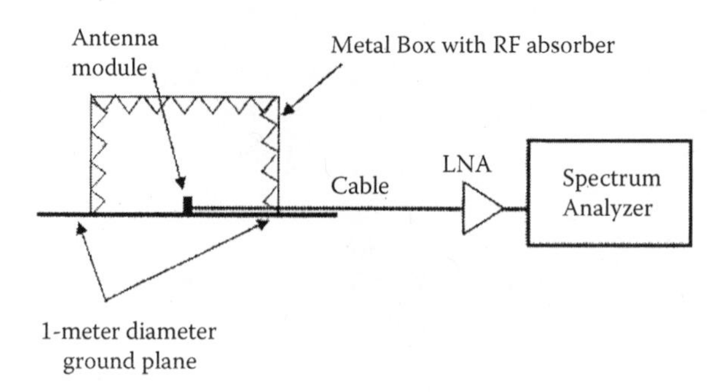

FIGURE 7.31
Antenna noise temperature measurement arrangement.

7.16.3 SDARS Antenna Noise Temperature

According to the procedure described in Reference [2], the antenna noise temperature measurement arrangement is shown in Figure 7.31. An external high gain LNA has a noise figure ~1 dB. The estimation of the active antenna noise temperature includes the following steps:

Step 1 — Place an SDARS active antenna module on a 1 m ground plane into an absorber box with metal outer walls. The spectrum analyzer indicates the power P_{hot} (dBm).

Step 2 — The antenna module is directed to the sky without absorber box, away from large buildings. The spectrum analyzer indicates the power P_{cold} (dBm).

Step 3 — Perform calculations according to the following equations. The Y factor is given by:

$$Y = P_{hot} - P_{cold}\ (\text{dB})$$

or

$$y = 10^{\left(\frac{Y}{10}\right)} \tag{7.17}$$

A typically measured y factor is 4 to 5 dB. The formula to calculate active antenna module noise temperature T_S is:

$$T_S = T_{cold} + \frac{T_0 - y \cdot T_{cold}}{y - 1} \tag{7.18}$$

where T_0 = ambient temperature at the measurement site (usually room temperature, 270 K) and T_{cold} = 35 K (sky noise temperature). The antenna noise temperature can be calculated as:

$$T_{ant} = T_S - (F - 1) \cdot T_0 \tag{7.19}$$

F is the noise figure of spectrum analyzer measured at the external LNA input. Assuming that the Y factor is equal to 4 dB (2.5), $T_{cold} = 35\ K$. Using Equations (7.18) and (7.19), we estimate: $T_S \approx 157\ K$ and $T_{ant} \approx 87\ K$.

7.17 Diversity Circuits

An antenna mounted on a car roof (an ideal location) uses the metal roof plane as the ground and its performance typically meets Sirius and XM requirements. For certain car configurations (e.g., convertible), an antenna cannot be mounted on the roof. As stated in previous sections, the radiation pattern of a trunk-mounted antenna shows increased deviations for various azimuth angles. The same phenomenon is typical for an antenna mounted inside a vehicle, for example, under the front panel or deck lid.

Another factor that causes fluctuations of signals received by an SDARS antenna is multipath fading of the received signal caused by tall buildings in an urban area or vegetation effects [20]. The diversity technique helps to reduce the signal fluctuations and also improves reception quality. This technique is commonly used for FM radios and cellular phones and can be applied to SDARS antennas.

Figure 7.32a shows a single SDARS antenna with diversity ports A1, A2, A3, and A4 [21]. This is one design option for this type of system. Another choice is to use two or more identical antennas mounted in the different parts of the vehicle [1,23]. Antenna units may be located (Figure 7.32b) under a trunk lid

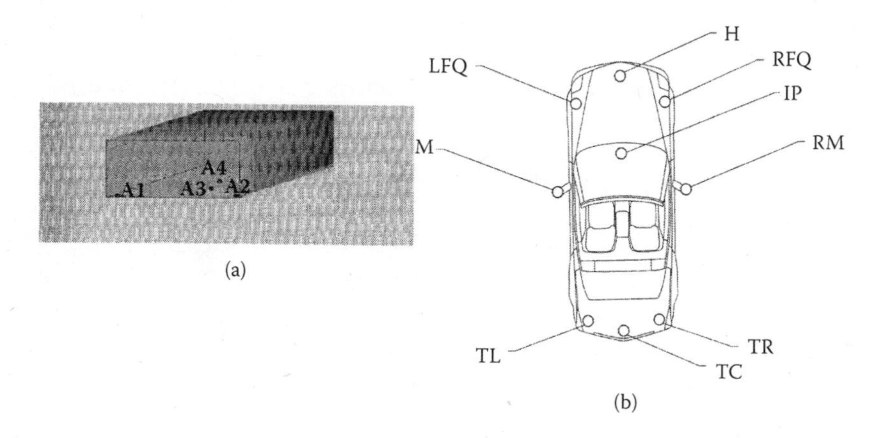

FIGURE 7.32
Diversity antenna design: (a) single antenna with four output ports (From H. Lindenmeier, *IEEE Antennas and Propagation Society International Symposium*, 2002, Copyright 2002. IEEE. With permission.); (b) design with a few diversity antennas (From K. Yegin, *IEEE Antennas and Propagation Society International Symposium*, IB, 2005, Copyright 2005. IEEE. With permission.)

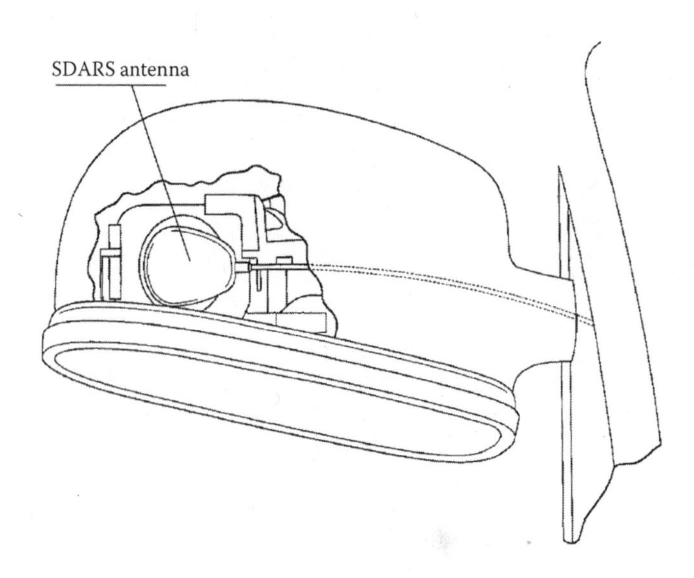

FIGURE 7.33
Antenna mounted in side mirror.

in a center location (TC), on the left (driver) side (TL), right (passenger) side, (TR), hood (H), left (driver) side front quarter panel (LFQ), right (passenger) side front quarter panel (RFQ), instrument panel (IP), left (driver) side mirror (LM), right (passenger) side mirror (RM), or any location a designer desires. Figure 7.33 shows an antenna mounted in a side mirror [23,24]. The driver or passenger mirror housing generally comprises a plastic cover that provides a hidden and aesthetically pleasing system. A side mirror-mounted system can include combined GPS/SDARS patch antenna elements and antennas intended for cellular frequency range. Two antennas operating in a diversity system can be mounted in such a way that one is in the left driver side mirror and second is positioned in the right passenger side mirror housing.

Figure 7.34 shows the location of a diversity system on a Cadillac XLR vehicle [1] and Table 7.4 presents the combined gain values of different elevation angles of a passive antenna system. Of course, the cost increase associated with the additional antenna module is a significant disadvantage. Figure 7.35 shows a diversity radiation pattern involving an overlap curve of the radiation patterns of the first and second antennas. Diversity efficiency can be estimated [21] as:

$$n = \frac{P_d}{P_S} \tag{7.20}$$

where $P_S = \dfrac{\Sigma_v \Delta t_{vS}}{t_{tot}}$, $P_d = \dfrac{\Sigma_v \Delta t_{vd}}{t_{tot}}$

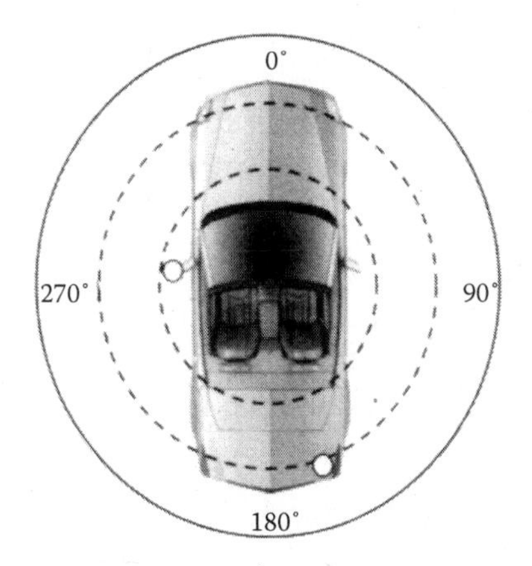

FIGURE 7.34
Hidden antenna locations on XLR. (From K. Yegin, *IEEE Antennas and Propagation Society International Symposium*, 1B, 2005. Copyright 2005. IEEE. With permission.)

TABLE 7.4

XLR Passive Antenna Diversity Patterns[a]

Elevation Angle	20	25	30	35	40	45	50	55	60
Minimum	–4.4	–4.7	–2	–1.5	–0.5	–2.4	–0.4	–0.6	–0.8
Average	2.3	2.6	2.7	2.9	2.9	3.1	3.1	2.8	2.9
Maximum	6.3	6.4	7.1	6.9	6.7	6.7	5.6	6.1	5.2

[a] Angle shown in degrees; gain shown in dBic.

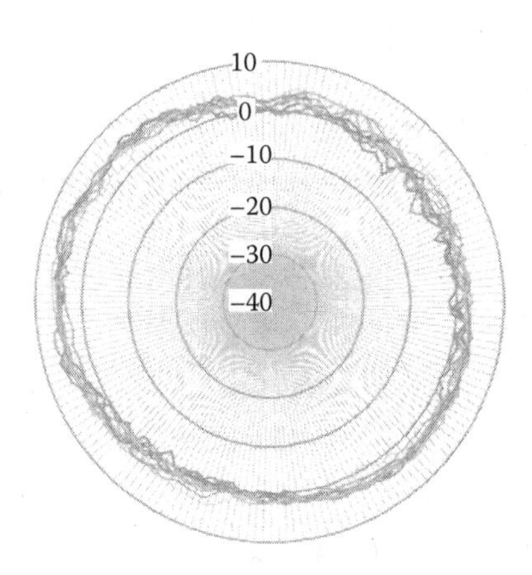

FIGURE 7.35
Diversity antenna radiation pattern. (From K. Yegin, *IEEE Antennas and Propagation Society International Symposium*, 1B, 2005. Copyright 2005. IEEE. With permission.)

Value P_S determines the ratio of the sum of time intervals during which the signal level is above a given threshold to the total time of observation for the single antenna. P_d is a similar ratio for a diversity system. The following derivation reveals the relationship between the signal-to-distortion ratio and the efficiency value n for a diversity system. If V is the voltage signal at the output of a single antenna, the probability distribution power function can be expressed as:

$$P(V^2) = \frac{1}{V_m^2} \cdot e^{-\frac{V^2}{V_{mm}^2}} \tag{7.21}$$

where V_m^2 = average power value. This probability density function corresponds to the integral probability function:

$$p_e = 1 - e^{-\frac{V_{\min}^2}{V_m^2}} \tag{7.22}$$

$V_{\min}$ represents the minimum required reviving voltage for symbol detection.

Defining the ratio $V_m/V_{\min}$ as the signal-to-distortion ratio SND for the single antenna:

$$SND = 20 \log\left(\frac{V_m}{V_{\min}}\right) \tag{7.23}$$

It is shown [21] that for n noncorrelated diversity branches, $SND_S(n)$ can be expressed as:

$$SND(n) \approx n \cdot SND \tag{7.24}$$

Equation (7.24) assumes that $SND >> 1$.

The following data show the diversity effectiveness n of the antenna presented in Figure 7.32a, in a Rayleigh multipath scenario, if the best signal is selected from the group of indicated ports:

A1, A2 $\rightarrow n = 1.9$; A1, A3 $\rightarrow n = 1.5$; A1, A4 $\rightarrow n = 1.99$; A2, A3 $\rightarrow n = 1.5$; A2, A4 $\rightarrow n = 1.99$; A3, A4 $\rightarrow n = 1.95$; A1, A2, A3, A4 $\rightarrow n = 3.1$.Clearly a considerable improvement in SDARS reception can be expected from diversity, based on utilizing one antenna with a few outputs or with a few antennas producing a single output.

References

1. K. Yegin, Satellite radio diversity antenna systems, *IEEE Antennas and Propagation Society International Symposium*, 1B, 2005, pp.72–75.
2. A. Petros, I. Zafar, and R. Pla, Antenna measurement techniques for SDARS antennas, *Antenna Measurement Techniques Association Proceedings*, PIDO58, Stone Mountain Park, GA, October 2004.

3. P. Bhartia et al., *Microstrip Antenna Design Handbook*, Boston: Artech, 2000.
4. Kin-Lu Wong, *Compact and Broadband Microstrip Autennas*, John Wiley & Sons, Inc., 2002, New York.
5. C. Balanis, *Antenna Theory: Analysis and Design*, 2nd Ed., New York: John Wiley & Sons, 1997, pp. 760–762.
6. IE3D software package.
7. R. Sainati, *CAD of Microstrip Antennas for Wireless Applications*, Boston: Artech, 1996.
8. M. Kara, Formulas for the computation of the physical properties of rectangular microstrip antenna elements with various substrate thicknesses, *Microwave and Optical Technology Letters*, 12, 234–239, 1996.
9. R. Bancroft, *Microstrip and Printed Antenna Design*, New York: SciTech Publishing, 2004, pp. 54–55.
10. Impedance, axial ratio and receiving power bandwidth of microstrip antennas, *IEEE Transactions on Antennas and Propagation*, 52, 2769–2774, 2004.
11. A. Duzdar et al., Radiation efficiency measurements of a microstrip antenna designed for the reception of XM satellite radio signals, *Society of Automotive Engineers World Congress*, Detroit, (2006-01-1354) April 2006.
12. K. Yegin et al., Patch Antenna with Parasitically Enhanced Perimeter, U.S. Patent 7,038,624, May 2006.
13. N. Yuan et al., Analysis of probe-fed conformal microstrip antennas on finite grounded substrate, *IEEE Transactions on Antennas and Propagation*, 54, Part 2, 554–563, 2006.
14. H. Kuboyama et al., Post loaded microstrip antenna for pocket size equipment at UHF, *ISAP Proceedings*, 1985, pp. 433–436.
15. M. Daginnus, R. Kronberger, and A. Stephan, Ground plane effects on the performance of SDARS antennas, *IEEE Antennas and Propagation Society International Symposium*, 2002, pp. 748–751.
16. R. Kronberger, G. Hassmann, and S. Schulz, Measurement and analysis of vehicle influences on the radiation pattern of SDARS antennas, *IEEE Antennas and Propagation Society International Symposium*, 2002 pp. 740–743.
17. Y. Hong, S-Band Dual-path dual-polarized antenna system for satellite digital audio radio service (SDARS) application, *IEEE Transactions on Microwave Theory and Techniques*, 54, Part 1, 1569–1575, 2006.
18. D. Sievenpiper, H. Hsu, and R. Riley, Low-profile cavity-backed crossed-slot antenna with a single-probe feed designed for 2.34 GHz satellite radio applications, *IEEE Transactions on Antennas and Propagation*, 52, 873–879, 2004.
19. K. Yegin et al., Vehicular Glass-Mount Antenna and System, U.S. Patent 7,190,316, March 2007.
20. S. Themistoklis and C. Philip, Propagation model for vegetation effects in terrestrial and satellite mobile systems, *IEEE Transactions on Antennas and Propagation*, 52, 1917–1920, 2004.
21. H. Lindenmeier, et al., A new design principle for a low profile SDARS antenna including the option for antenna diversity and multiband application, *Society of Automotive Engineers World Congress*, Detroit, March 2002, Technical Paper 2002-01-0122.

22. H. Lindenmeier et al., Vehicle-conformal SDARS diversity antenna system by a fast operating diversity algorithm, *IEEE Antennas and Propagation Society International Symposium*, 1b, 2005 pp. 76–79.
23. K. Yegin et al., Directional Patch Antenna, U.S. Patent 7,132,988, November 2006.
24. K. Yegin et al., Vehicle Mirror Housing Antenna Assembly, U.S. Patent 7,248,225, July 2007.

8

GPS Antennas

8.1 Typical GPS Antenna Parameters

GPS systems are very popular among civilian users for map positioning, scientific uses, and timing services. One of the most important components of a GPS is an antenna to provide stable communication between a satellite transmitter and a receiver. Modern GPS systems utilize three carrier frequency options: 1.575 GHz (L1 band), 1.227 GHz (L2 band), and 1176 MHz (L5 band). GPS antennas for the Ecall (European) and OnStar (U.S.) safety systems are used with cellular phone antennas to supply rapid assistance to a driver involved in an emergency situation. The system simultaneously allows a driver to make a request and conveys information about the vehicle location. A variety of GPS antennas are currently available commercially.

The most common design for vehicle application is an active microstrip single-feed patch antenna. Unlike a SDARS antenna, the microstrip GPS patch operates in right hand circular polarization mode (RHCP). The design is very attractive for vehicle applications because it has a low profile and it can be integrated into a single package with other car antennas. Typical passive patch parameters in the L1 frequency band are as follows [1]:

- Operation Frequency: 1575 MHz
- Polarization: RHCP
- θ coverage: 0 to 90 degrees
- Azimuth (φ) coverage: 360 degrees
- Gain: 3 dBic at zenith ($\theta = 0$ degrees); 1 dBic at $\theta = 50$ degrees; –10 dBic at horizon
- Axial ratio: <4 dB
- Output impedance: 50 ohms
- VSWR: 1.5:1

As a rule, all measurements are carried out for a patch mounted on a 1 m metal ground plane. Typical patch dimensions are $25 \times 25 \times 4$ mm^3, but a variety of designs with smaller dimensions are available on the market.

TABLE 8.1

Parameters of Low Noise Amplifier for Active GPS Antenna Design

Amplifier Gain without Cable	Amplifier Gain with Cable Loss	Typical Input P1 dB Range	Noise Figure	Out-of-Band Filter Attenuation	DC Power Current	DC Power Voltage	Connector Type
28 dB	20 dB	–6 to –2 dBm	2 dB	–25 dB at ±50 MHz	30 mA	5 V through RF cable	TNC

A smaller antenna has lower gain and wider beamwidth. For example, TOKO suggests a 20 × 20 mm antenna (DAG 1575CR3T) that has a gain of about 2 dBic (measured with a 40 mm square ground plane). A TOKO 18 × 18 mm patch antenna DAX1575MS63T exhibits a zenith gain of about 0 dBic. Table 8.1 lists parameters of a low noise amplifier for an active GPS antenna design. Values can vary slightly, depending on supplier.

8.2 Patch Parameters and Ground Plane Size

Patch antenna performance is strongly coupled to the dimensions of the ground plane on which it is mounted [2]. Figure 8.1 shows examples of measured and calculated RHCP directionality curves [3] for a GPS square patch antenna mounted on a circular dielectric (ε_r = 9.8) grounded circuit board at 1575 MHz frequency. Solid lines in Figure 8.1a through Figure 8.1d show measured elevation plane directionality cuts (RHCP) for different diameters D of the circular ground plane. Note that the cross polarization level (LHCP) of the radiation pattern is 10 to 25 dB less than that of the RHCP component (depending on ground size). Figure 8.1e demonstrates the calculated radiation patterns for a 6 inch (15.24 cm) ground circular diameter.

A popular GPS passive patch with dimensions of 25 × 25 × 4 mm^3, built on a ceramic substrate, with dielectric constant $\varepsilon_r \approx 20$, dielectric thickness h = 4 mm (TOKO's DAK1575MS50), and measured with a 70 mm square ground plane passive antenna has the following parameters:

- Center frequency without redome: 1580.5 MHz
- Center frequency without case is 1580.5 MHz, with case 1575.42 MHz ± 1.023 MHz
- Bandwidth: 9 MHz
- Gain at zenith: +5 dBic
- Gain at 10 degrees elevation: –1 dBic
- Return lost at resonant frequency: less than –25 dB
- Impedance: 50 ohms

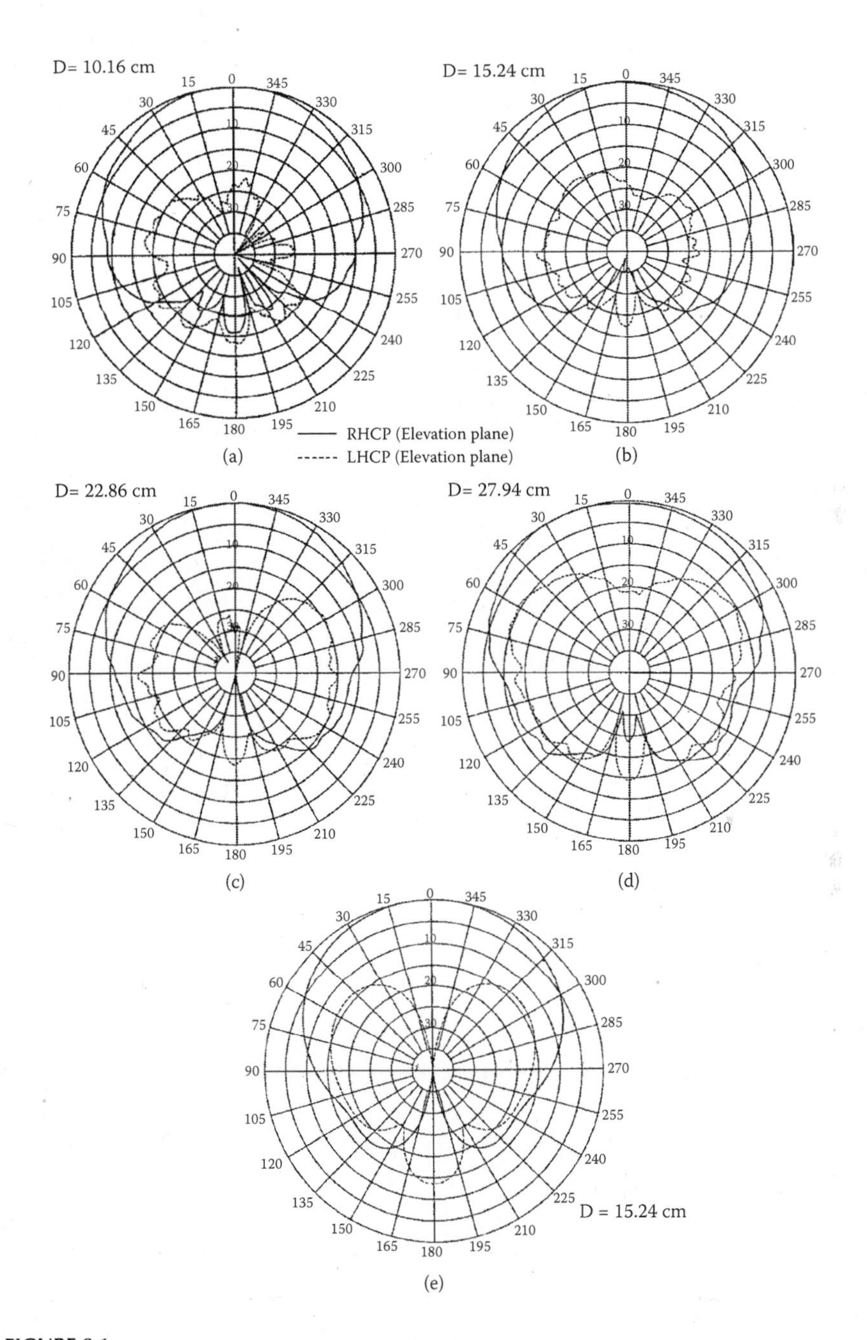

FIGURE 8.1
Radiation patterns of GPS antenna for RHCP plotted against different diameter size *D* of ground plane: (a) to (d) depict measurement results. (e) depicts simulation results. (From T. Milligan and P. Kelly, *Second IEEE Antennas and Propagation Symposium*, July 1996. Copyright 1996 IEEE. With permission.)

TABLE 8.2A

Antenna Zenith Gain (dBic) versus Size and Square Ground Plane with Side Length A (mm)

A	40	45	50	60	70	80	90	100
Gain	1	2	2.6	4.3	5.1	5.3	5.3	5.3

TABLE 8.2B

Frequency Shift (MHz) of Dielectric Patch Antenna versus Ground Plane Size *A* (mm)

A	40	45	50	60	70	80	90	100
F	1576.5	1577.3	1578	1579.5	1580.5	1581.6	1582.4	1583.3

Table 8.2a shows the approximate values of zenith gain *Gain* in dBic for the antenna compared with the size of a square ground plane with side length A in mm. The antenna is measured without a case at 1580.5 MHz. The gain varies from 1 dBic (ground plane side length equals side length of patch) to 5.3 dBic when the side length of the ground plane is about 100 mm. The gain does not significantly change when the side length of the ground plane varies from 70 to 100 mm. The resonance frequency shift in MHz of a dielectric patch antenna by ground plane size is shown in Table 8.2b. The maximum shift is about 10 MHz when the ground side length changes from the patch side length to 100 mm. An antenna mounted on a 1 m circular plane has directionality values equal to –1.5 dB at 30 degrees from zenith (in comparison with maximum value at zenith), –3 dB at 60 degrees, and –8 dB at a horizontal direction.

8.3 On-Vehicle GPS Antenna Measurements

Of course, the parameters of a GPS antenna mounted on a vehicle differ from parameters measured on the ground plane. Figure 8.2 shows a ceramic square (dielectric constant 35, 20 × 20 mm) passive patch antenna [4] used for two series of measurements, one in an anechoic chamber on a 1 m circular ground plane and the other installed at different locations on a vehicle. The circular ground plane had rolling edges to minimize the gain measurement data ripples. Table 8.3 presents the measurement results of the average gain (over 360 degrees) G_{avg} (dBic), and gain ripple G_{max}/G_{min} (dB) in the azimuth plane (azimuth angle changed from 0 to 360 degrees) for different elevation angles (anechoic chamber).

The second series of measurements related to the antenna mounted at different vehicle positions designated RF (4 cm from sunroof), RC (4 cm from back light), RR (rear roof), and TC (8 cm from back light and trunk edge) as shown in Figure 8.3. The measured antenna gain values are summarized in

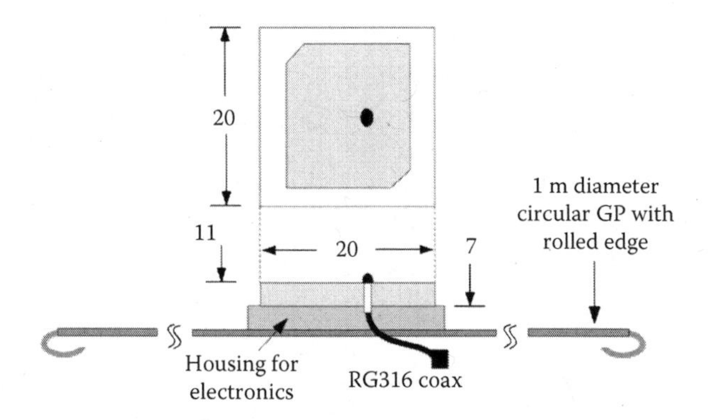

FIGURE 8.2
Typical dimensions of RHCP ceramic patch antenna on 1 m square ground plane. (From K. Yegin, *IEEE Antennas and Propagation*, 6, 2007. Copyright 2007 IEEE. With permission.)

TABLE 8.3
Average Gain and Gain Ripples in Azimuth Plane for Different Elevation Angles (Anechoic Chamber)

Angle θ	0	10	20	30	40	50	60	70	80	90
G_{avg}	3.69	2.76	2.84	2.41	1.21	1.63	1.35	−0.56	−3.4	−6.25
Ripple	0	1.33	0.68	1.24	1.1	1.51	1.9	2.78	3	4.24

$\theta = 0°$ corresponds to zenith angle direction.
Source: K. Yegin, *IEEE Antennas and Propagation*, 6, 2007. Copyright 2007 IEEE. With permission.

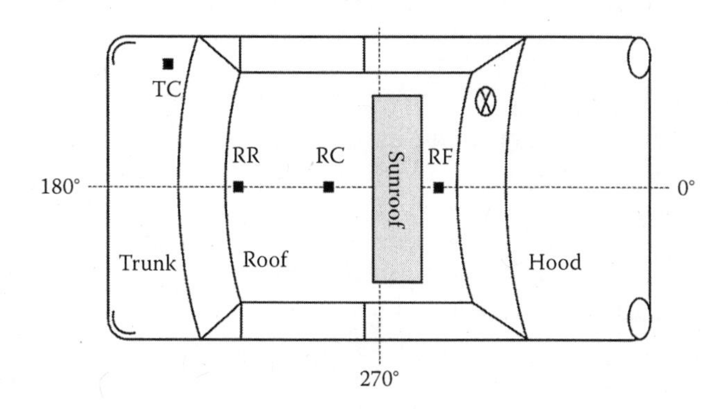

FIGURE 8.3
Vehicle antenna positions. (From K. Yegin, *IEEE Antennas and Propagation*, 6, 2007. Copyright 2007 IEEE. With permission.)

TABLE 8.4

Measured Antenna Gain Values for Different Antenna Locations on the Car

Angle θ	0	10	20	30	40	50	60	70	80
RC	2.98/ 0.83	2.96/ 3.08	2.10/ 2.53	1.47/ 4.09	0.64/ 3.64	0.51/ 6.32	−0.67/ 5.02	−1.79/ 8.67	−3.69/ 8.75
RF	2.43/ 1.05	2.17/ 3.48	1.64/ 5.39	0.81/ 7.18	0.90/ 5.18	0.36/ 4.7	−0.86/ 9.34	−2.08/ 12.1	−3.42/ 11.5
RR	2.68/ 0.84	0.75/ 3.23	1.66/ 4.57	1.09/ 4.75	1.24/ 6.15	0.43/ 4.91	0.35/ 8.96	−1.5/ 9.73	−2.91/ 10.7
TC	1.48/ 0.86	1.24/ 3.64	1.31/ 4.14	1.32/ 4.65	0.48/ 7.7	-0.8/ 9.6	−0.98/ 13.6	−1.45/ 33.1	−3.02/ 30.2

$\theta = 0°$ corresponds to zenith angle direction.
Source: K. Yegin, *IEEE Antennas and Propagation*, 6, 2007. Copyright 2007 IEEE. With permission.

Table 8.4. Every cell of the table has two values: the first is the average gain (dBic) and the second is the gain ripple (dB). For example, 2.98/0.83 indicates an average gain of 2.98 dBic and ripple equal to 0.83 dB.

For θ angles between 0 (zenith angle direction) and 30 degrees, the roof center performs the best, as expected. When the elevation angle θ_{el} ($\theta_{el} = 90° - \theta$) is lowered, the ripple grows. Although an antenna at the trunk corner mounting location seems to show reasonable gain, it exhibits a large ripple at low elevation angles due to higher cross polarization and blockage by the roof.

8.4 Circular Annular Ring Microstrip Antenna

Figure 8.4 shows a microstrip antenna [5] designed with an annular ring RHCP topology for L band applications. The center resonance frequency point is about 1526 MHz, which differs from the GPS center frequency of 1575 MHz. However, this annular ring compact CP antenna can meet GPS specifications with little geometry modification.

The antenna has an outer radius diameter b of 20 mm and an inner radius a of 13 mm. The FR-4 dielectric substrate has a thickness of 1.6 mm and relative permittivity of 4.4. The impedance transformer has a length l of 31 mm and a width W of 0.68 mm. The insert slit length l_S is 2.0 mm and width W_S is 1.0 mm. The distance between ground plane and annular ring patch h is 1.6 mm. Circular polarization is provided by specific slit locations and feed point topology. Figure 8.5a shows the measured input impedance. The axial ratio in the broadside direction versus frequency is demonstrated in Figure 8.5b. The measured CP bandwidth (3 dB axial ratio) is determined by the value equal to 12 MHz. Measured radiation patterns in two orthogonal planes at the center frequency point are plotted in Figure 8.6. Good RHCP radiation is achieved.

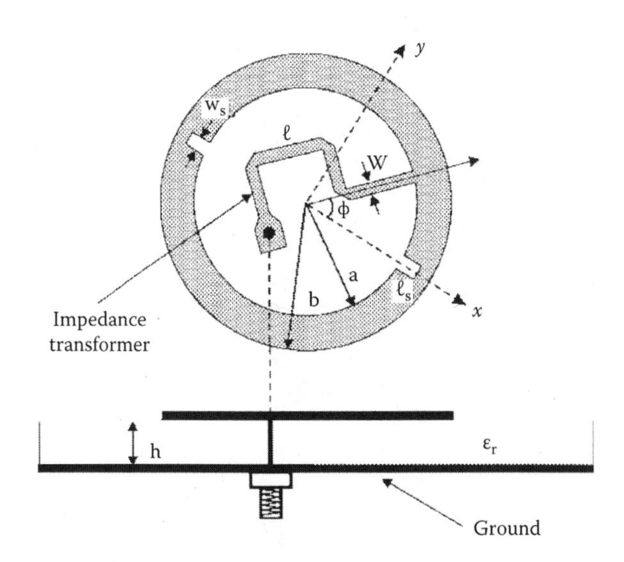

FIGURE 8.4
Annular ring RHCP topology. (From H. Chen and K. Wong, *IEEE Transactions on Antennas and Propagation*, 47, 1999. Copyright 1999 IEEE. With permission.)

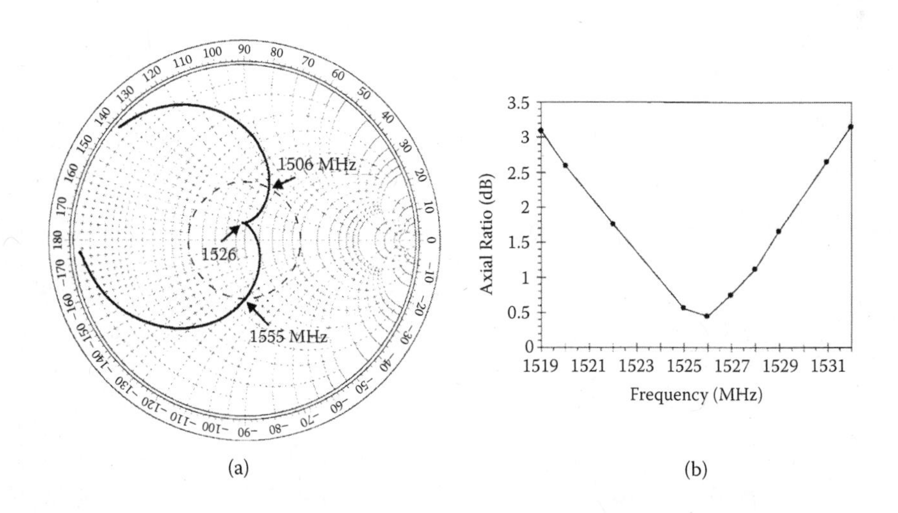

FIGURE 8.5
Measured parameters of annular ring antenna: (a) impedance; (b) axial ratio. (From H. Chen and K. Wong, *IEEE Transactions on Antennas and Propagation*, 47, 1999. Copyright 1999 IEEE. With permission.)

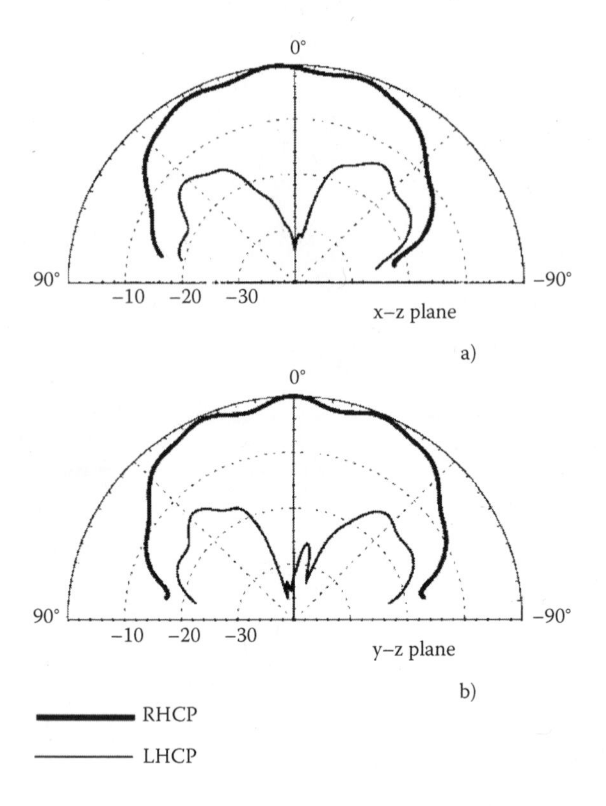

FIGURE 8.6
Measured elevation cuts of RHCP radiation patterns for GPS antenna: (a) X-Z plane (φ = 0 degrees); (b) Y-Z plane (φ = 90 degrees). (From H. Chen and K. Wong, *IEEE Transactions on Antennas and Propagation*, 47, 1999. Copyright 1999 IEEE. With permission.)

8.5 Dual-Band GPS Antenna

In general, GPS receivers operate in the 1575 MHz (L1) band. However, for some applications [6], an antenna must cover both L1 and L2 (1227 MHz) bands. Figure 8.7 shows two geometries of a simple corner-truncated square CP single feed microstrip patch antenna [7] that operates in both frequency bands.

The design in Figure 8.7a uses two stacked patches to achieve CP in two frequency bands. The total height is less than 4 mm, the side length of the upper square is 60 mm, the side length of the lower square is 59 mm, and substrate FR-4 has a dielectric constant of 4.4. The probe feed excites the upper patch via a hole in the lower patch. A thin air layer exists between the two patches. Measurement results show that the return loss at 1227 MHz does not exceed –25 dB and at 1575 MHz is about –15 dB. The axial ratio at both frequency points in the broadside direction is around 1. The impedance

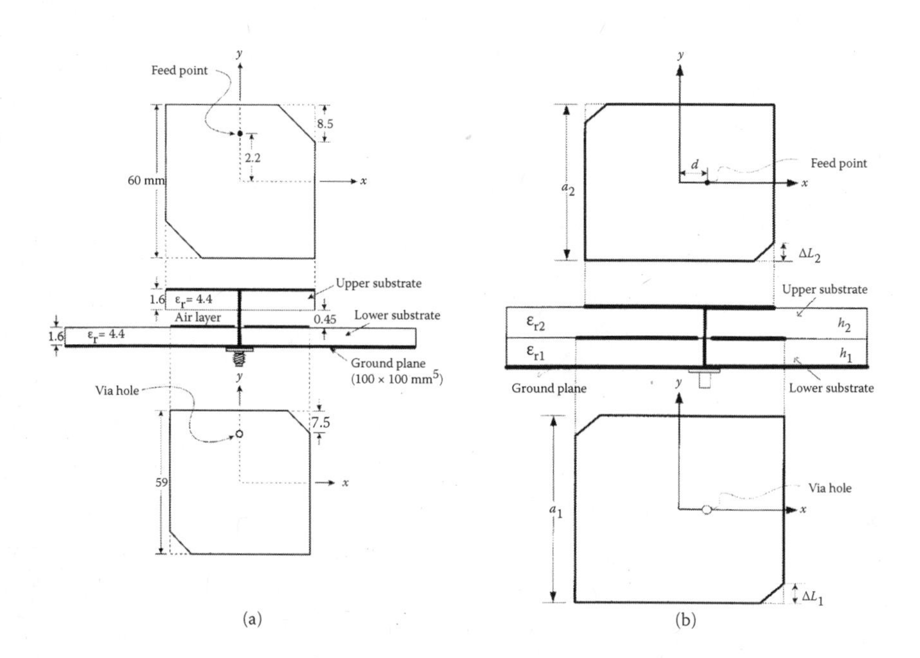

FIGURE 8.7
Dual-band microstrip antennas for GPS application: (a) option with air layer (From C. Su and K. Wong, *Microwave and Optical Technology Letters,* 33, 2002. Copyright 2002 *Microwave and Optical Technology Letters.* Reprinted with permission of Wiley-Blackwell, Inc.); (b) option without air layer (From X. Peng et al., *Microwave and Optical Technology Letters,* 44, 2005. Copyright 2005 *Microwave and Optical Technology Letters.* Reprinted with permission of Wiley-Blackwell, Inc.)

bandwidth, determined from the 10 dB return loss, is about 53 MHz for 1227 MHz and 44 MHz for 1575 MHz. Gain values are more than 1.5 dBic at the lower frequency band and more than 4.5 dBic at the upper band.

A similar geometry with a high density dielectric constant and without an air layer is presented in Reference [8]. The antenna (Figure 8.7b) dimensions are $h_1 = 4$ mm, $h_2 = 2.5$ mm, $\varepsilon_{r1} = 12$, $\varepsilon_{r2} = 9.2$, $a_1 = 31.5$ mm, and $a_2 = 31$ mm. The results show that the frequency bandwidth that corresponds to the 10 dB return loss level is about 26 MHz for both bands. The peak gain values are 2.4 and 4.5 dBic for the lower and upper frequencies, respectively.

Figure 8.8 depicts dual-band annular patch CP antenna geometry [9] for GPS application. Two shorted elliptical annular patches are concentrically printed on two stacked substrates separated by an air gap as shown in Figure 8.8a and Figure 8.8b. Both substrates have the same dielectric constant ε_R of 2.55 and height h of 3.2 mm. The major and minor axes of the outer boundaries are indicated as a_{up}, b_{up}, a_{down}, and b_{down} for the upper and lower patches, respectively. The inner border of the antenna with the two rings is shorted to the ground plane and its minor axes are a_{inner} and b_{inner}.

The stacked antenna is fed using a single coaxial probe electrically connected to the upper ring through a hole of radius $r = 2.13$ mm in the lower

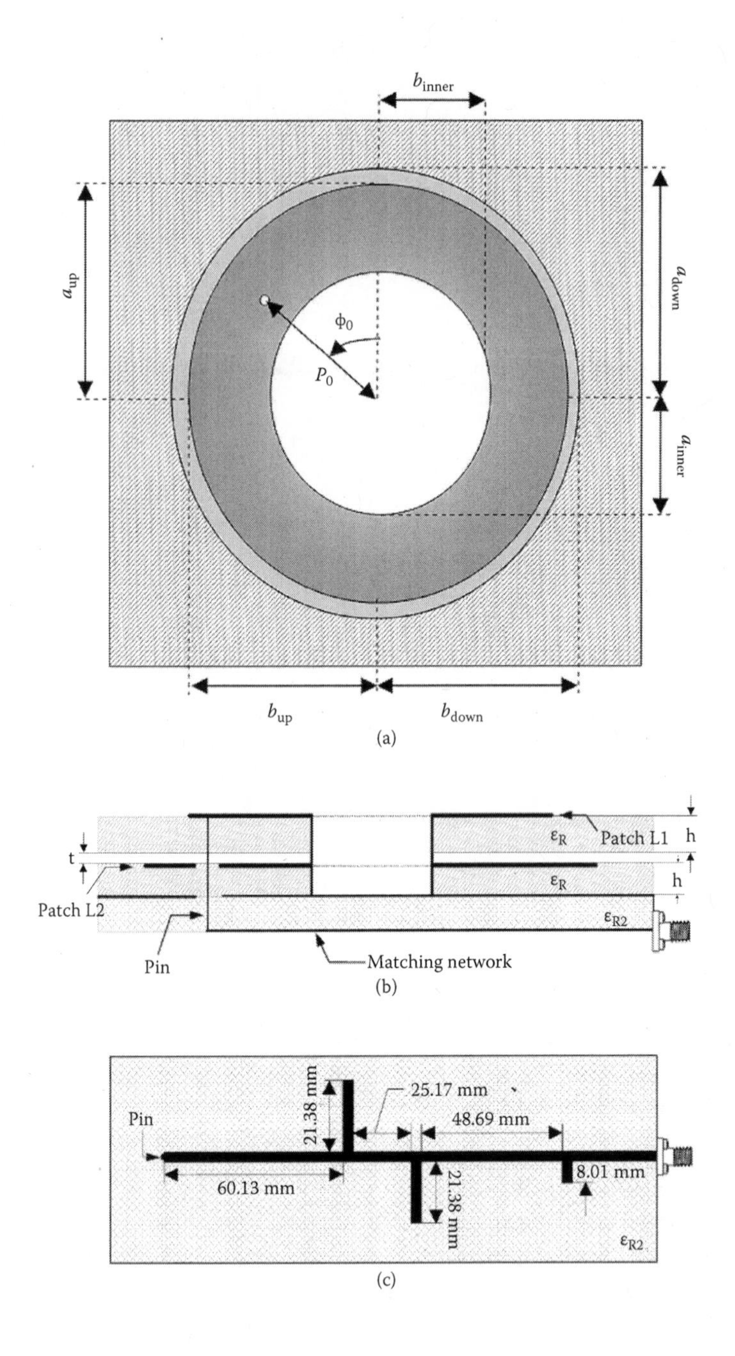

FIGURE 8.8
Stacked shorted annular elliptical patch antenna geometry: (a) top view; (b) side view; (c) matching network. (From L. Boccia, G. Amenodola, and G. Massa, *IEEE Antennas and Wireless Propagation Letters*, 3, 157–160, 2004. Copyright 2004 IEEE. With permission.)

patch. The position of the feed is defined by polar coordinates ρ_0 and ϕ_0, with respect to the antenna center. These ring sizes were chosen to make the upper and lower patches resonate at frequency bands L1 (1557 to 1595 MHz) and L2 (1212 to 1244 MHz). The input impedance was matched by tuning both the probe position and the air gap thickness ($t = 1$ mm and $\rho_0 = 35$ mm). Optimized values of the antenna dimensions were $\phi_0 = 135$ degrees, $a_{inner} =$ 23.5 mm, $b_{inner}/a_{inner} = 0.975$, $b_{up}/a_{up} = 0.972$, $b_{down}/a_{down} = 0.97$, $a_{up} = 52.8$ mm, and $a_{down} = 57.23$ mm. Additional matching is provided by the matching network strip line shown in Figure 8.8c (substrate parameters are $\varepsilon_{R2} = 2.33$ and thickness = 0.762 mm).

The impedance matching is implemented by three open-ended stubs whose lengths and positions are optimized to effectively compensate the feed inductance at both resonant frequencies of the stacked patch. Figure 8.9 shows the reflection coefficient (solid line), RHCP antenna gain (dotted line), and axial ratio of the design. The measured and simulated cuts of radiation pattern (for the antenna in Figure 8.8) in elevation plane are demonstrated in Figure 8.10. Figure 8.10a reflects the results for the L1 frequency band and Figure 8.10b presents the results for the L2 band. The gain of the antenna is 8.2 dBic and 8.9 dBic at the L2 and L1 bands, respectively. The proposed design has at both frequencies an amplitude roll-off from bore sight to horizon of about 20 dB while the measured front to back ratio is around –30 dB. The axial ratio of the prototype is below 3 dB for elevation angles above 30° at 1.575 GHz and within the entire hemispherical coverage at the lower frequency.

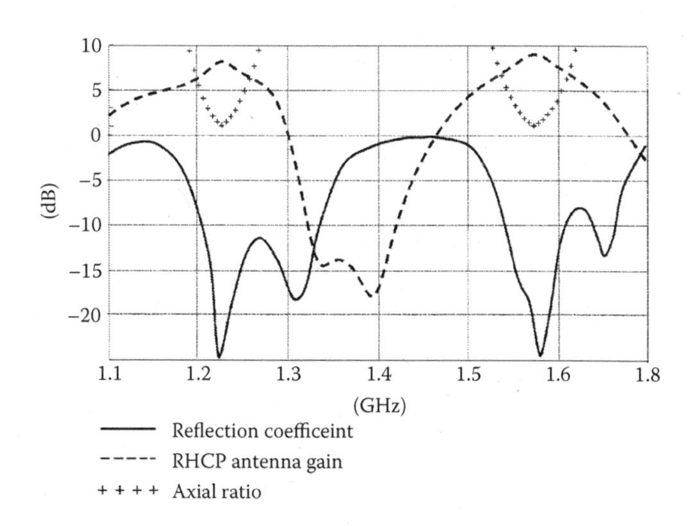

FIGURE 8.9
Reflection coefficient, antenna gain, and axial ratio. (From L. Boccia, G. Amenodola, and G. Massa, *IEEE Antennas and Wireless Propagation Letters*, 3, 157–160, 2004. Copyright 2004 IEEE. With permission.)

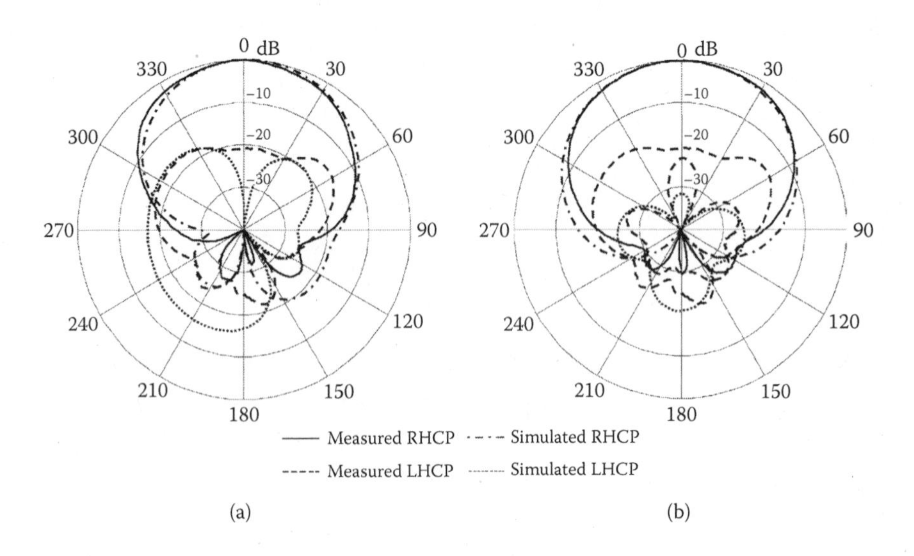

FIGURE 8.10
Measured and simulated radiation pattern in elevation plane: (a) 1575 MHz; (b) 1227 MHz. (From L. Boccia, G. Amenodola, and G. Massa, *IEEE Antennas and Wireless Propagation Letters*, 3, 157–160, 2004. Copyright 2004 IEEE. With permission.)

8.6 Tri-Band Applications

Modification of an existing GPS system includes the addition of a signal at 1176.45 MHz (L5) [10]. The design concept investigated in [10] consists of the development of a small antenna that can cover all three GPS bands: L1 (1575 MHz), L2 (1227 MHz), and L5 (1176 MHz). Each bandwidth is about 24 MHz with a gain of about 0 dBic (RHCP) and 50 ohm impedance. The geometry of the proposed antenna is shown in Figure 8.11 (a = 28 mm, h_1 = 6 mm, h_2 = 6 mm, l_1 = 22 mm, l_2 = 18 mm, l_h = 4 mm, and l_v = 10 mm).

The antenna presents a stacked patch known to allow for dual-band operation. Two metallic patches are placed on top of stacked dielectric substrates. The L-shaped probes connected to the 50 ohm coaxial cables were placed near the top patch. Copper tape is placed on the top of the lower substrate to form the lower patch. Epoxy (ε_r = 3.5 and tan δ = 0.03) was used to glue the upper and lower layers. To excite the RHCP, two-port quadrature phase feeding is provided by a hybrid that can be printed on the back of the ground plane. Instead of the common direct coaxial probe feeding [7], these authors introduced proximity coupled feeding [11]. The measured and simulated data for antenna gain are shown in Figure 8.12; the gain exceeded 2 dBic for all three bands.

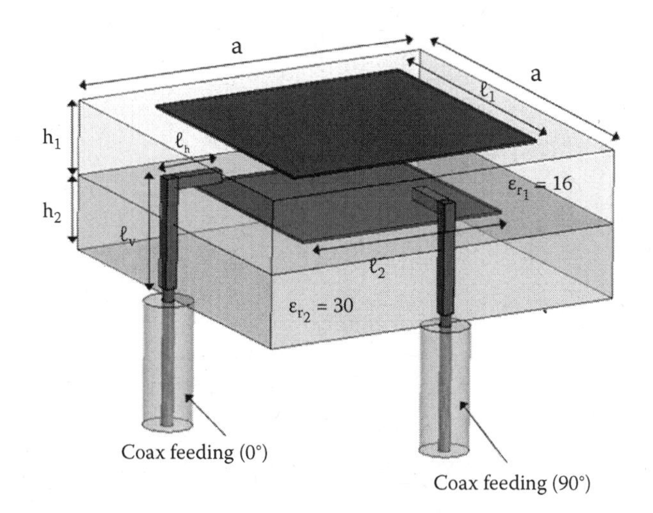

FIGURE 8.11
Geometry of tri-band antenna covering 1575, 1227, and 1176 MHz frequencies. (From Y. Zhou, C. Chen, and J. Volakis, *IEEE Transactions on Antennas and Propagation*, 55, 2007. Copyright 2007 IEEE. With permission.)

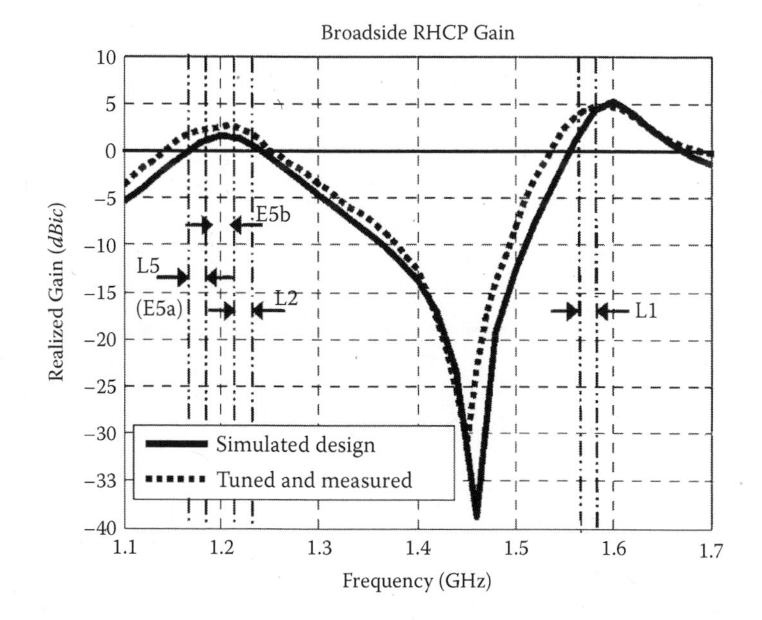

FIGURE 8.12
Measured and simulated broadside RHCP tri-band antenna gain. (From Y. Zhou, C. Chen, and J. Volakis, *IEEE Transactions on Antennas and Propagation*, 55, 2007. Copyright 2007 IEEE. With permission.)

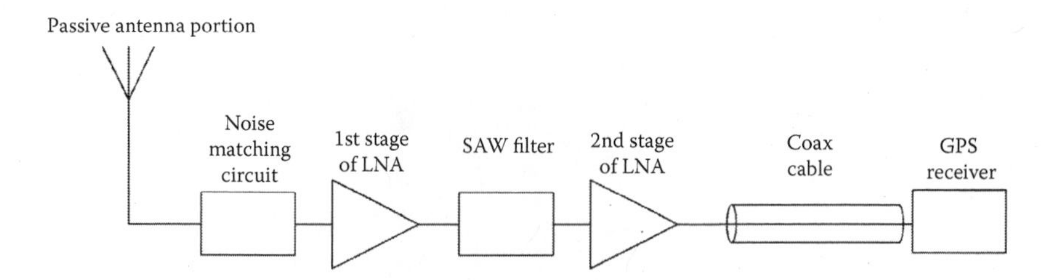

FIGURE 8.13
Block diagram for GPS active antenna amplifier.

8.7 Amplifier Circuit for Active Design

Figure 8.13 illustrates amplifier topology for an active GPS antenna. The amplifier includes a noise L-C matching passive circuit and two active stages surrounding a band pass filter. A narrow band SAW filter is a compact element that provides a high level of out-of-band interference rejection. Figure 8.14 shows an example of a SAW filter frequency response. The gain loss of this type of filter does not exceed 1.5 to 2 dB in the operating frequency range. The out-of-band response value for the filter is less than –30 dB. An integrated

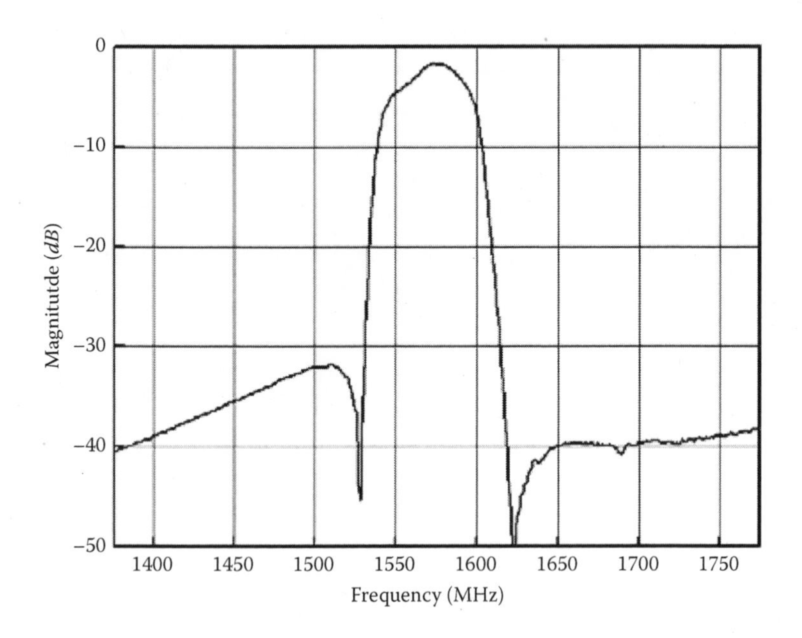

FIGURE 8.14
SAW filter frequency response.

monolithic amplifier element specially designed for GPS application can serve as the first and second stages.

The UPC8211TK element from California Eastern Laboratories is a good candidate for this application. Its noise figure is abut 1.3 dB, typical gain is about 18.5 dB, and P1 compression point output power is about –4 dBm. Tyco Electronics introduced the MAALSS0027 GPS low noise amplifier. This monolithic two-stage design with 50 ohm input and output impedance minimizes external components. Optimized for 1575 MHz frequency, the amplifier exhibits 1.5 dB noise figure a 20 dB gain, and 8.5 dBm output 1 dB compression power point.

Avago Technologies (formerly Agilent) markets the MMIC amplifier element ALM-1106 that uses a GaAs enhancement mode PHEMT process to achieve high gain operation with very low noise value. It has a 15 dB gain and about 0.9 dB noise figure at 1575 MHz. A design based on discrete transistors is also widely used for LNA designs. Typically, discrete elements are cheaper in comparison with an integrated circuit, but demand more in-front and output passive components to meet gain and noise requirements. For example, Infineon Technologies has demonstrated a SiGe BPF740F transistor for GPS applications. Measurement results appear on the Infineon website (application note LWR SD00051 LNA P). Briefly, minimum gain is 17 dB, maximum noise figure is 0.8 dB, and input 1 dB compression power point is –18 dBm at 1575 MHz.

8.8 Combined GPS and Cellular Phone Systems

Today's GPS technology joined with cellular phone equipment is used in the automotive industry to determine accurate vehicle locations. For example, OnStar Corporation (a General Motors subsidiary) provides the following services to U.S. and Canada subscribers who own GM vehicles:

- Driving directions
- Help to police in tracking down stolen vehicles
- Contacting emergency medical services in case of an accident
- Unlocking doors for drivers (after verifying authorization by phone)
- Roadside assistance

8.8.1 Typical Antenna System

A variety of combined cell and GPS antenna designs are available now for the automotive market. Figure 8.15 presents the topology for a combined GPS and cellular antenna. Typically, a microstrip patch with an amplifier mounted under the ground serves as a GPS antenna and printed on a dielectric substrate or helical monopole, with a reduced height, is used as

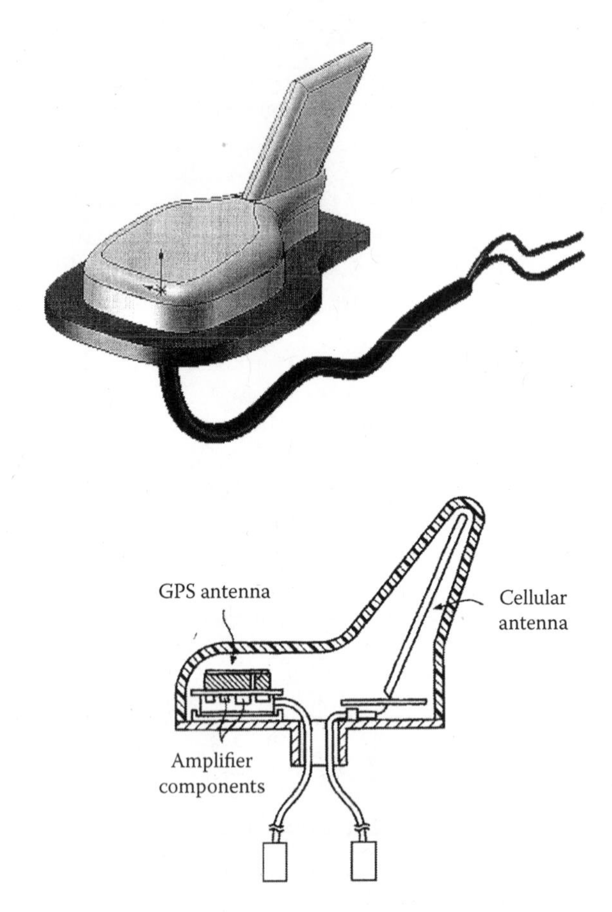

FIGURE 8.15
Geometry of combined GPS/cellular phone antenna.

cellular phone antenna. A multiband cellular antenna must operate in a few frequency bands. Table 8.5 shows the parameters of commercially available combined antennas presented by a few vendors. The gain of the LNA of an active RHCP antenna shown in Table 8.5 corresponds to the 1575 MHz band, the passive cellular antennas function on linear vertical polarization, and the output impedance is 50 ohms at all frequency bands. The dimensions are presented as length × width × depth.

8.8.2 Isolation of Collocated Cellular and GPS Antennas

When GPS and cellular antennas are integrated into one package, the coupling effect of antennas becomes critical [1,12]. Integration of GPS and cellular antennas increases the risk of interference from a cellular transmitter

TABLE 8.5

Commercially Available Systems Combining GPS and Cellular Phone Antennas

Company	Part Number	Cell Bands (MHz)	Operational Cell Bands	Gain for Cell Bands (dBi)	Dimensions (mm)
Panorama Antennas (GPS LNA gain = 26 dB)	GPSF-FF	890 to 960	GSM900	2	68 × 48 × 40
		1710 to 1880	GSM1800		
		1850 to 1990	PCS		
		1900 to 2170	3G UMTS		
Larsen (GPS LNA gain = 28 dB)	GPSCWCP	824 to 960	AMPS	2	39 × 57 × 82
		1710 to 2170	DCS/PCS/UMTS		
Laird (GPS Passive gain = 2.5 dBic)	GPSQ	824 to 896	AMPS	2.3	127 × 34 × 6.4
		890 to 960	GSM900	2.3	
		1710 to 1880	DCS	2.1	
		1850 to 1990	PCS	2.5	
TDC (GPS LNA gain = 26 dB)	GPS FIN	890 to 960	GSM 900	2	63 × 53 × 75
		1710 to 1880	GSM 1800		
PCTEL (GPS LNA gain = 25 dB)	GPSGSM	824 to 896	AMPS	2	72 × 62 × 14
		1710 to 1990	DCS/PCS	1	

to a receiver. This interference then impacts the GPS antenna amplifier. For example, for cellular AMPS mode, maximum transmitting power is about +34 dBm. A collocated active GPS antenna amplifier that operates during the transmission of high power from the cellular antenna must have a protection circuit to provide proper GPS operation.

The combined cellular and GPS antenna topology [1] optimized for minimum coupling interference is presented in Figure 8.16a. A 25 × 25 × 4 mm dielectric patch with a dielectric constant of 20 receives the GPS RHCP signal. The patch is rotated 45 degrees in the transverse plane to minimize wire-to-patch coupling, especially at low elevation angles. This orientation is based on experimental investigation. An antenna mounted with the cellular helix is measured on a 1 m ground plane for antenna-to-antenna isolation. The $VSWR_1$ for the helix antenna, $VSWR_2$ for the GPS antenna, and coupling coefficient S_{21} between ports 1 and 2 are shown in Figure 8.16b. Table 8.6 lists passive antenna isolation measurements, transmission power, and input 1 dB power compression point for LNA. Note that at least 10 dB more filtering at the AMPS band is required. It is also important that the low noise characteristics of the amplifier are maintained during cellular phone transmission.

8.8.3 Gain Measurements of Passive Cell/GPS Elements

Passive gain measurements of the antenna shown in Figure 8.16a are performed on a 1 m diameter circular ground plane with rolled edges. Azimuth cuts of the radiation pattern at different elevation angles: $\theta = 90$ degrees (elevation angle direction along horizon, $\theta_{el} = 0$), $\theta = 85$ degrees, and $\theta = 80$ degrees for

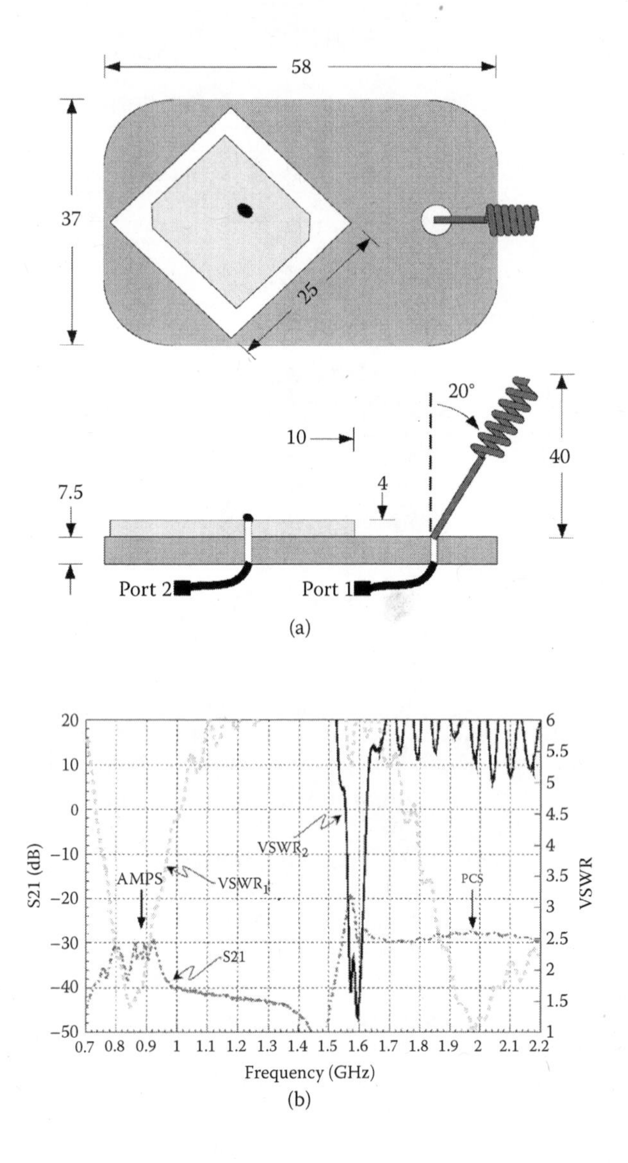

FIGURE 8.16
(a) Optimized topology of GPS/cellular phone antenna. (b) Plots of coupling coefficients of GPS and cellular phone portions and VSWR. (From K. Yegin, *IEEE Antennas and Propagation*, 6, 2007. Copyright 2007 IEEE. With permission.)

859 MHz and 1920 MHz are shown in Figure 8.17a and b. The elevation and azimuth cuts for the GPS patch are displayed in Figure 8.17c and d ($\theta = 0$ degrees corresponds to zenith angle direction). The ripples on the elevation GPS radiation pattern cut arise from finite screen size. The axial ratio of the GPS antenna portion is about 2.1 dB at $\theta = 0$ degrees (zenith direction).

TABLE 8.6

Passive Antenna Isolation Measurements, Transmission Power, and Input 1 dB Compression Point for LNA

	AMPS (824 to 894 MHz)	PCS (1850 to 1990 MHz)
Peak transmission power	+34 dBm	+28 dBm
Measured isolation (GPS patch to helix)	30 dB	28 dB
Peak input power at GPS LNA	+4 dBm	0 dBm
Typical input P1 dB range for GPS LNA	−6 to −2 dBm	0 to +2 dBm

Source: K. Yegin, *IEEE Antennas and Propagation*, 6, 2007. Copyright 2007 IEEE. With permission.

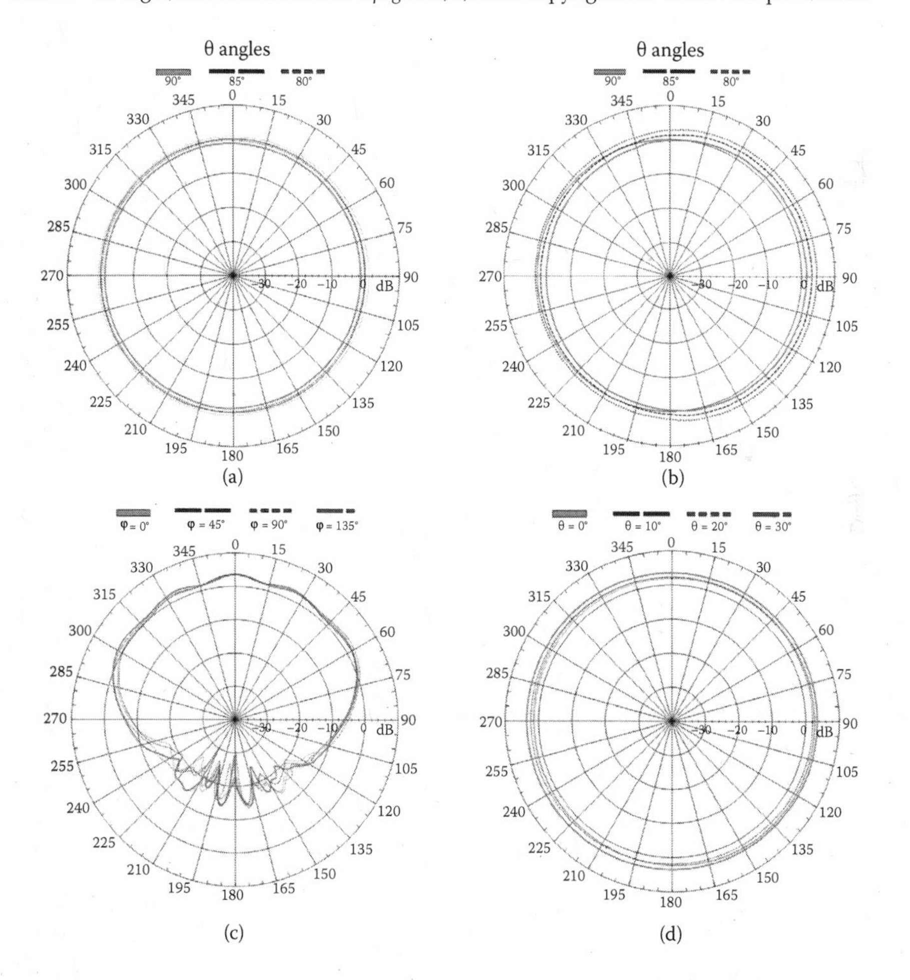

FIGURE 8.17
(a) Cell antenna, 859 MHz, azimuth cuts for Θ different angles. (b) Cell antenna, 1920 MHz, azimuth cuts for different θ angles. (c) GPS antenna, elevation cuts for different φ angles. (d) GPS patch antenna, azimuth cuts. (From K. Yegin, *IEEE Antennas and Propagation*, 6, 2007. Copyright 2007 IEEE. With permission.)

TABLE 8.7
Average Gain Values in Azimuth Cuts for Different Angles θ

Average Gain/Angle θ	0	10	20	30	40	50	60	70	80	85	90
GPS	3.5	2.15	2.06	2.27	1.07	1.23	1.27	−0.3	−3.3	–	–
AMPS	–	–	–	–	–	–	–	–	0.85	−0.2	−1.2
PCS	–	–	–	–	–	–	–	–	2.5	1.04	−0.5
Target GPS	+3	–	–	–		+1	–	–	−10	–	–
Target AMPS	–	–	–	–	–	–	–	–	−1 to −2	−1 to −2	−1 to −2
Target PCS	–	–	–	–	–	–	–	–	0 to−2	0 to −2	0 to −2

The measured averaged azimuth gain values are summarized in Table 8.7. The gain in GPS frequency band is measured in dBic; for AMPS and PCS, measurement is in dBi units. Target values were missed only by 0.24 db at the AMPS band and 0.48 dB at the PCS band. The minimum antenna gain target values were met, except for the 0 degree elevation PCS gain of −2.61 dBi.

8.8.4 Active Antenna and Filter Topology

Two different geometries of the active antenna demonstrated in Figure 8.18a and 8.18b were investigated [1]. The first option includes a front ceramic SAW filter and two amplifier stages. This type of configuration increases the noise figure of circuit at a value equal to loss of the SAW filter (2 to 3 dB) and, as a result decreases the sensitivity of the receiver system.

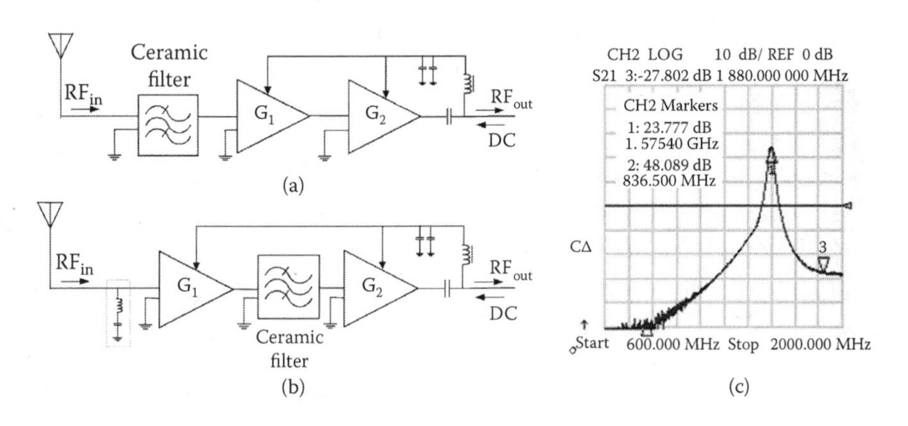

FIGURE 8.18
Low noise amplifier for GPS antenna: (a) first block is a ceramic filter; (b) ceramic filter placed between two amplifier stages; (c) GPS LNA gain measurement. (From K. Yegin, *IEEE Antennas and Propagation*, 6, 2007. Copyright 2007 IEEE. With permission.)

The second and preferable option, (Figure 8.18b) similar to the design shown in Figure 8.13, has a ceramic filter between the stages and a notch filter in front of an amplifier. The notch filter loss is not more than 0.2 to 0.4 dB, depending on the inductor quality. The overall noise of the system increases insignificantly. The gain response of the amplifier versus frequency is shown in Figure 8.18c. Isolation from GPS to AMPS and PCS are improved to 65 dB and 45 dB, respectively.

8.9 Combined GPS and SDARS Antenna System

8.9.1 Stacked Patch Design

Combined SDARS and GPS antenna [13,4] topologies with single feed pins are shown in Figure 8.19. Figure 8.19a presents an antenna that utilizes a single plane metallization surface to receive SDARS and GPS signals. The antenna [13] includes inner and outer patches installed on the top of the dielectric substrate. The ground is built onto the bottom side of the substrate. The single feed extends through the substrate and is connected to the inner patch.

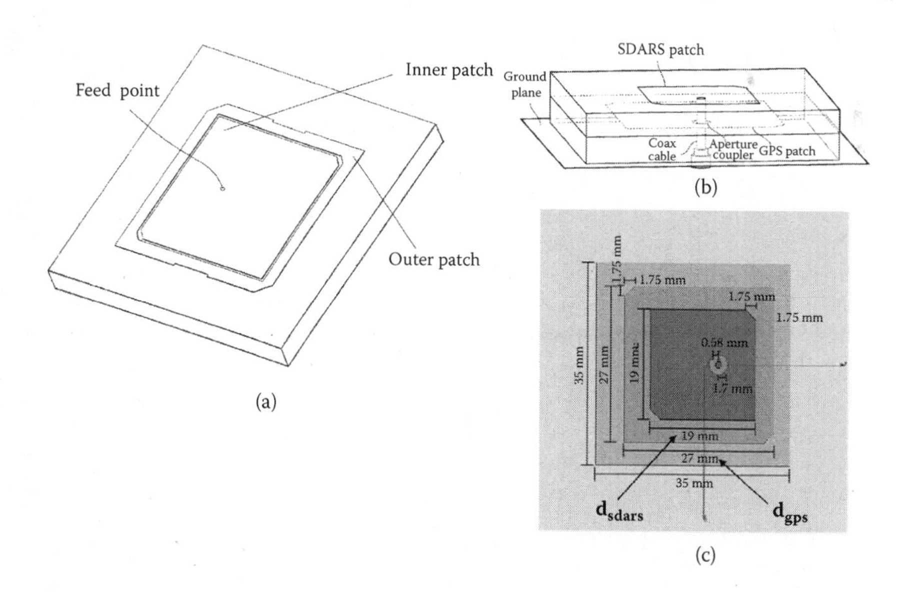

FIGURE 8.19
Combined GPS and SDARS antenna with single feed pin: (a) GPS and SDARS antenna portions are placed on the same plane, (b) GPS and SDARS antennas arranged in two-floor configuration, (c) top-down view of (b) (From K. Geary et al., *IEEE Vehicular Technology Conference*, 2008. Copyright 2008 IEEE. With permission.)

The inner patch is a square with two truncated corners cut at 45 degree angles to provide the LHCP antenna solution. The outer patch is shaped like a picture frame around the inner patch. The patch frame also has two outer trunk corners. The inner patch receives LHCP signals associated with satellite radio and the inner patch along with the outer frame patch receives RHCP associated with GPS signals. The inner patch size, outer patch frame size, and gap between the inner patch and outer frame are chosen to provide good reception quality for both SDARS and GPS frequency ranges.

Measurement results [13] show that return losses at a GPS frequency of 1575 MHz are about –20 dB, and at a XM frequency of 2340 MHz are about –10 dB. The maximum zenith gain for a RHCP GPS antenna is about 7.5 dBic at 1576 MHz. The value of the LHCP signal at the same frequency point in the zenith direction is –10 dBic. At 2338MHz, the zenith signal value for LHCP is about 6 dBic (30 degrees = 4.5 dBic, 60 degrees = 3.8 dBic) and for the cross polarization (RHCP), the signal value is about –11 dBic. The pin feed output is connected to a low noise amplifier that amplifies both GPS and SDARS signals. The amplifier output through the splitter with a dual-pass band filter communicates with both receivers (Figure 8.22a). The splitter receives a signal from the amplifier and divides it into a signal centered at the GPS frequency and a second signal centered in the SDARS band so that the correct signals are received by the GPS and SDARS receivers.

A microwave splitter circuit [17] is demonstrated in Figure 8.22b. It includes a dielectric board with etched transmission lines that have specificity for GPS and SDARS band lengths and widths. It has a single combined GPS/SDARS input and two decoupling outputs. Tuning stub lines provide proper decoupling of splitter outputs. Good results are obtained when the impedance of the transmission lines is 50 ohms and the stub line has an impedance of 120 ohms. Experiments show that insertion losses between the input and each output of this design are about 0.22 dB and isolation exceeds 20 dB.

Figure 8.19b and 8.19c show a "two-floor" SDARS/GPS combined antenna design [14] with a single output. The GPS component consists of a square metal patch on the top of a grounded dielectric substrate. The SDARS component consists of a slightly smaller square metal patch on the top of an ungrounded substrate. The SDARS patch is directly fed by the center conducting pin of the coaxial cable and the GPS patch is fed via a circular aperture coupler etched in the GPS metal patch. Figure 8.19c shows a top-down schematic of the stacked patch antenna in Figure 8.19b. The dielectric constant of the top and bottom substrate is 9.8. The appropriately trunked corners of the upper and lower patches along with feed point location provide LHCP for the SDARS antenna and RHCP for the GPS antenna.

The simulated reflection coefficient for this antenna system (Figure 8.20a) within two frequency bands (15 MHz for the L1 GPS band and 70 MHz for the SDARS band) is less than –10 dB. Figure 8.20 b and c demonstrates measured

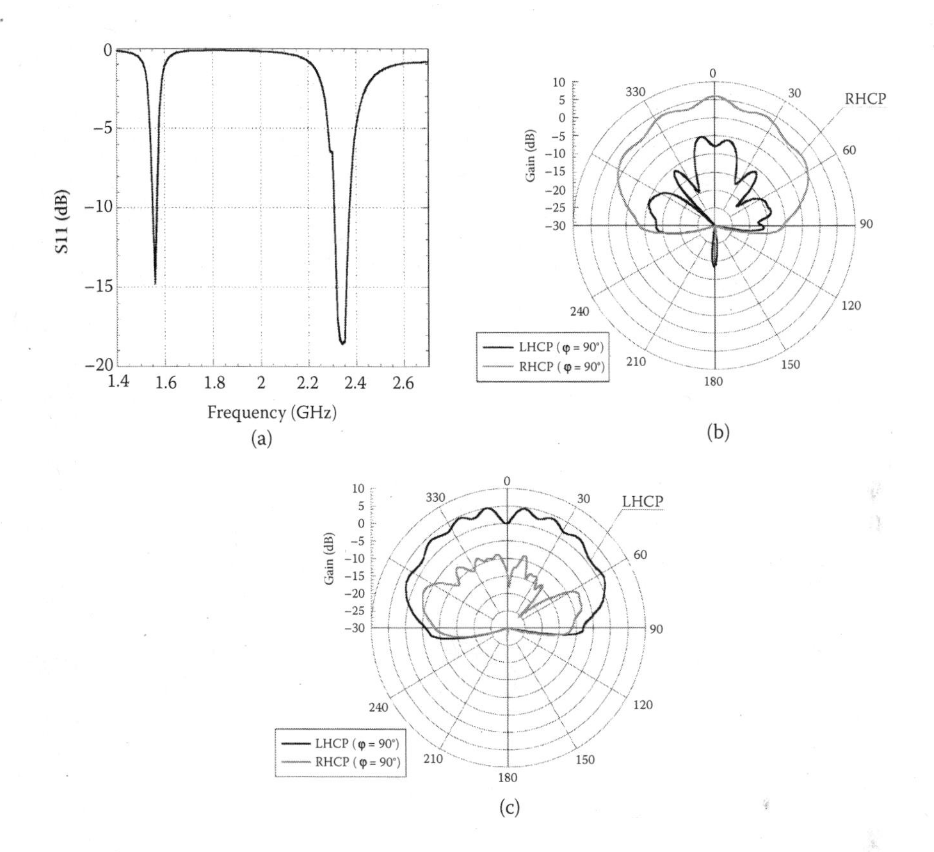

FIGURE 8.20
Parameters of combined GPS and SDARS antennas in two-floor configuration: (a) reflection coefficient, (b) measured RHCP and LHCP elevation gain patterns (φ = 90 degrees) at 1.543 GHz, (c) measured RHCP and LHCP elevation gain patterns (φ = 90 degrees) at 2.39 GHz. (From K. Geary et al., IEEE Vehicular Technology Conference, 2008. Copyright 2008. IEEE. With Permission.)

elevation radiation patterns (φ = 90 degrees) of the antenna mounted on a 1 m diameter circular ground plane for RHCP and LHCP. The ripples in the measurement results are due to the finite ground plane size used. Results also show improvements in gain for low elevation angles for linear polarization of about 2 dB in comparison with the commercially available SDARS patch antenna.

Combined GPS/SDARS antenna designs with two separated outputs [15,16] are shown in Figure 8.21. The antenna shown in Figure 8.21a and b includes an upper metallization element, an intermediate metallization element, and a rectangular bottom metallization element (ground plane). Based on Figure 8.21a, the upper element includes opposing cut corners, resulting in an LHCP antenna element and an intermediate metallization element that

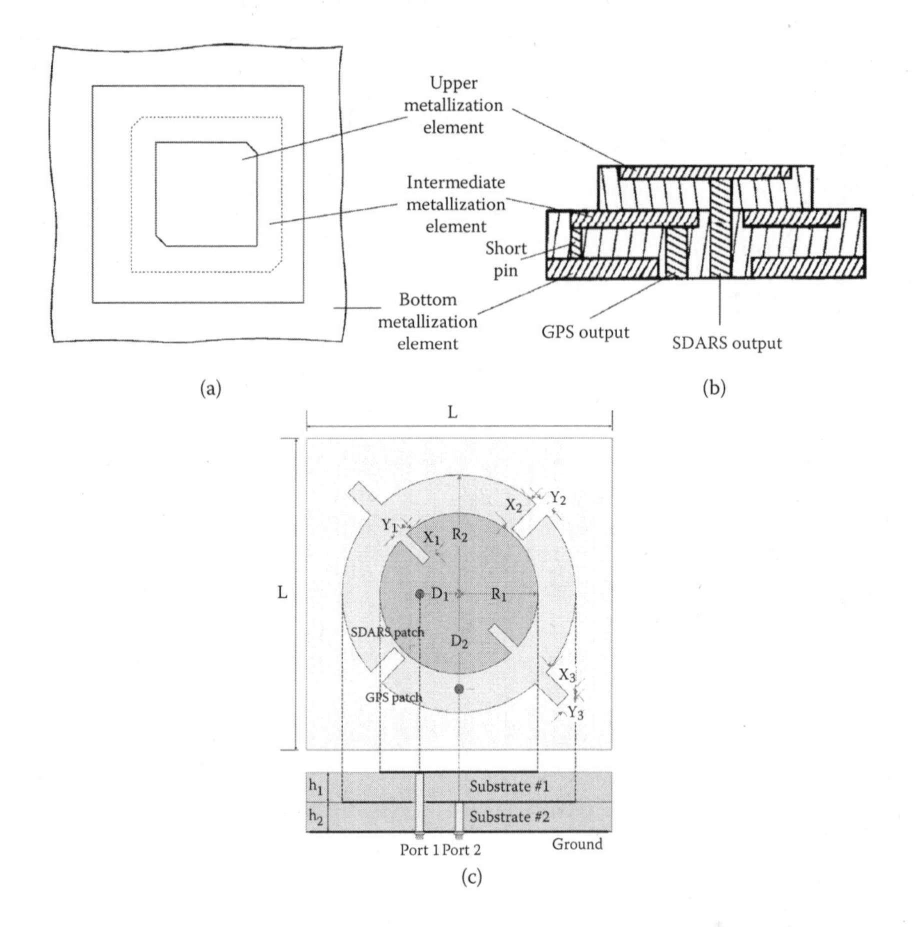

FIGURE 8.21
Combined GPS/SDARS antenna with two outputs: (a) Top view of square antenna; (b) side view of square antenna; (c) ring patch dual frequency antenna geometry. (From J. Oh, IEEE. Antennas Propagation Society International Symposium, 2008. Copyright 2008. IEEE. With Permission.)

also includes opposing cut corners remote to the non-square corners of the upper patch and therefore provide RHCP. The upper metallization element is dispersed over the top surface of a top dielectric material, while the intermediate metallization material is dispersed over the top surface of a lower dielectric material as shown in Figure 8.20b. This antenna system has two separate feed pins and one shorting pin.

The upper metallization antenna element is resonant at SDARS frequencies while the intermediate metallization antenna element resonates at GPS frequencies. This antenna has two outputs, one for receiving GPS signals and another for receiving SDARS signals. A similar topology with two outputs but without a short pin [16] appears in Figure 8.21c.

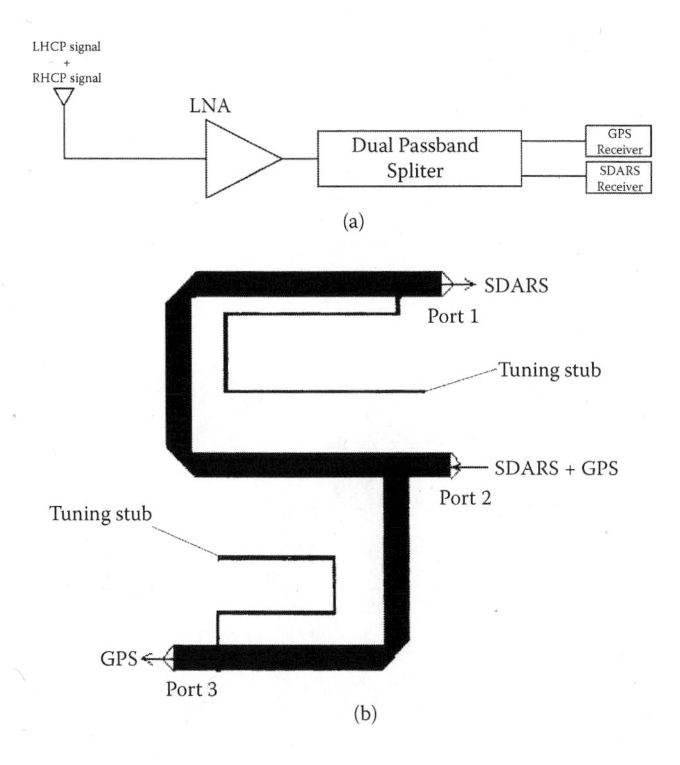

FIGURE 8.22
(a) Block diagram of active combined GPS/SDARS antenna with single feed pin output; (b) splitter combiner circuit for GPS/SDARS antenna.

8.9.2 Two Antennas in One Package

An antenna assembly that includes two separate patches, one for GPS and another for SDARS applications, is shown in Figure 8.23 [18]. Both are mounted on the same metal chassis with total dimensions of about 8 cm × 4 cm. Each antenna has its own low noise amplifier placed on the bottom side of the dielectric board. The system has two output cables and each is connected to a corresponding receiver.

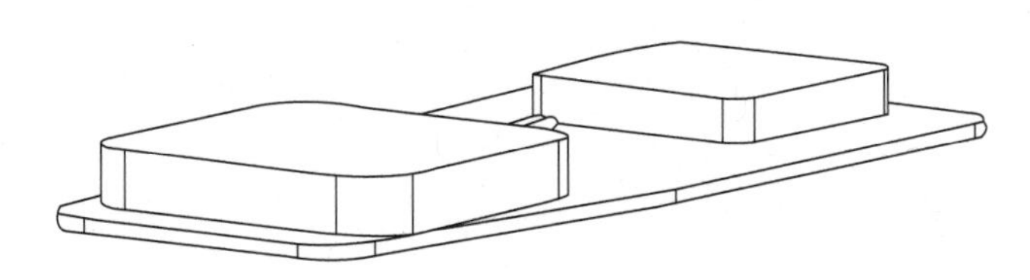

FIGURE 8.23
Separate GPS and SDARS antennas in single package.

8.10 Microstrip Antenna for GPS and DCS Applications

The antenna [19] consists of two radiation elements: a truncated square patch with RHCP radiation and annular ring patch intended for DCS (1705 to 1880 MHz) application. Both patched antennas are centered above a slotted ground plane (Figure 8.24). The annular ring patch with outer radius R_1 and

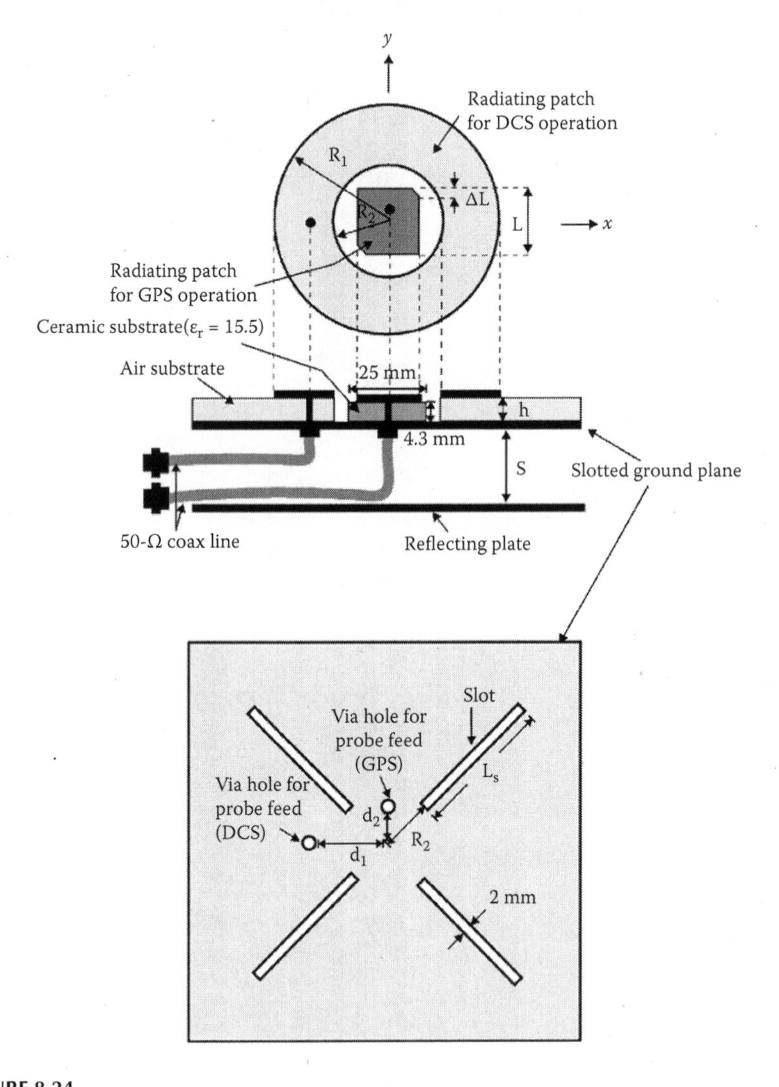

FIGURE 8.24
Configuration of compact antenna for GPS and DCS applications. (From S. Lin and K. Huang, *IEEE Transactions on Antennas and Propagation*, 53, 2005. Copyright 2005 IEEE. With permission.)

inner radius R_2 has resonant frequency calculated as [19]:

$$\frac{w}{R} = 0.4 \quad (8.1)$$

$$k \cdot R = 2 \quad (8.2)$$

where $k = \frac{2 \cdot \pi}{\lambda}$; $w = \frac{R_1 - R_2}{2}$; and $R = \frac{R_1 + R_2}{2}$. This antenna has four radial slots about 2 mm wide embedded on the square ground plane. This type of configuration allows a reduction in size of the annular patch. The resonant frequency f_c decreases approximately linearly with increasing slot length L_S:

$$f_c \approx f_0 - 19.4 \cdot L_S \quad (8.3)$$

where f_0 = resonant frequency of the patch with regular ground plane. An additional reflected ground plane shown in Figure 8.24 blocks background radiation from the slots. Figure 8.25 shows the measured impedance of the GPS and DCS antenna portions for a prototype. The geometric dimensions are $L = 21.4$ mm, $\Delta L = 2.2$ mm, $R_1 = 52.5$ mm, $R_2 = 22$ mm, $h = 6$ mm, $S = 12$ mm, $L_S = 60$ mm, $d_1 = 38$ mm, $d_2 = 3.38$ mm, and slotted ground plane = $180 \cdot 180$ mm^2. The impedance bandwidth for VSWR 2:1 is about 24 MHz (1568 to 1592 MHz) and the circular polarization bandwidth (axial ratio <3 dB) is about 8 MHz. The impedance bandwidth for DCS is about 175 MHz. Figure 8.26 demonstrates the measured isolation of the two feeding ports. Figure 8.27 shows the investigated radiation patterns.

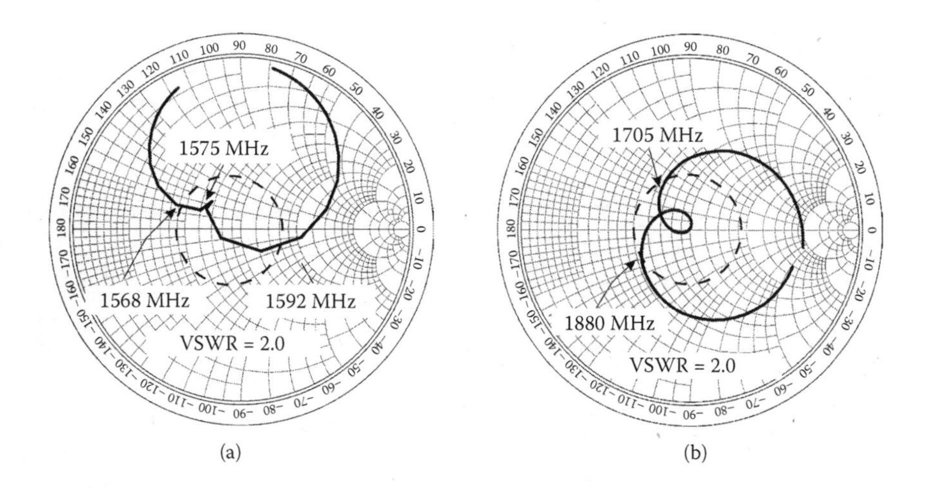

FIGURE 8.25
(a) Measured input impedance for GPS antenna; (b) measured input impedance for DCS antenna. (From S. Lin and K. Huang, *IEEE Transactions on Antennas and Propagation*, 53, 2005. Copyright 2005 IEEE. With permission.)

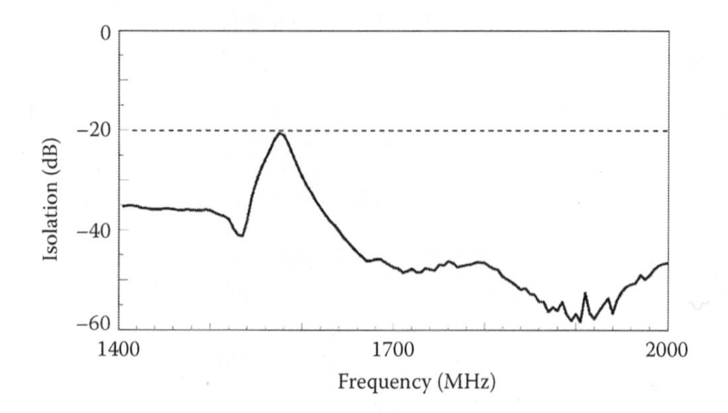

FIGURE 8.26
Measured isolation between feeding points of GPS and DCS antennas. (From S. Lin and K. Huang, *IEEE Transactions on Antennas and Propagation*, 53, 2005. Copyright 2005 IEEE. With permission.)

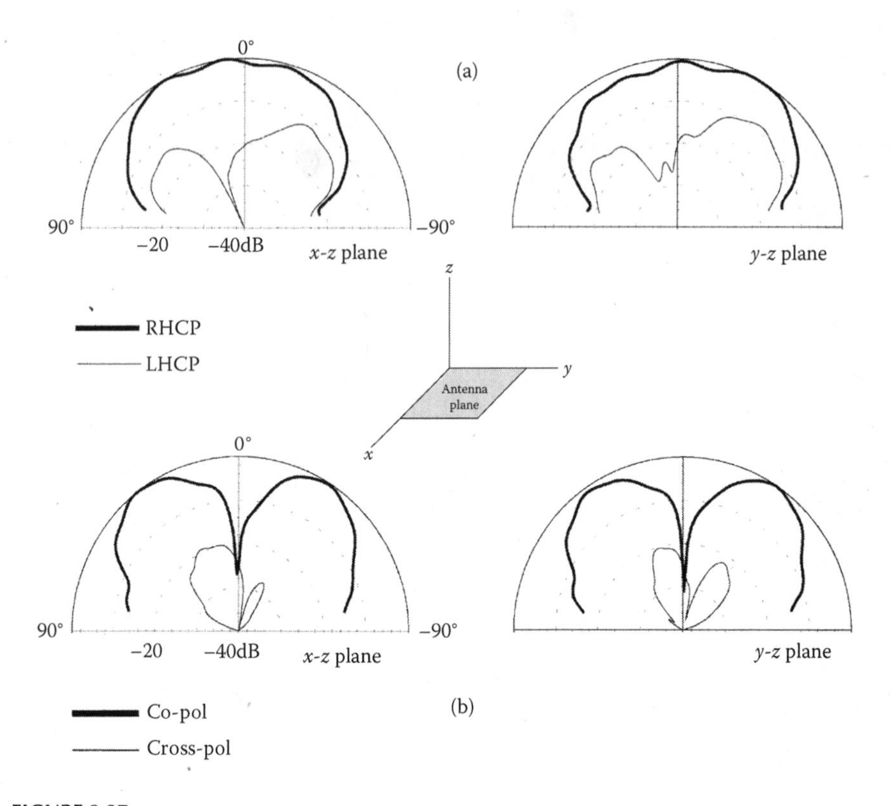

FIGURE 8.27
Measured radiation patterns: (a) 1575 MHz; (b) 1800 MHz. (From S. Lin and K. Huang, *IEEE Transactions on Antennas and Propagation*, 53, 2005. Copyright 2005 IEEE. With permission.)

8.11 Integrated GPS/PCS/RKES System

Section 5.11 of Chapter 5 examined the cellular antenna—one component of an integrated GPS/PCS/RKES antenna system. Another component of this system is the GPS portion. All the antenna segments are mounted on the same base; the distance between the GPS and PCS/RKES portions equals 20 mm (Figure 8.28). The GPS antenna represents a corner-truncated microstrip

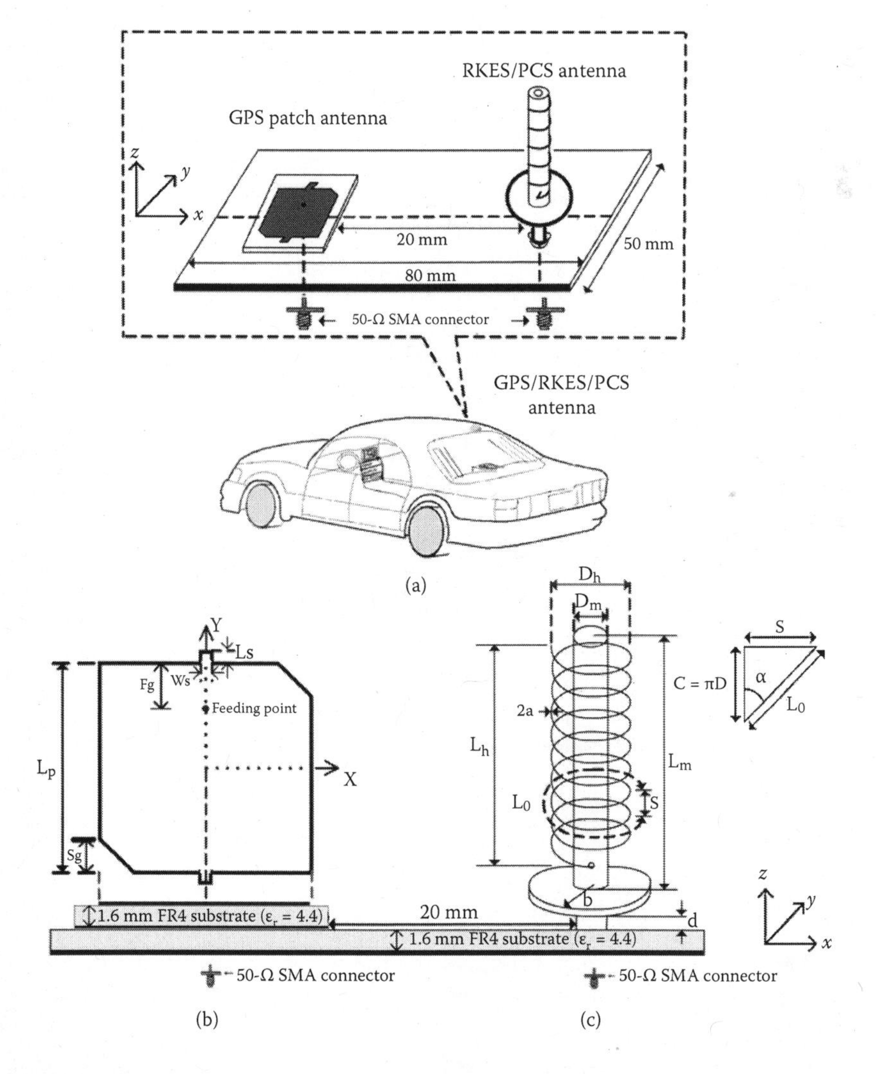

FIGURE 8.28
GPS/RKES/PCS antenna geometry. (From K. Oh, B. Kim, and J. Choi, *IEEE Microwave and Wireless Component Letters*, 15, 2005. Copyright 2005 IEEE. With permission.)

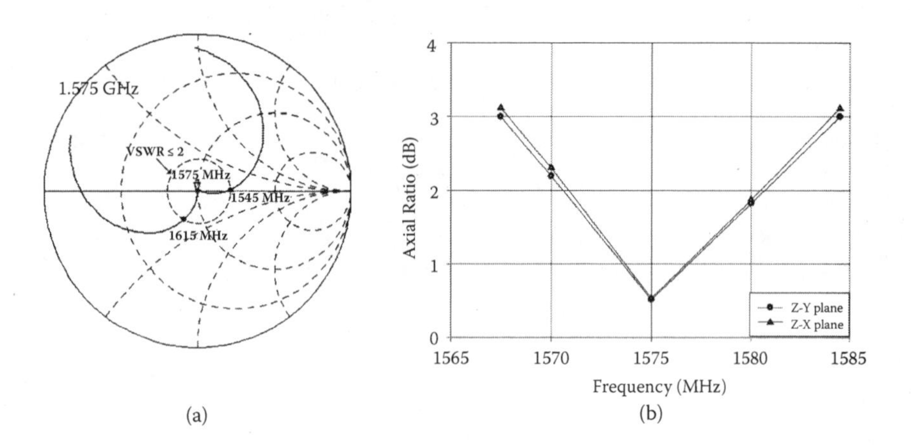

FIGURE 8.29
Measured input impedance (a) and axial ratio (b) in boresight direction for GPS antenna. (From K. Oh, B. Kim, and J. Choi, *IEEE Microwave and Wireless Component Letters*, 15, 2005. Copyright 2005 IEEE. With permission.)

patch with a tuning stub. These stubs placed at the centers of boundaries allow independent adjustment of input impedance for fixed coaxial feed.

Figure 8.29 demonstrates the measured input impedance and axial ratio in the boresight direction of the GPS antenna portion designed and built as a prototype sample. The antenna parameters are $L_p = 46$ mm, $S_g =$ 5.3 mm, $L_S = 1.5$ mm, $W_S = 4$ mm; the dielectric parameters are $\varepsilon_r = 4.4$, $h = 1.6$ mm, and ground plane = 50 · 80 mm^2. The measured bandwidths of the proposed GPS antenna (for VSWR <2 and axial ratio in 0 to 3 dB range) are 70 MHz and 17 MHz, respectively. Simulated and measured gain values at 1575 MHz frequency show a peak antenna gain of about 4.1 dBic (RHCP), with a small variation (<0.2 dB) in the GPS frequency range.

8.12 Car Location Options

In general, a GPS antenna combined with other antenna systems is mounted on a car roof. This location provides the maximum visibility for incoming satellite signals and the best performances because the antenna is sited on a big roof ground plane. However, as mentioned in previous chapters, some car models do not have enough space for roof antenna mounting (e.g., a convertible). Therefore alternative locations need to be considered for GPS antenna mounting. A trough glass GPS antenna configuration similar to the

FIGURE 8.30
Covert double band GPS/cellular phone antenna for vehicle interior application.

trough glass SDARS antenna described in a previous chapter is available commercially (for example, the glass-mounted GPS/GSM antenna from Radiall-Larsen Antenna Technologies).

Another mounting described in a U.S. patent application [20] presents a GPS/SDARS antenna assembly in an externally located vehicle mirror housing that includes a cable exit passage that passes an antenna cable extending from the antenna circuit board into the vehicle. A commercially available covert double band GPS antenna (Figure 8.30) can be mounted inside a vehicle instrument panel (IP). The antenna operates in the GPS band (1575 MHz) and cellular band (806 to 960 MHz, 2 dBic gain). The GPS antenna has a gain of 5 dBic (zenith) and the amplifier gain is about 26 dB. Dimensions of the antenna case are width = 41 mm, length = 61 mm, and height = 14 mm. The antenna with wings has a linear size of about 114 mm.

According to the investigation [21], a passive GPS patch installed in an IP (Figure 8.31) exhibits acceptable gain only for the θ angles from a zenith of about 50 degrees. At θ angles that correspond to the horizon angle, the average gain and gain ripples do not meet typical antenna requirements.

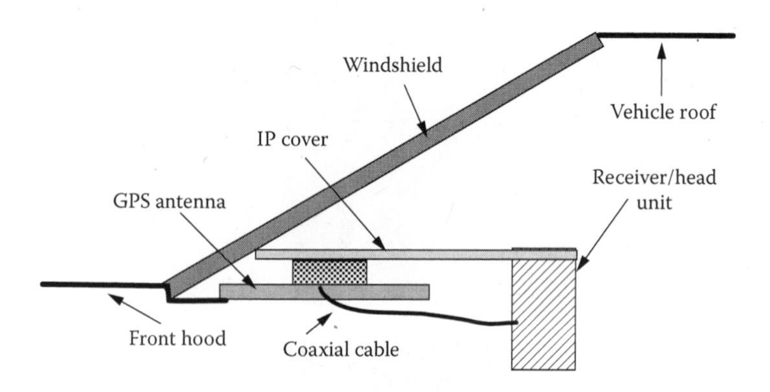

FIGURE 8.31
GPS antenna mounted into instrument panel (IP). (From K. Yegin, *Microwave and Optical Technology Letters*, 49, 2007. Copyright 2007 IEEE. With permission.)

Hidden GPS antenna topology printed on transparent film that can be attached to car glass [22] is shown in Figure 8.32. This antenna forms a metallic loop conductor placed on a transparent plastic film and a wire parasitic element independent from the loop. The measured average zenith gain is about 0 dBic, a $\theta = 30$ degree angle corresponds to a –2 dBic gain, $\theta = 60$ degrees to –12 dBic, and $\theta = 90$ degrees to –20 dBic. This antenna has a tiny amplifier with a gain of about 25 dB and a cable length of 5 m. This type of antenna is commercially available, for example, from SPK Electronics Co. Ltd. (www.spkecl.com).

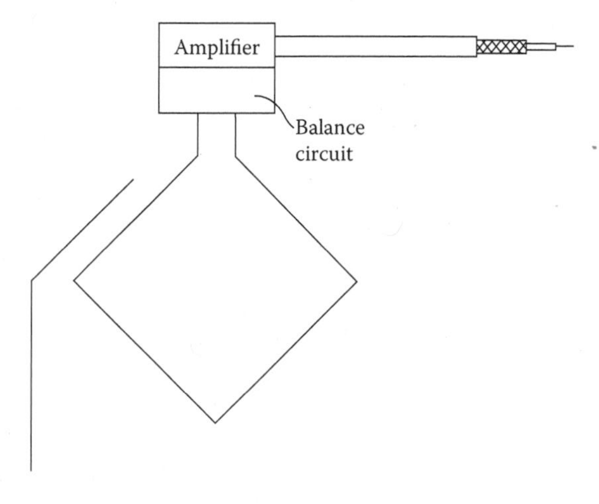

FIGURE 8.32
Hidden GPS antenna printed on transparent film.

8.13 Diversity Antennas for GPS Applications

In urban environments or where an antenna is mounted in the interior of a car, a GPS radio wave propagates through complicated reflection or scattering processes. The GPS receiver begins to encounter link problems when it goes into non-line-of-sight environments because of signal blocking from multipath fading. When the GPS signal is reflected by, for example, buildings, trees, or a car body, right hand circular polarized (RHCP) GPS signals may be changed to left hand circular polarization (LHCP), elliptical polarization, or linear polarization [23,24]. As a result, signal fading is becoming a major source of error in GPS positioning.

Experiments [25] have shown that the advantages of circular polarization disappear when it is used in a heavy multipath environment. Therefore, the diversity technique that effectively improves the signal-to-noise ratio (SNR) can be successful in GPS systems. In the simplest case, the diversity technique uses two receiving antennas spaced at a distance that provides a correlation coefficient between them of less than 0.5.

Antennas can be located, for example, at the same positions as the diversity antennas used for satellite radio (Figure 7.32b). A second option is based on the utilization of two diversity antennas in one package. One is the RHCP unit and second one is a VP monopole. Figure 8.33 shows an example. The vertically polarized antenna can be built as a helix or a meander monopole printed on a dielectric board, similar to the monopole used for SDARS terrestrial application shown in Figure 7.2. The height of this antenna is 1 to 1.5 cm. The annular ring patch intended for DCS (1705 to 1880 MHz) application shown in Figure 8.27 also can be used as a GPS diversity system. In this case, the ring patch must be tuned to the 1575 MHz band.

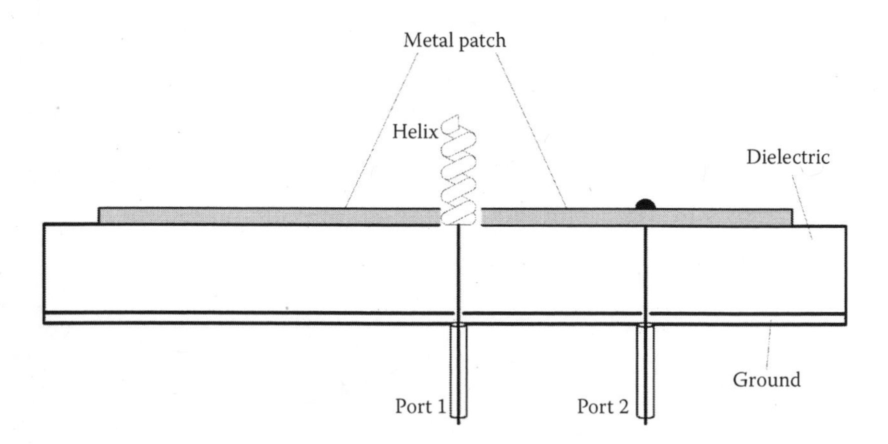

FIGURE 8.33
Combined diversity RHCP/VP GPS antenna.

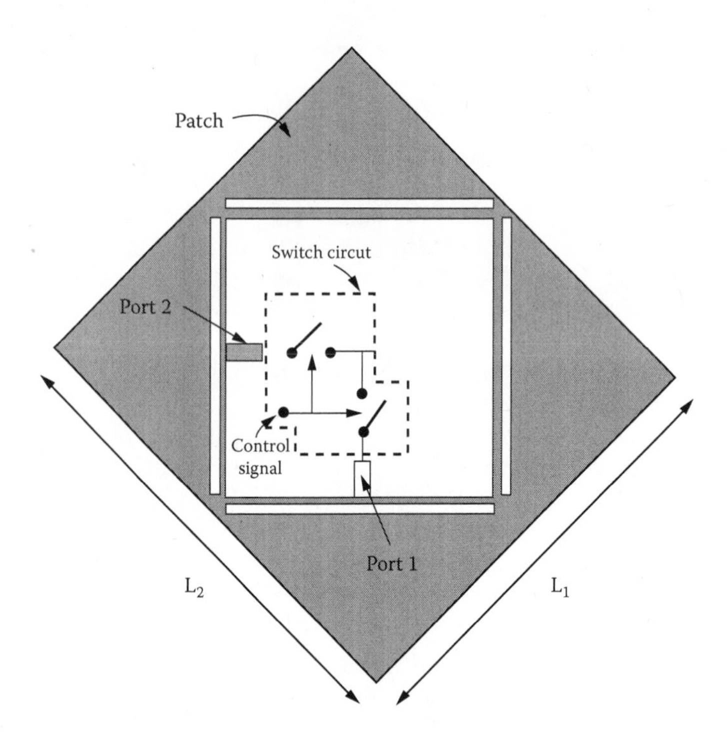

FIGURE 8.34
Diversity design with electronic switch.

A third option for a GPS diversity system is a microstrip antenna with circular polarization diversity [24]. This antenna presents a reconfigurable design with an electronically controllable switch. An antenna that operates on RHCP or LHCP waves is shown in Figure 8.34. The rectangular ring patch has two probe feeding structures. The outside length L_2 is slightly greater than L_1, and RHCP at port 1 and LHCP at port 2 are obtained, respectively. Four identical narrow slots are embedded along the inside edge of the patch. These have equal length L and narrow width W. Active circuitry is mounted in the square opening of the radiator.

All the parts are assembled on the top layer of the FR-4 dielectric board. The switch device was from Analog Devices (ADG918). A plastic HEMT transistor (Agilent ATF-55143) with low noise parameters was chosen for this application. The fabricated antenna size was 33.5×35 mm^2 (including the active circuit). The measured impedance bandwidth for RHCP and LHCP, determined by 2:1 VSWR, was about 83 MHz. Bandwidth determined from 3 db axial ratio for RHCP is about 29 Mhz and 34 Mhz for LHCP. The simulation and measurement results indicate that the proposed antenna system successfully performs CP diversity.

References

1. K. Yegin, AMPS/PCS/GPS active antenna for emergency call systems, *IEEE Antennas and Propagation*, 6, 255–258, 2007.
2. D. Pozar and D. Schaubert, Microstrip antennas, in *The Analysis and Design of Microstrip Antennas and Arrays*, New York: IEEE Press, 1995, p.100.
3. T. Milligan and P. Kelly, Optimization of ground plane for improved GPS antenna performance, *Second IEEE Antennas and Propagation Symposium*, July 1996, pp. 1250–1253.
4. K. Yegin, On-vehicle GPS antenna measurements, *IEEE Antennas and Propagation*, 6, 488–491, 2007.
5. H. Chen and K. Wong, On the circular polarization operation of annular ring microstrip antennas, *IEEE Transactions on Antennas and Propagation*, 47, 1289–1292, 1999.
6. D. Pozar and S. Duffy, A dual-band CP aperture-coupled stacked microstrip antenna for GPS, *IEEE Antennas and Propagation*, 45, 1618–1625, 1997.
7. C. Su and K. Wong, A dual band GPS microstrip antenna, *Microwave and Optical Technology Letters*, 33, 238–240, 2002.
8. X. Peng et al., Compact dual-band GPS microstrip antenna, *Microwave and Optical Technology Letters*, 44, 58–61, 2005.
9. L. Boccia, G. Amenodola, and G. Massa, A dual frequency microstrip patch antenna for high precision GPS applications, *IEEE Antennas and Wireless Propagation Letters*, 3, 157–160, 2004.
10. Y. Zhou, C. Chen, and J. Volakis, Dual-band proximity-fed stacked patch antenna for tri-band GPS applications, *IEEE Transactions on Antennas and Propagation*, 55, 220–223, 2007.
11. K. Luk et al., Broadband microstrip patch antenna, *Electronics Letters*, 34, 1442–1443, 1998.
12. D. Aloi and M. Alsiety, A methodology to determine the isolation requirements between collocated cellular and GPS antennas, *IEEE Antennas and Wireless Propagation Letters*, 6, 1–4, 2007.
13. D. Duzdar et al., Single-feed multi-frequency multi-polarization antenna, U.S. Patent 7,405,700, July 2008.
14. K. Geary et al., Single-feed dual-band stacked patch antenna for orthogonal circularly polarized GPS and SDARS applications, *IEEE Vehicular Technology Conference*, 2008.
15. K. Yegin et al., Integrated GPS and SDARS antenna, U.S. Patent 7,253,770, August 2007.
16. J. Oh, et al., Dual circularly polarized stacked patch antenna for GPS/SDMB, *Antennas and Propagation Society International Symposium*, 2008.
17. C. Callewaert and C. Maynord, Splitter/Combiner Circuit, U.S. Patent Application 20070216495, September 2007.
18. L. Ralf et al., Modular Antenna Assembly for Automotive Vehicle, U.S. Patent Application 20080111752, May 2008.
19. S. Lin and K. Huang, A compact microstrip antenna for GPS and DCS application, *IEEE Transactions on Antennas and Propagation*, 53, 1227–1229, 2005.

20. Y. Korkut et al., Vehicle Mirror Housing Antenna Assembly, U.S. Patent Application 20080001834, January 2008.
21. K. Yegin, Instrumental panel mount GPS antenna, *Microwave and Optical Technology Letters*, 49, 1979–1981, 2007.
22. K. Ogino et al., Circular Polarization Antenna and Composite Antenna, U.S. Patent 7,286,098, October 2007.
23. D. Manadhar and R. Shibasaki, Possibility analysis of polarization diversity scheme for multipath mitigation in GPS receivers, *International Symposium on GNSS/GPS*, Sydney, December 2004.
24. G. Yun, Compact active integrated microstrip antennas with circular polarization diversity, *Microwave Antennas and Propagation*, 2, 82–87, 2008.
25. V. Pathak et al., Mobile handset system performance comparison of a linearly polarized GPS internal antenna with a circularly polarized antenna, *IEEE Antennas and Propagation Symposium*, 2003, pp. 666–669.

9

Antennas for Short Range Communication

9.1 Introduction

Typical short range automotive communication systems include remote key entry (RKE), remote start engine (RSE), tire pressure monitoring system (TPS), electronic toll collection (ETC), radio frequency identification (RFID), long range radar (LRR) and short range radar (SRR) antenna components. More than 70% of modern vehicles are equipped with standard or optional RKE systems [1–3]. These remote controllable (RK) security devices commonly use 315 MHz frequency in the United States, Canada, and Japan; Europe utilizes 433.9 MHz and 868 MHz bands. This chapter focuses on antennas intended for remote control security devices.

The simplified vehicle mounted RK receiving module shown in Figure 9.1 includes a receiving antenna, RF receiver, digital microprocessor, and control module. The RF receiver captures the RF signal, demodulates it, and sends the data stream to the microprocessor that decodes it and sends a command to the module responsible for starting the car engine, opening and closing the doors, or sensing tire pressures. The antenna is a significant component [1] of such systems because it determines the communication range between the transmitter (e.g., key fob) and RK device. A better antenna allows an extended communication range.

Today's two main types of commercial antennas are an internal antenna mounted on a board along with electronic components of the receiving module and an external antenna separate from the receiving module. Different types of external antennas are used with automotive RK modules and include the dipole or "pigtail" coaxial antenna, the printed-on-glass version, and the compact type printed on a dielectric circuit board.

Figure 1.8 in Chapter 1 depicts typical internal antennas mounted on a board with electronic components. The integration of RF and digital electronic components with receiving antennas reduces the number of wires and connectors and thus reduces system cost. However, such designs present one important disadvantage: parasitic emissions from electronic components on the circuit board can significantly reduce device sensitivity and also the communication range of the system. The typical communication range of an

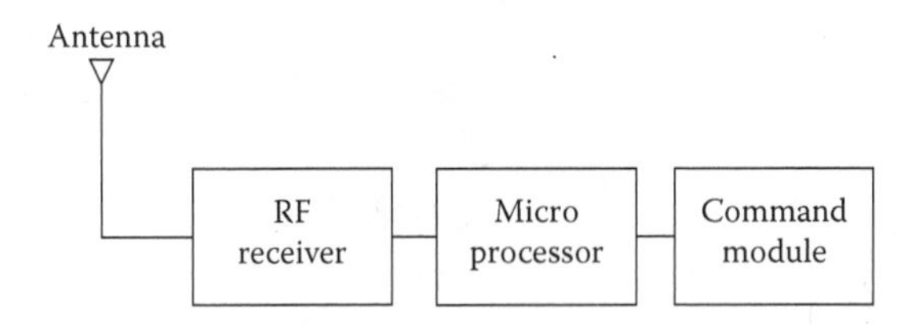

FIGURE 9.1
RKE module architecture.

RK system with an integrated internal antenna does not exceed 20 to 40 m. These values are determined by receiver sensitivity, antenna gain, noise of electronic components, and effective radiation power of the transmitter. For example, according to the U.S. Federal Communications Commission (FCC) requirements, maximum effective radiation power cannot exceed –15 dBm; Japan prohibits radiation above –43 dBm in the 315 MHz band.

External dipole antennas [4] can extend the range to 100 m, but this solution is inconvenient for interior vehicle applications: the size of an antenna for 315 MHz frequency is about 15 cm. The "pigtail" coaxial antenna [5,6] avoids some of the problems of dipoles and is thus more convenient for automotive interior applications. The pigtail is made by simply stripping off the outer conductor of the coaxial cable to extend the inner conductor by a quarter wavelength. The cable thus becomes a part of the antenna. The problem with the pigtail, however, is that the antenna is positioned very near the vehicle body as a part of the cable harness. Due to metal shadows from the vehicle body, the pigtail exhibits very small gain that in turn reduces the communication range. Window glass antennas for short range communication systems [7] have increased gain and thus provide extended range. However the printed-on-glass design depends on the available free car window space and must be tuned for each car model.

An alternative antenna type for vehicular use is a compact device printed on a circuit board [8], located separate from the control module in any convenient interior part of the vehicle, and connected with the RF receiver via an RF cable. Such an antenna can easily be packaged with a low noise amplifier (LNA) if hidden inside a vehicle.

It is known [9,10] that as the size of an antenna decreases, the radiation efficiency and gain decrease. Moreover, the impedance of small antennas has a significant reactive component and such antennas must be tuned very carefully to the proper impedance value. The main parameters that determine the quality of an antenna include:

- Radiation and loss resistance
- Radiation efficiency
- Bandwidth
- Directionality and gain

- Matching technique allowing tuning for 50 ohm impedance
- RF cable effect (of external antenna with receiver)

9.2 Compact Planar Passive Antenna

9.2.1 Geometries

The investigated planar antenna geometries [11] are shown in Figure 9.2. All antennas use a ground pad for the amplifier ground. The maximum antenna dimensions are 50 × 70 mm—less than wavelength. The planar printed loop antenna in Figure 9.2a, was examined because it is a popular configuration

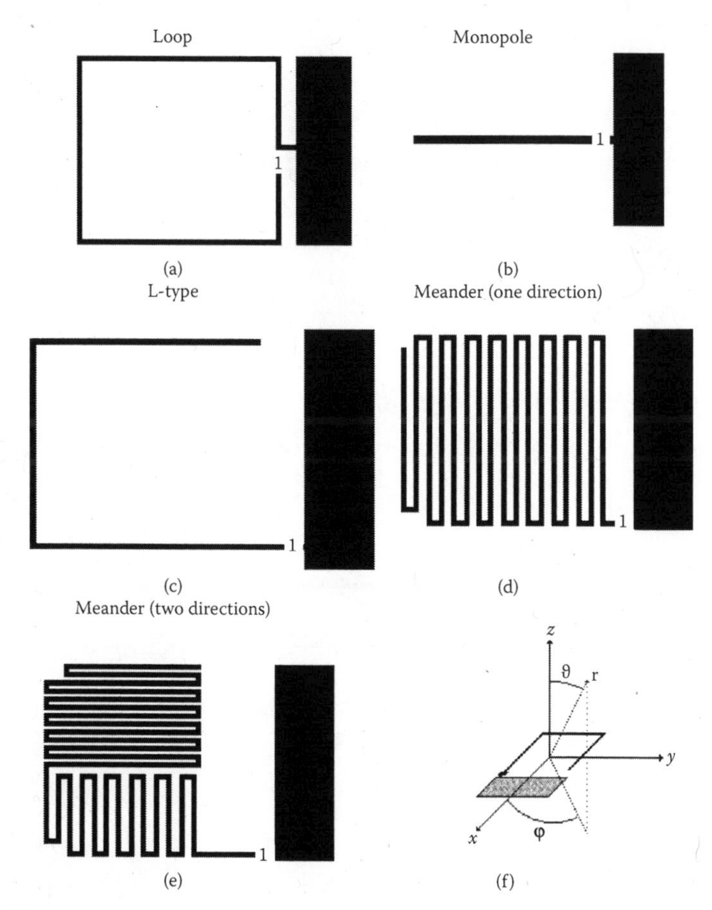

FIGURE 9.2
Investigated planar antenna geometries. (From V. Rabinovich et al., *IEEE Transactions on Vehicular Technology*, 55, 1425–1435, 2006. Copyright 2006 IEEE. With permission.)

for the keyless entry application. A second type is the printed monopole shown in Figure 9.2b. Other types are an L-inverted antenna (Figure 9.2c) and the meander line antennas shown in Figure 9.2d and e.

The antennas are printed on the front of a 1.6 mm thick FR-4 substrate (dielectric constant 4.5). The width of the printed antenna trace lines is 1 mm. Ground pads are used as a second "arm" on each antenna and serve concomitantly as low noise amplifier grounds.

Printed antenna parameters are often very difficult to compute using analytical expressions, but they can be estimated with the help of electromagnetic software or determined experimentally. We used IE3D software for antenna parameter simulation analysis [12] and computed three groups of parameters for every antenna type: (1) input impedance, radiation resistance and loss resistance, radiation efficiency and bandwidth; (2) matching to 50 ohm circuits and matching stability of voltage standing wave ratio (VSWR) when lumped elements were within tolerance values; (3) directionality and gain. Point 1 in Figure 9.2 indicates the antenna feed.

9.2.2 Simulation Results

9.2.2.1 *Impedance, Efficiency, and Directivity*

The simulated real *Re(Z)* and imaginary *Im(Z)* parts of antenna input impedance, radiation resistance, loss resistance, and radiation efficiency are shown in Table 9.1. The table also shows directivity D_{xy} in the x-y plane for the electromagnetic field component $E_{\varphi}(\theta = 90$ degrees). As shown in Table 9.1 all antennas have approximately the same directivity in the x-y plane.

9.2.2.2 *Parameters for Matching Circuits*

Matching circuits for these antenna types are shown in Figure 9.3. The circuits were designed with the help of Eagleware's Genesys software [13]. Inductor

TABLE 9.1

Input Impedance, Radiation Resistance, Loss Resistance, and Directivity of Printed Planar Antennas

Antenna Type	Re(Z) (Ohms)	Im(Z) (Ohms)	R_{rad} Ohms	R_{loss} Ohms	$\frac{R_{rad}}{R_{rad}+R_{loss}}$ (%)	D_{xy} (dBi)
Loop	23.96	911.4	3.57	20.39	14.9	1.8
Monopole	5.61	–548.7	0.95	4.66	16.9	1.9
L type	4.64	–324.4	1.25	3.39	27.0	1.8
One-direction meander	17.03	19.5	5.11	11.92	30.0	1.8
Two-direction meander	14.60	17.59	2.63	11.97	18.0	1.8

Source: V. Rabinovich et al. Compact planar antennas for short-range wireless automotive communication, *IEEE Transactions on Vehicular Technology*, 55, 2006. Copyright 2006 IEEE. With permission.)

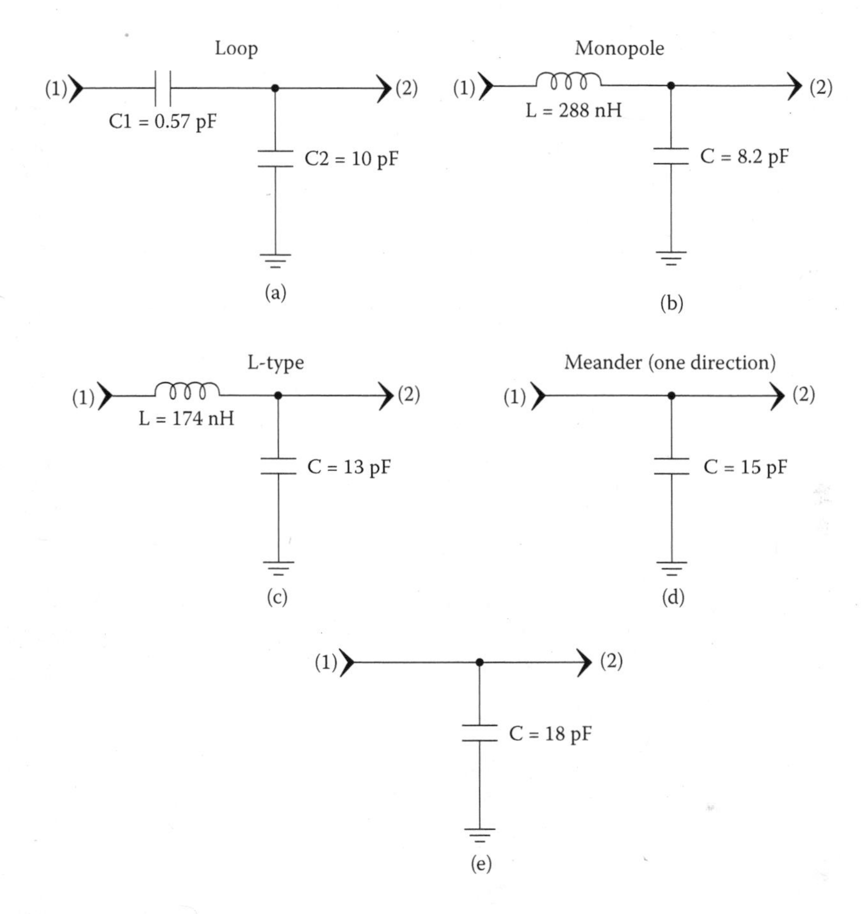

FIGURE 9.3
Matching to 50 ohm impedance circuits. (From V. Rabinovich et al., *IEEE Transactions on Vehicular Technology*, 55, 1425–1435, 2006. Copyright 2006 IEEE. With permission.)

components have a quality factor value of 25; for capacitor components the factor is 100. Table 9.2 shows the main parameters of the matched-to-50-ohm antennas. The efficiency η_s (ratio 2.28) takes into account matching network losses including heat ohmic loss within the network components and mismatching losses. Matching circuits for monopole and L-type antennas incorporate inductors. Due to the low quality factors of inductors (Q = 25), those matched to 50 ohm antennas exhibit reduced gain. The gains of both meander line antennas as seen in Table 9.2 exceed the gains of the loop, monopole, and L types. The bandwidths of the designs under investigation are not critical for this application. Gain G_{xy} for horizontal polarization is presented in the horizontal x-y plane.

To understand the variations of VSWR values due to the within-tolerance variation of the lumped matching elements from their nominal values, we

TABLE 9.2

Parameters of Planar Antennas with Output Matching 50 Ohm Impedance

Antenna Type	Re(Z) (Ohms)	Im(Z) (Ohms)	R_{rad} (Ohms)	R_{loss} (Ohms)	η_S (%)	*VSWR*	*BW* (%)	G_{xy} (dBi)
Loop	49.8	2.0	7.3	42.5	14.7	1.1	2	–6.5
Monopole	44.4	1.3	1.5	42.9	3.3	1.1	10	–12.9
L type	39.8	2.1	2.7	37.1	6.8	1.3	8	–9.9
One-direction meander	39.2	–0.9	11.7	27.5	29.9	1.3	2.5	–3.4
Two-direction meander	35.5	–2.5	6.4	29.1	17.9	1.4	2.5	–5.7

Source: V. Rabinovich et al. Compact planar antennas for short-range wireless automotive communication, *IEEE Transactions on Vehicular Technology*, 55, 2006. Copyright 2006 IEEE. With permission.)

used the Monte Carlo analysis option included in the Genesys software. Monte Carlo analysis evaluates circuit behavior with a random distribution of component values within specific limits. It is a statistical process that reveals whether production results will fall within acceptable limits. Monte Carlo supplies multiple responses, each generated with a pseudo-random set of component values based on the specified distribution and tolerances. Results for uniform distribution values with component tolerances of ±5% are shown in Figure 9.4.

We can see from these results that both one- and two-directional meander antennas produce the most stable VSWRs when matching components are within a tolerance of ±5%. The meander line antenna with its inherent matching of 50 ohms has fewer components. In reality, the meander line is physically small and electrically large. Therefore, the matching circuit for a meander line is not as critical as for the other antenna types in Figure 9.2. The antennas in Figures 9.2a through c exhibited large reactance values before matching. Therefore the ±5% tolerance for the matching components drops the VSWR to unacceptable values.

The engineering estimation of antenna gain using Equation (2.30) in Chapter 2 shows value of about 0 dBi. The discrepancy between estimation value and calculation is due to the fact that losses are not taken into account in estimating the upper limit.

9.3 Symmetrical and Asymmetrical 315 MHz Meander Line Printed on Dielectric Board

9.3.1 Introduction

An important point related to the asymmetrical antenna discussed in the previous section is the significant current flow in the outer conductor of the RF cable that connects the antenna to the RK control module. Essentially,

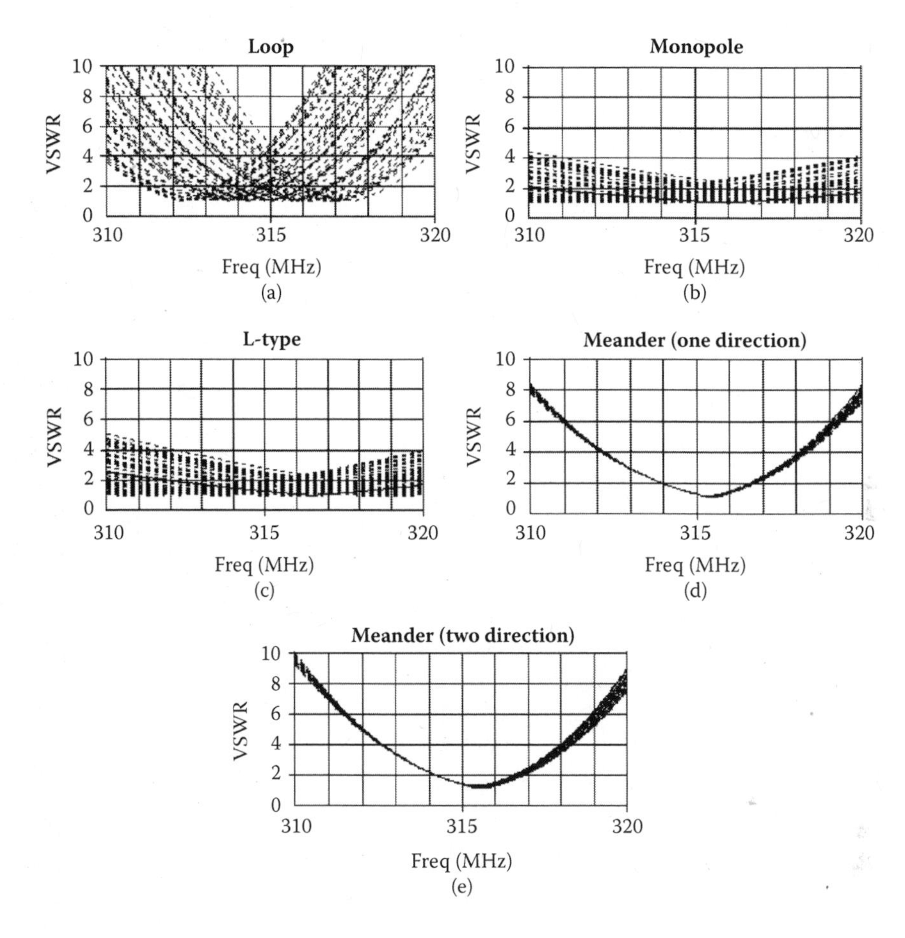

FIGURE 9.4
Monte Carlo analysis results for VSWRs of matching circuits with component tolerances within ±5%. (From V. Rabinovich et al., *IEEE Transactions on Vehicular Technology*, 55, 1425–1435, 2006. Copyright 2006 IEEE. With permission.)

the cable functions as an important radiating part of the antenna and can increase or reduce the communication range of the RK system.

Modern vehicles are equipped with many electronic devices including heaters, air conditioning with automatic temperature control, audio amplifiers, heated seats, power controls, and sunroof modules. Parasitic emissions from the electronic devices near the routing path of the external antenna RF cable reduce the communication range of an RKE system. To gain a feel for how an RK system can be affected by "real life" interference, assume that a nominal communication range for an RKE system is 100 m in the absence of parasitic emissions. Experimental measurements show that the noise received by an RF cable can exceed the noise floor of the receiver by 20 dB.

According to simulation graphs [14], such a noise level reduces the system range to 20 m or less. On the other hand when the parasitic emission effect of electronic components on an RF cable is negligible, an asymmetrical antenna with RF cable as described in this chapter can significantly increase the communication range.

The investigation presented in the next section had two goals. The first goal is to examine a meandered dipole antenna with reduced linear size that appeared to be a good candidate for 315 MHz and 433 MHz automotive applications. The second goal is to numerically and experimentally estimate the effect of an RF cable without a balun [15] on the parameters of both symmetrical and asymmetrical antennas.

9.3.2 Antenna Geometry

Figure 9.5 [16] shows meander line antennas of several length L, width *W*, and other significant geometric parameters. The width of the printed antenna trace lines is 1 mm. All values of *L* are below 1/10 wavelength. The asymmetrical meander line antenna [8] shown in Figure 9.5d has the same linear *L* and *W* as the antenna shown in Figure 9.2e. The *W*/*L* ratio for each antenna is less than 1. All antennas are printed on an FR-4 substrate, with a thickness of 1.6 mm and relative permittivity of 4.5. The antennas shown in Figures 9.5a to d are printed on one side of the dielectric board. Figure 9.5e shows a double-sided printed antenna. Black lines are drawn on the top of the dielectric; grey lines with a ground spot are located on the bottom of the dielectric.

The antennas presented in Figures 9.5d and e have a spot that can be used as a ground for the amplifier circuit when an active receiving antenna design is used. Figure 9.5e shows the geometry of the assembly including the RF cable. The total printed line length and number of bends for each antenna were chosen to provide 50 ohm input impedance.

Accurate impedance tuning for designs shown in figure 9.5a, b, c, e was achieved experimentally by insertion of an inductor between the positive and negative dipole arms and additional cutting of the edge meander antenna copper trace bends. (Circles in Figure 9.5a show these antenna components.) The tuning of the meander asymmetrical antenna impedance to 50 ohms was accomplished by inserting an additional capacitor between the meander line and a ground spot near the input antenna port. All antennas were operated as external devices connected via a control RKE module through the RF cable.

9.3.3 Numerical Results

Table 9.3 presents radiation efficiency η_s simulation results for linear antennas (frequency 315 MHz; tuning for 50 ohms; calculations performed by IE3D software). As the table reveals, the meander asymmetrical antenna with the

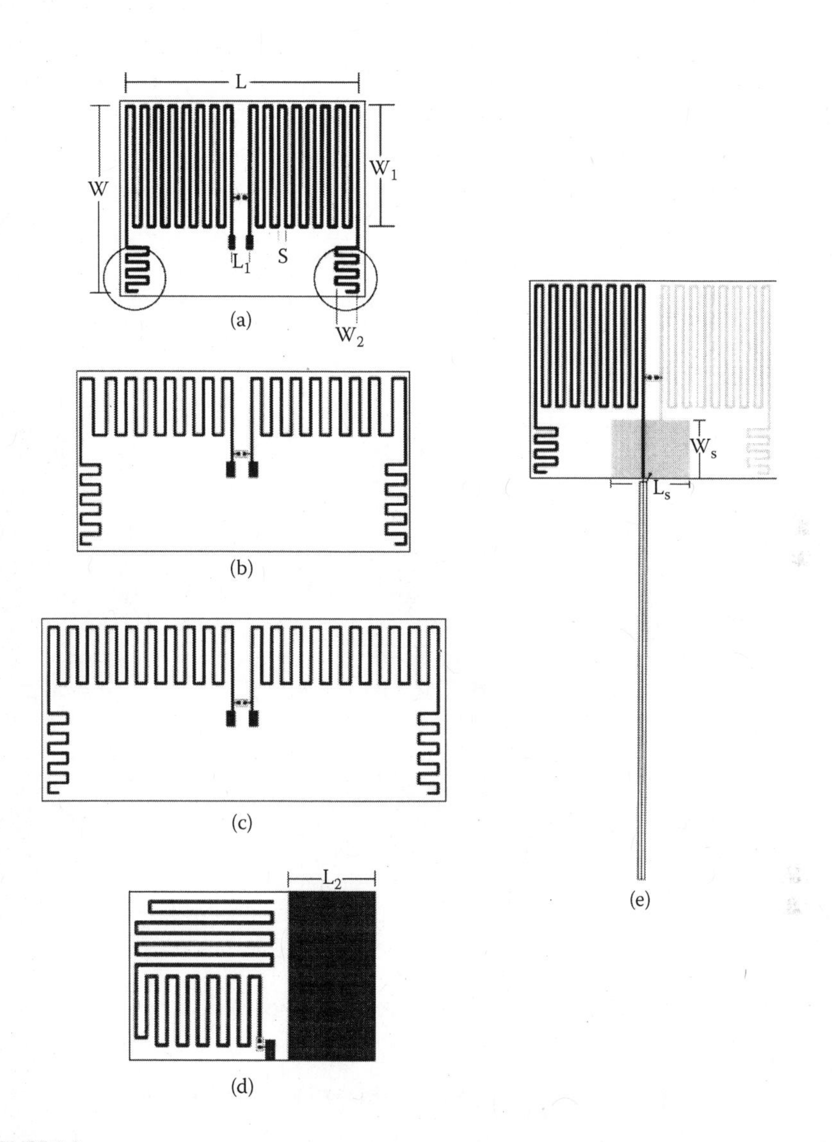

FIGURE 9.5
Investigated antenna geometries (mm): (a) L = 70, L_1 = 5, W =54, W_1 = 33, W_2 = 6 S = 1; (b) L = 10, L_1 = 6, W = 54, W_1 = 17, W_2 = 6, S = 3; (c) L = 120, L_1 = 6, W = 54, W_1 = 17 W_2 = 6, S = 3; (d) L = 70, L_2 = 24, W = 54; (e) , L = 70, L_1 = 5, L_s = 24, W = 54, W_1 = 33, W_2 = 6, W_s = 12, S = 1. V. (From Rabinovich et al., *Microwave and Optical Technology Letters,* 48, 2006. Copyright 2006 *Microwave and Optical Technology Letters. Reprinted with* permission of Wiley-Blackwell, Inc.)

dimensions shown in Figure 9.5d and without an RF cable exhibited the lowest efficiency: 0.12 (–9.2 dB). In comparison, the symmetrical meander dipole was 1.9 times more efficient. Note that the asymmetrical meander antenna (L = 70 mm) with an RF cable exhibited the same efficiency as the 100 mm meandered dipole without an RF cable. This indicates that the cable serves as a

TABLE 9.3

Radiation Efficiency Simulation Results for Different Linear Antenna Sizes L(W = 54 mm)

Parameter	Printed Meandered Dipole Antenna				Printed Asymmetrical Meander Line	Wire Half Wave Dipole
Length (mm)	70	100	120	70 + ground spot	70	475
Efficiency η_S without cable	0.23	0.42	0.52	0.21	0.12	0.98
Efficiency η_S with 1 m cable	0.28	0.45	0.54	0.33	0.45	0.98

Source: V. Rabinovich et al., *Microwave and Optical Technology Letters*, 47, 2005. Copyright 2005 *Microwave and Optical Technology Letters*. Reprinted with permission of Wiley-Blackwell, Inc.)

significant enhancement to an asymmetrical meander antenna. Such antennas may therefore be effective in vehicle applications in which electronic components near the cable do not radiate interference at the operating frequency.

However, the difference in the efficiency of a symmetrical antenna with or without an RF cable is insignificant. Further numerical and experimental directionality data were examined for both the symmetrical (Figure 9.5a) and asymmetrical (Figure 9.5d) antennas. Figure 9.6 plots simulated radiation patterns of symmetrical and asymmetrical meander line printed antennas (horizontal polarization, X-Y plane) with two different RG 174 RF cable lengths L_c (cable location is identical to shown in Figure 9.5e). Antenna orientations with regard to directionality angles are shown in Figure 9.2. The asymmetrical antenna (with RF cable) reveals a multi-lobed radiation pattern similar to the multilobed pattern of the dipole antenna with length about equal to the length of the asymmetrical meander with RF cable.

Figure 9.7 shows the calculated efficiency η_s and cable length (cm) for the asymmetrical meander antenna in Figure 9.5d. Efficiency (dB) was normalized to the half wave dipole efficiency (0 dB). As shown in Figure 9.7, the asymmetrical antenna with a cable approximately 25 cm long exhibits efficiency almost equal to that of the half wave dipole and similar to that of a coaxial antenna with an inner conductor length equal to one quarter of the wavelength, as described in [6]. Instead of the inner conductor of the coaxial antenna, we used a meander line shorter than one quarter wavelength with a total trace longer than a quarter wavelength. We can introduce a mean square error ε [16], averaged over 360 degrees that numerically estimates the "similarity" of two power directionality curves: the first when $F(\varphi)$ corresponds to the antenna without a cable, and the second when $F_1(\varphi)$ corresponds to the antenna with RF cable:

$$\varepsilon = \frac{\int_0^{360^\circ} (F(\varphi) - F_1(\varphi))^2 d\varphi}{\int_0^{360^\circ} F^2(\varphi) d\varphi} \tag{9.1}$$

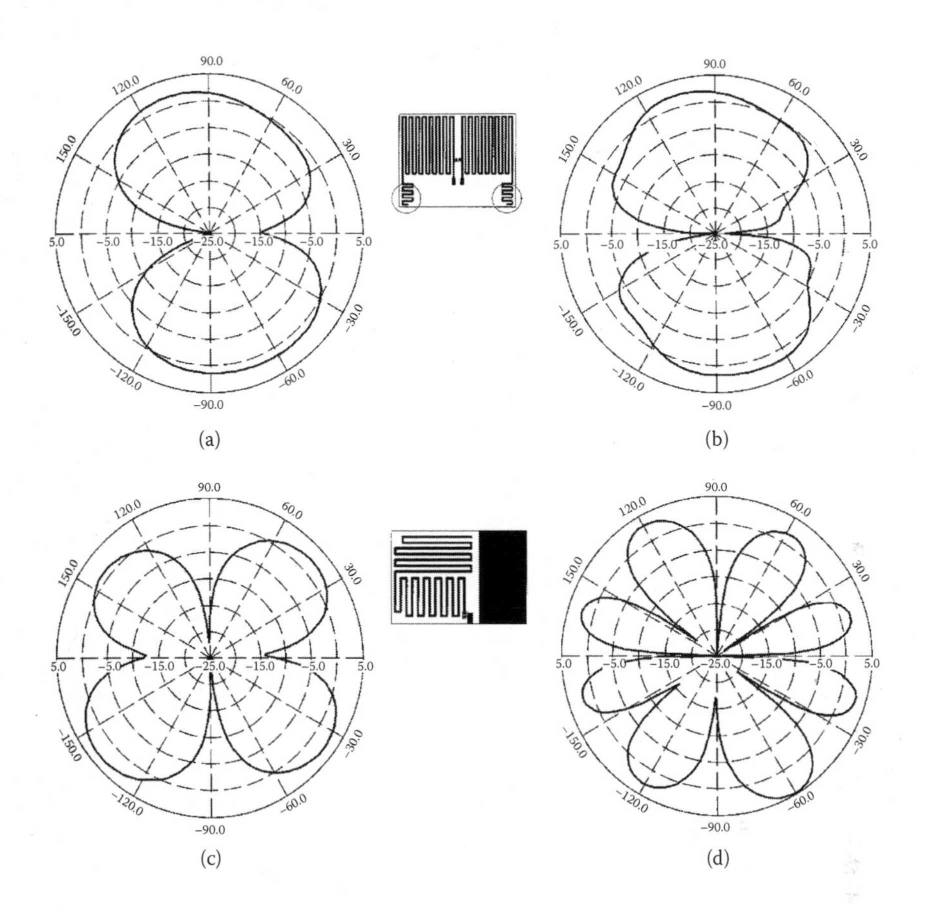

FIGURE 9.6
Simulated results: (a) symmetrical meander line dipole, L_c = 65 cm; (b) symmetrical meander line dipole, L_c = 160 cm; (c) asymmetrical meander line antenna, L_c = 65 cm; (d) asymmetrical meander line, L_c = 160 cm. (From Rabinovich et al., *Microwave and Optical Technology Letters,* 48, 2006. Copyright 2006 *Microwave and Optical Technology Letters. Reprinted with* permission of Wiley-Blackwell, Inc.)

Table 9.4 shows the calculated results. The asymmetrical antenna exhibits a maximum error value; it benefits from the largest increase in gain due to the added effect of the cable, but can suffer from interference effects from parasitic interference sources near the RF cable route in a car. The meander symmetrical dipole shows the smallest error ε; thus this antenna obtains minimal benefit from the addition of a cable, but also minimal possible interference effects. These results indicate that an asymmetrical antenna design with careful RF cable routing can increase communication range in a car that does not have electronic components radiating parasitic emissions at 315 MHz. However, if electronic components radiate parasitic emissions near the cable route, a symmetrical dipole antenna is a better candidate

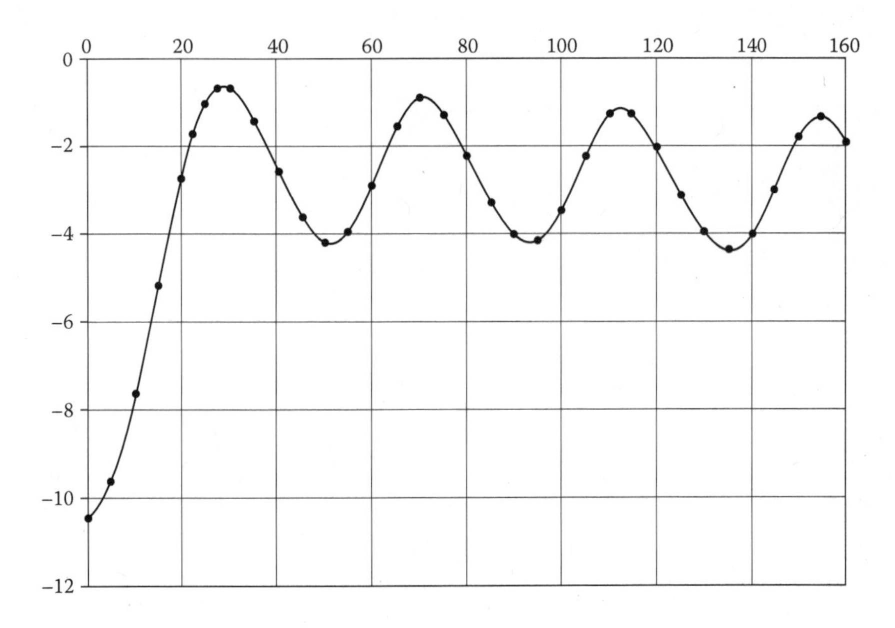

FIGURE 9.7
Efficiency as a function of cable length for asymmetrical meander line antenna. (From Rabinovich et al., *Microwave and Optical Technology Letters,* 48, 2006. Copyright 2006 *Microwave and Optical Technology Letters. Reprinted with* permission of Wiley-Blackwell, Inc.)

for RKE applications. The first design step for an RKE installation should always be investigation of the noise environment in the car.

9.3.4 Measurement Results

A passive meander line dipole antenna printed on an FR-4 dielectric substrate with RF cable was placed horizontally on a turntable (substrate board plane parallel to floor plane). The antenna was activated in a transmitting mode. A

TABLE 9.4

Calculated Square Error Parameter (315 MHz)

Antenna Type	Length (mm)	Mean Square Error ε
Printed meander dipole	70	0.3
	100	0.16
	120	0.15
	70 + ground spot	0.74
Printed asymmetrical meander line	70	0.81

Source: V. Rabinovich et al., *Microwave and Optical Technology Letters,* 47, 2005. Copyright 2005 *Microwave and Optical Technology Letters. Reprinted with* permission of Wiley-Blackwell, Inc.)

horizontally polarized receiving Yagi antenna operating at 300 to 1000 MHz frequency was positioned in the far zone of the antenna assembly (passive antenna with RF cable located in the antenna plane and perpendicular to the L side direction). The resulting directionality measurements covered 360 degrees in the horizontal plane for HP. We used RG 174 cable for the measurements, with losses equal to 0.5 dB/m in the 315 MHz frequency band.

Figure 9.8 presents the measurement results for the symmetrical and asymmetrical antennas shown in Figures 9.5a to d. Figure 9.8a shows the directionality of a symmetrical dipole with cable length L_c equal to 65 cm. Figure 9.8b corresponds to a cable length of 160 cm. Figures 9.8c and d show

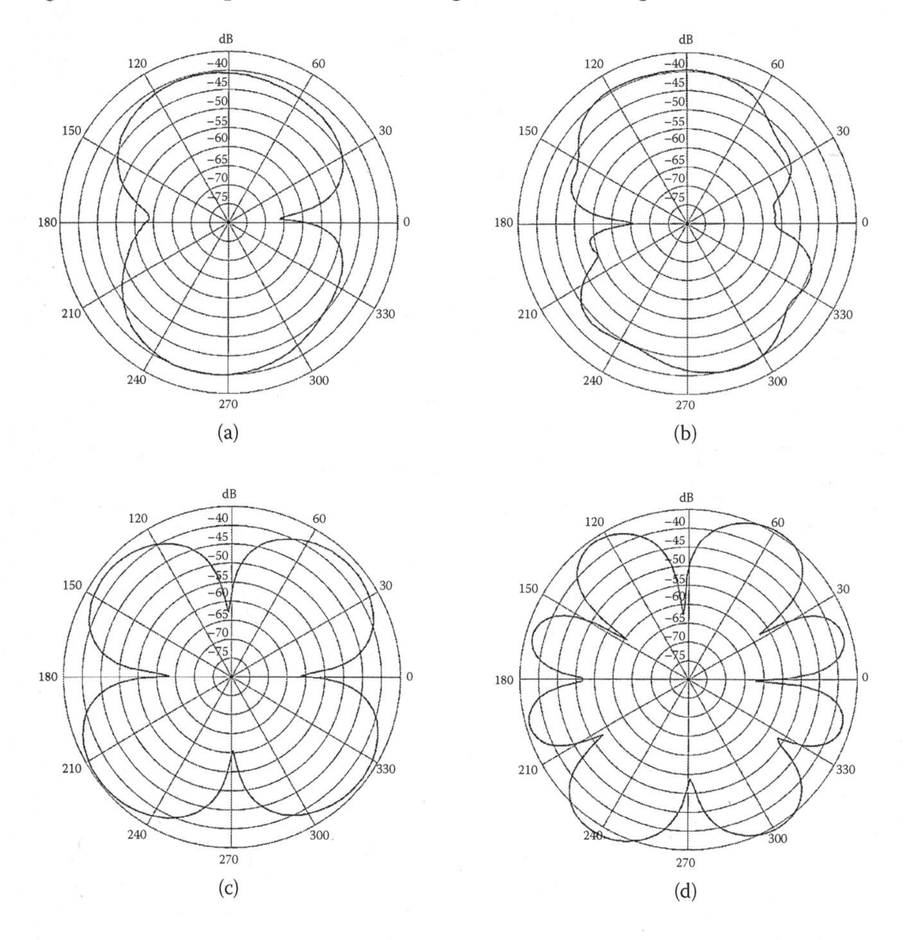

FIGURE 9.8
Measurement results: (a) symmetrical meander line dipole, L_c = 65 cm; (b) symmetrical meander line dipole, L_c = 160 cm; (c) asymmetrical meander line, L_c = 65 cm; (d) asymmetrical meander line, L_c = 160 cm. (From Rabinovich et al., *Microwave and Optical Technology Letters*, 48, 2006. Copyright 2006 *Microwave and Optical Technology Letters. Reprinted with* permission of Wiley-Blackwell, Inc.)

the horizontally polarized directionality plots in the azimuth plane for an assembly consisting of an asymmetrical meander line antenna with an RF cable. The measurement results confirm the findings of the numerical simulation: the cable effect is not very significant to the performance of a symmetrical antenna and very significant to the performance of the asymmetrical design.

9.4 Considerations for Small Antennas

As noted in previous sections, the radiation of a small printed antenna element with asymmetrical (unbalanced) geometry leads to the flow of current on the outer conductor of the coaxial cable connecting the antenna with the control module (Figure 9.9). Therefore, the parameters of a main antenna without a cable such as impedance, gain, and radiation pattern are measured with errors depending on the cable length and orientation relative to the antenna. For example, Reference [17] demonstrates that improper measurement set-up (RF cable connected directly to antenna) produced gain error measurements up to 10 dBi. Complete elimination of cable radiation is not easy, but a significant reduction of the influence of an RF cable in small antenna measurements is achieved by using special methods described in the literature [18–22].

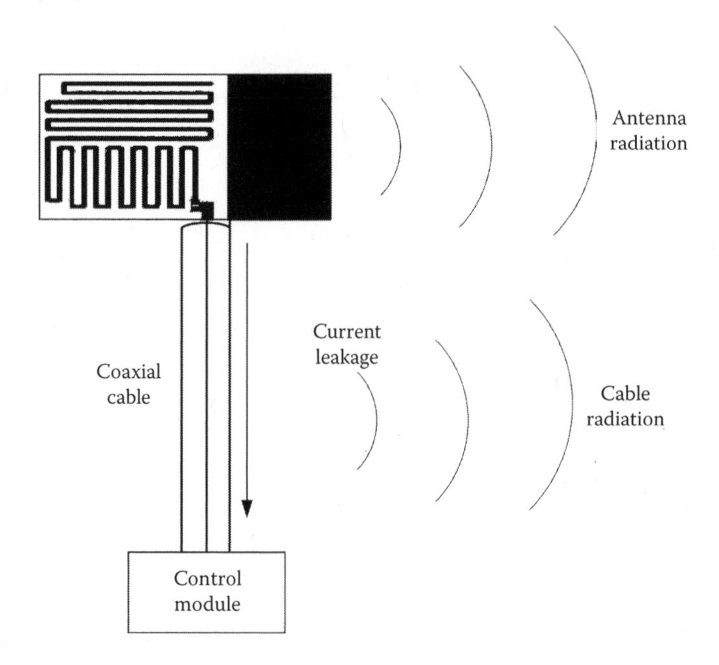

FIGURE 9.9
Influence of RF cable on radiated fields.

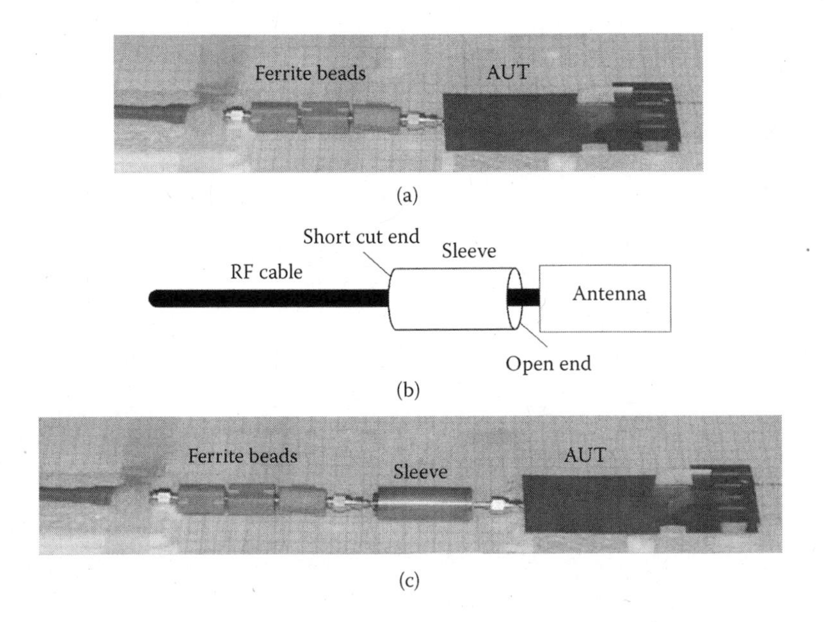

FIGURE 9.10
(a) Ferrite bead rings inserted into RF cable; (b) installation of sleeve balun on RF cable attached to antenna; (c) combined balun and ferrite bead attachment. AUT = antenna under test. (From C. Icheln et al., *IEEE Transactions on Instrumentation and Measurements*, 53, 2004. Copyright 2004 IEEE. With permission.)

Figure 9.10 shows a few techniques that may stop undesired current from flowing in a coaxial cable. The simplest variant is to place ferrite bead rings on the coaxial cable as shown in Figure 9.10a. The positioning of the beads plays an important part in determining radiation parameters of an antenna. The common practice [20] is to employ them on the cable as near as possible to the feed point of the antenna. When using ferrite beads, their performance at the frequencies of interest should be determined to ensure that current suppression is adequate. Another method involves installation of a sleeve choke between the antenna element and RF cable as shown in Figure 9.10b. A typical balun consists of a metal sleeve of quarter wavelength that covers the exterior of the outer conductor of the coaxial cable with its end shorted to the outer conductor. The quarter wavelength structure allows the stoppage of current flow on the outside conductor of the cable. A combination of balun plus ferrite ring beads [19] appears in Figure 9.10c.

Another simple way to reduce unbalanced current flowing on a feed cable is to use a choke coil [22] formed by winding a thin semi rigid cable around the feed cable. To avoid disturbances in radiation pattern measurements caused by cable radiation, a small battery operated transmitter can be connected with a passive antenna, thereby eliminating the connection of cable to antenna. In this case, the antenna under test (AUT) is used as a transmitting unit for pattern measurement. When measuring a radiation pattern in a specific plane, it

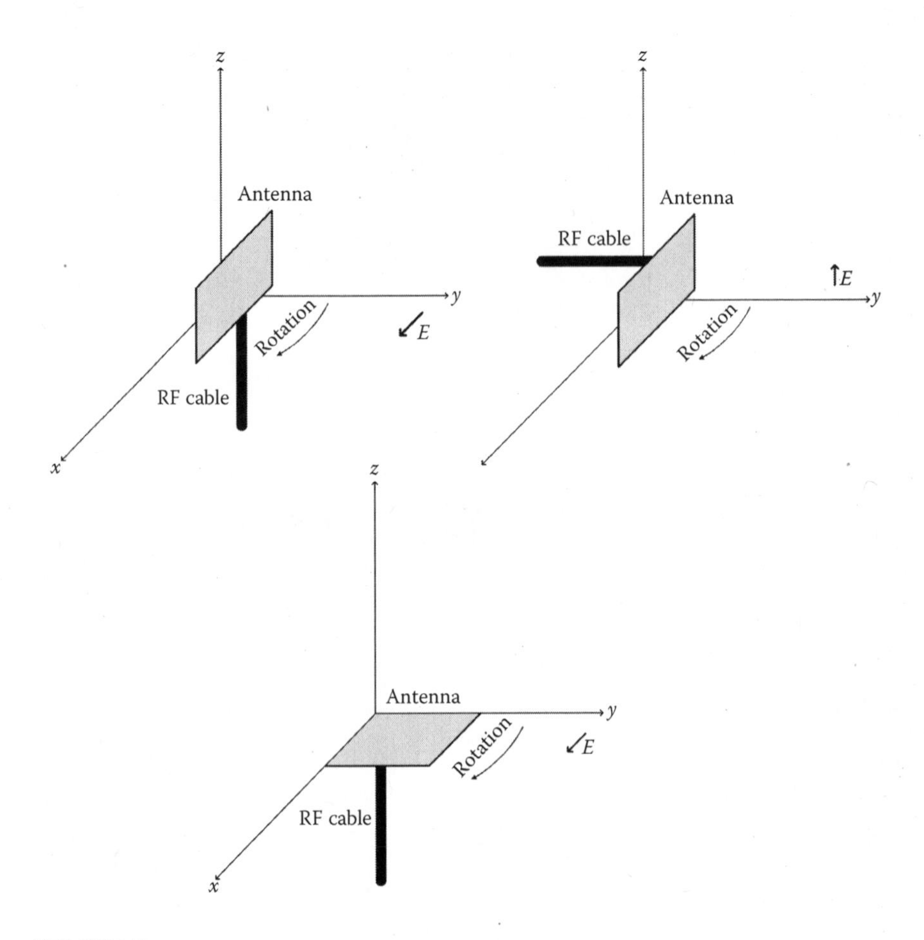

FIGURE 9.11
Set-up for radiation pattern measurements with RF cable mounted perpendicular to rotation plane.

is desirable (during rotation) to place the cable perpendicular to the polarization direction E of the transmitting antenna (Figure 9.11). A rough estimation can be achieved by using a near-field probe attached to a spectrum analyzer to measure radiation emissions from a cable connected to an antenna.

9.5 Active Meander Line Antenna Implementation for 315 MHz

9.5.1 Antenna and Amplifier Parameters

To study antenna performance in detail, an asymmetrical active meander prototype (Figure 9.12a; [8,37]) was printed on the front of an FR-4 substrate (thickness 1.6 mm, relative permittivity 4.5). The area occupied by the active

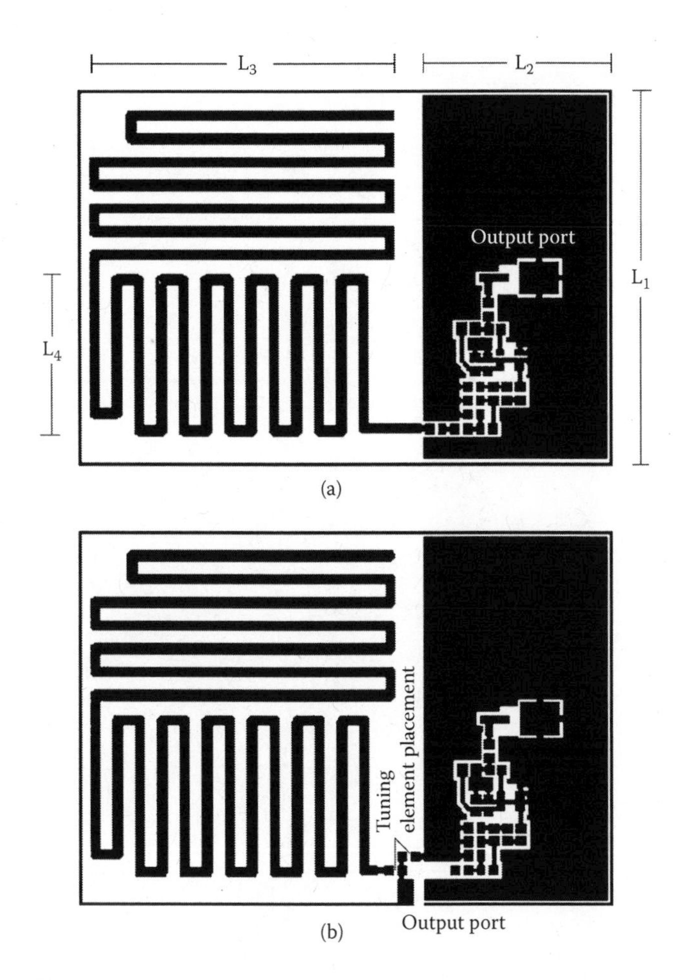

FIGURE 9.12
Geometry of printed meander line antenna, 315 MHz: (a) active topology; (b) passive portion with output port placement. (From V. Rabinovich, B. Al-Khateeb, and B. Oakley, *Microwave and Optical Technology Letters*, 43, 2004. Copyright 2004 *Microwave and Optical Technology Letters*. Reprinted with permission of Wiley-Blackwell, Inc.)

and passive electronic components was L_1 = 47.2 mm, L_2 = 24.7 mm. The antenna ground plane was printed on the bottom side under the electronic components and partially on the top of the substrate. The length of each horizontal line of the meandered arm L_3 was 40.8 mm; each vertical line L_4 was 20.4 mm long; and line thickness was 1.2 mm. The total dimensions of antenna and amplifier were 70.4 mm × 47.2 mm.

Figure 9.13 presents the measured output impedance and VSWR of the passive antenna element shown in Figure 9.12b matched with tuning elements (marker 3 corresponds to 315 MHz). The measured passive portion gain maximum in

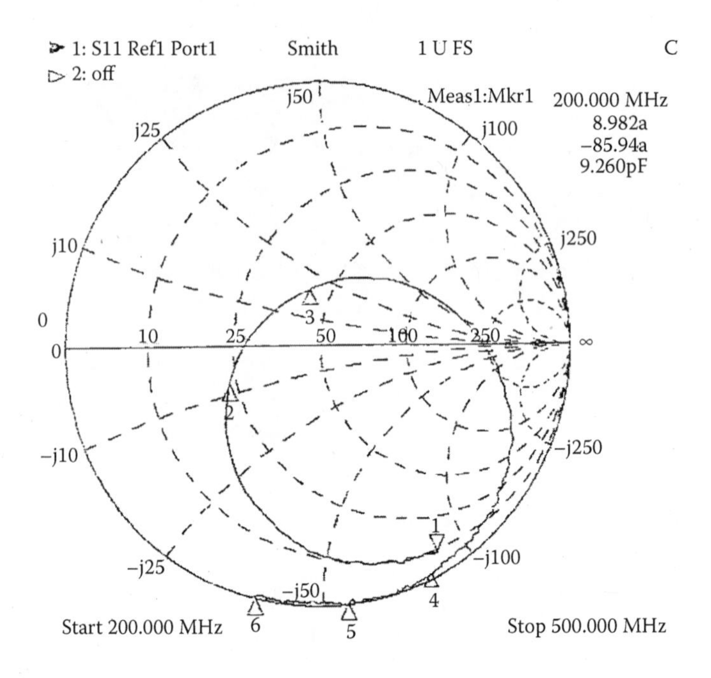

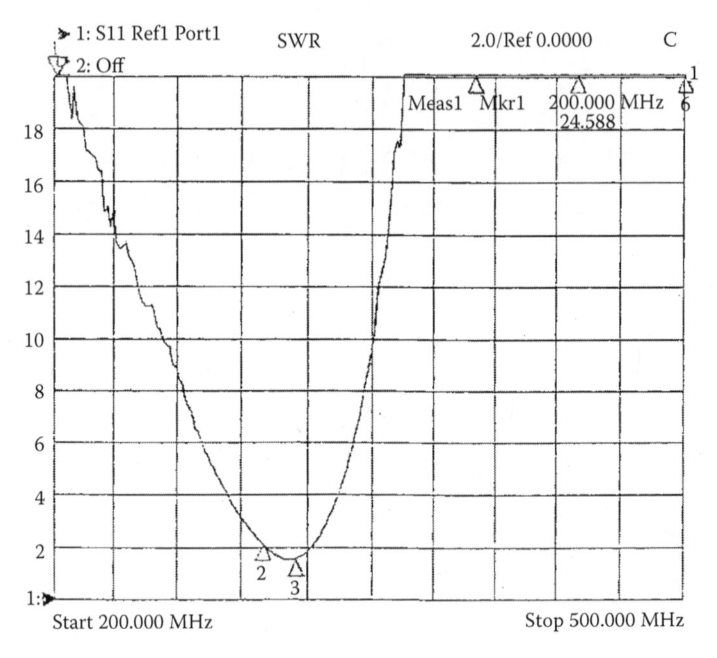

FIGURE 9.13
Smith chart and VSWR of passive antenna portion. (From V. Rabinovich, B. Al-Khateeb, and B. Oakley, *Microwave and Optical Technology Letters*, 43, 2004. Copyright 2004 *Microwave and Optical Technology Letters. Reprinted with* permission of Wiley-Blackwell, Inc.)

the X-Y plane was –8.5 dBi. The cable lay-out was carefully planned for measuring antenna gain. First, to ensure accurate measurements of the pattern, ferrite beads (Figure 9.10) were used. The beads eliminate stray current and undesired radiation from the coaxial cable. Second, the cable with ferrite beads was set perpendicular to the antenna plane.

The amplifier for the antenna was designed using computer software from Eagleware Corporation. Amplifier properties were optimized to achieve noise impedance matching between the antenna and transistor stage and power impedance matching between the amplifier and the 50 ohm load. The amplifier consisted of a single-stage NE662M04 transistor from California Eastern Laboratories coupled to a passive input circuit that provided low noise and an output matching circuit for maximum amplifier gain. The experimentally amplifier gain (measured separately from the antenna) in the 315 MHz band as a function of frequency is shown in Figure 9.14. The noise figure of the amplifier was about 2 dB.

9.5.2 Meander Line Antenna in Plastic Case

The flat printed antenna intended for remote control vehicle applications has a plastic case (Figure 9.15b) built of radio-transparent material [11]. The case is fixed to the car body interior and protects the antenna from damage. Of course, the radio-transparent dielectric material changes the antenna resonance frequency. Using IE3D software, we computed the influence of the infinite dielectric plates (dielectric constant $\varepsilon = 3.4$) on the VSWR, with the meander line antenna sandwiched between the plates (Figure 9.15a). Mathematical simulation results (Figure 9.16) show that the resonance frequency of such an antenna shifts into a lower frequency range than that of an antenna without such radio-transparent material (value of shifted frequency is about 15 MHz). Simply changing the matching series tuning capacitor a few picofarads provides the best matching results when the antenna is placed between infinite plastic plates.

9.5.3 On-Vehicle Radiation Pattern Measurements.

Outside radiation pattern measurements were obtained for the designed antenna system installed in a 2003 GM Yukon vehicle placed on an automatically controlled turntable. An HP Yagi transmitting antenna was located in the far zone of the receiving system consisting of the antenna assembly and vehicle. The antenna assembly included the meander antenna shown in Figure 9.12 and an RF cable about 1 m long. Ferrite beads and a sleeve balun were inserted between the antenna assembly and the spectrum analyzer.

Figure 9.17 shows the radiation pattern for the passive antenna assembly mounted directly under the front dashboard (the front of the vehicle is at

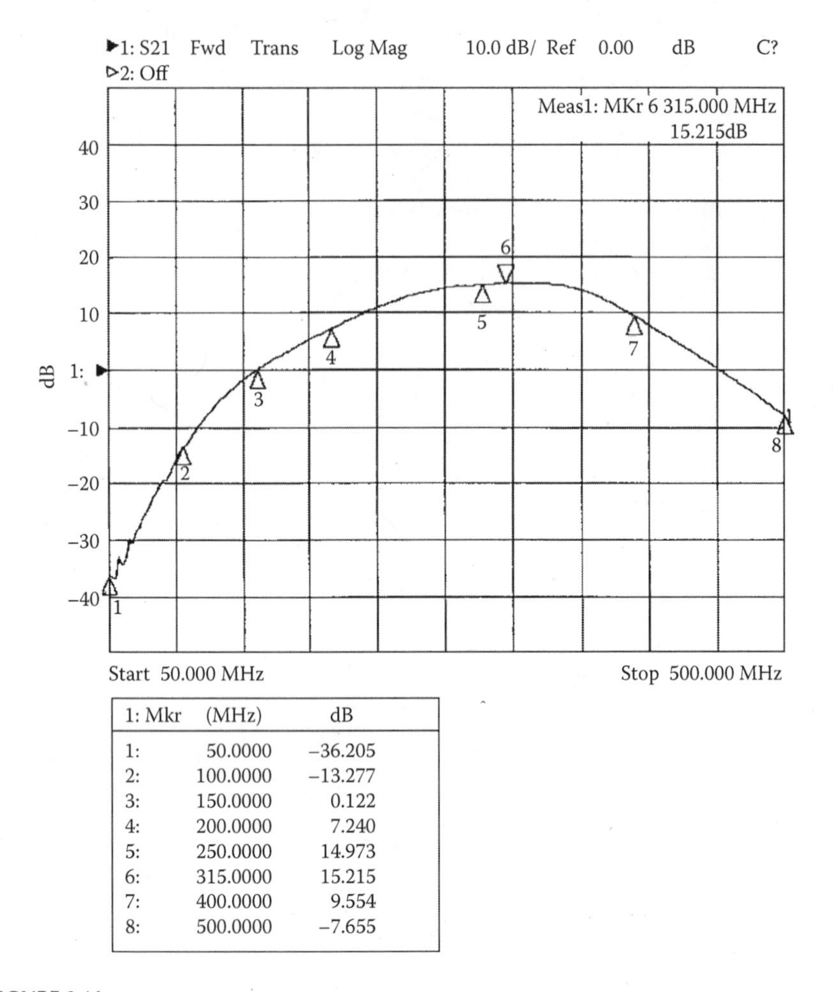

1: Mkr	(MHz)	dB
1:	50.0000	−36.205
2:	100.0000	−13.277
3:	150.0000	0.122
4:	200.0000	7.240
5:	250.0000	14.973
6:	315.0000	15.215
7:	400.0000	9.554
8:	500.0000	−7.655

FIGURE 9.14
Amplifier gain. (From V. Rabinovich, B. Al-Khateeb, and B. Oakley, *Microwave and Optical Technology Letters*, 43, 2004. Copyright 2004 *Microwave and Optical Technology Letters*. Reprinted with permission of Wiley-Blackwell, Inc.)

320 degrees). The gain value G_a averaged over 360 degrees in azimuth is about –7.4 dBi, with a standard deviation σ of 4.4 dB. The radiation pattern of a half wave dipole antenna located at the same place is shown in Figure 9.18 for comparison. The dipole gain value calculated from the chart shown in Figure 9.11 is –2.8 dBi, with σ of 5.1 dB. In this example, the car body reduced the dipole gain (from a theoretical 2.1 dBi to –2.8 dBi).

Figure 9.19 reveals the radiation pattern of the antenna assembly from measurements taken from installations in different parts of the car: Figure 9.19a and b shows the radiation pattern of the antenna mounted on the right front and rear roof support pillars. Figure 9.19c demonstrates

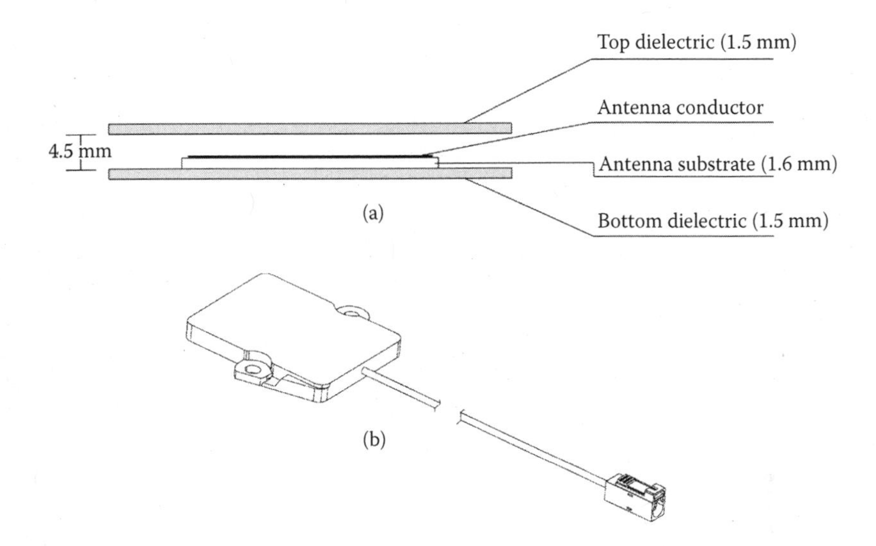

FIGURE 9.15
(a) Meander line antenna between two infinite plastic plates (ABS, = 3.4); (b) antenna in plastic case for vehicle application. (From V. Rabinovich et al. Compact planar antennas for short-range wireless automotive communication, *IEEE Transactions on Vehicular Technology*, 55, 2006. Copyright 2006 IEEE. With permission.)

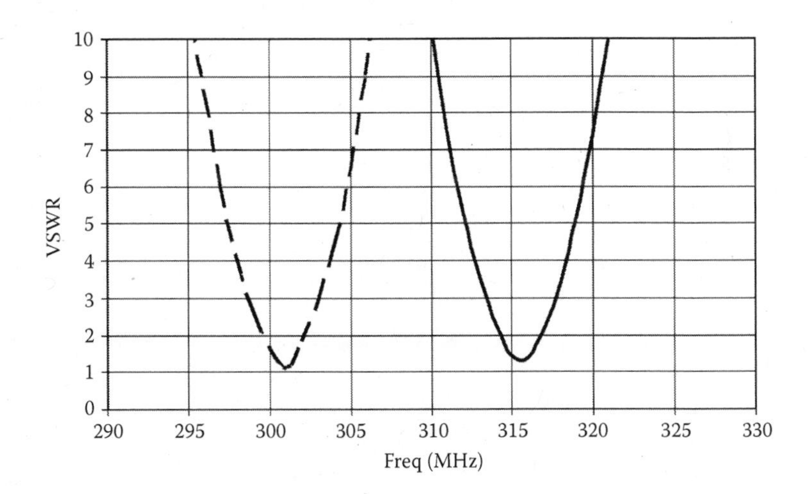

FIGURE 9.16
VSWR simulation results for antenna between plastic plates. Solid curve indicates tuning to 50 ohms; dashed curve indicates no tuning to 50 ohms. (From V. Rabinovich et al. Compact planar antennas for short-range wireless automotive communication, *IEEE Transactions on Vehicular Technology*, 55, 2006. Copyright 2006 IEEE. With permission.)

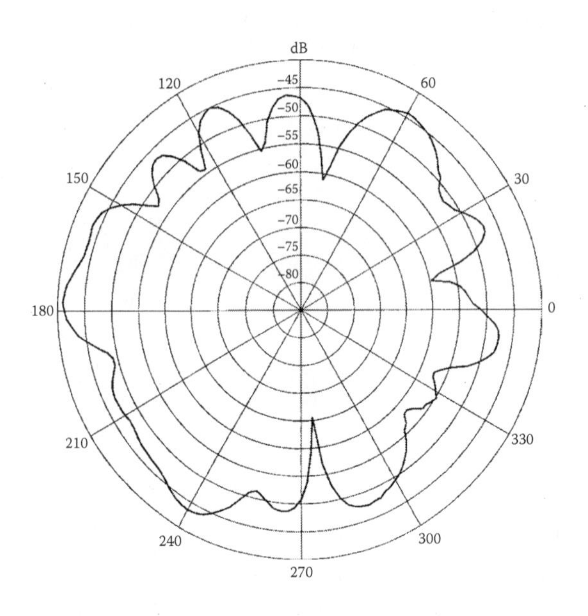

FIGURE 9.17
Radiation pattern of passive meander line antenna $G_a = -7.4$ dBi and $\sigma = 4.4$ dB). (From V. Rabinovich, B. Al-Khateeb, and B. Oakley, *Microwave and Optical Technology Letters*, 43, 2004. Copyright 2004 *Microwave and Optical Technology Letters*. Reprinted with permission of Wiley-Blackwell, Inc.)

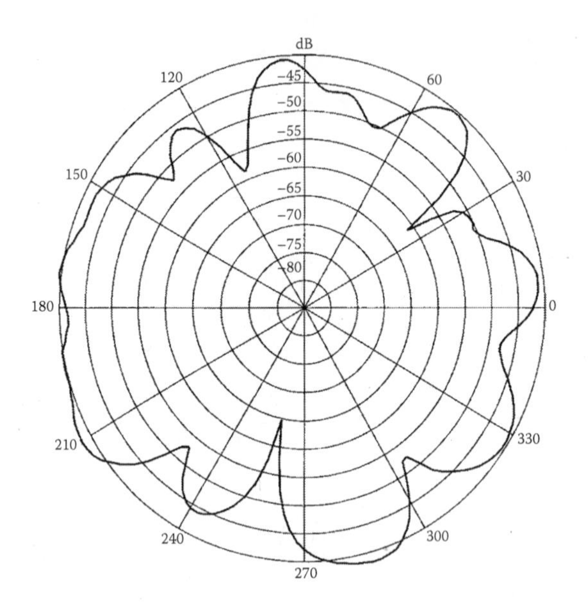

FIGURE 9.18
Radiation pattern of half wave dipole $G_a = -2.8$ dBi and $\sigma = 5.1$ dB). (From V. Rabinovich, B. Al-Khateeb, and B. Oakley, *Microwave and Optical Technology Letters*, 43, 2004. Copyright 2004 *Microwave and Optical Technology Letters*. Reprinted with permission of Wiley-Blackwell, Inc.)

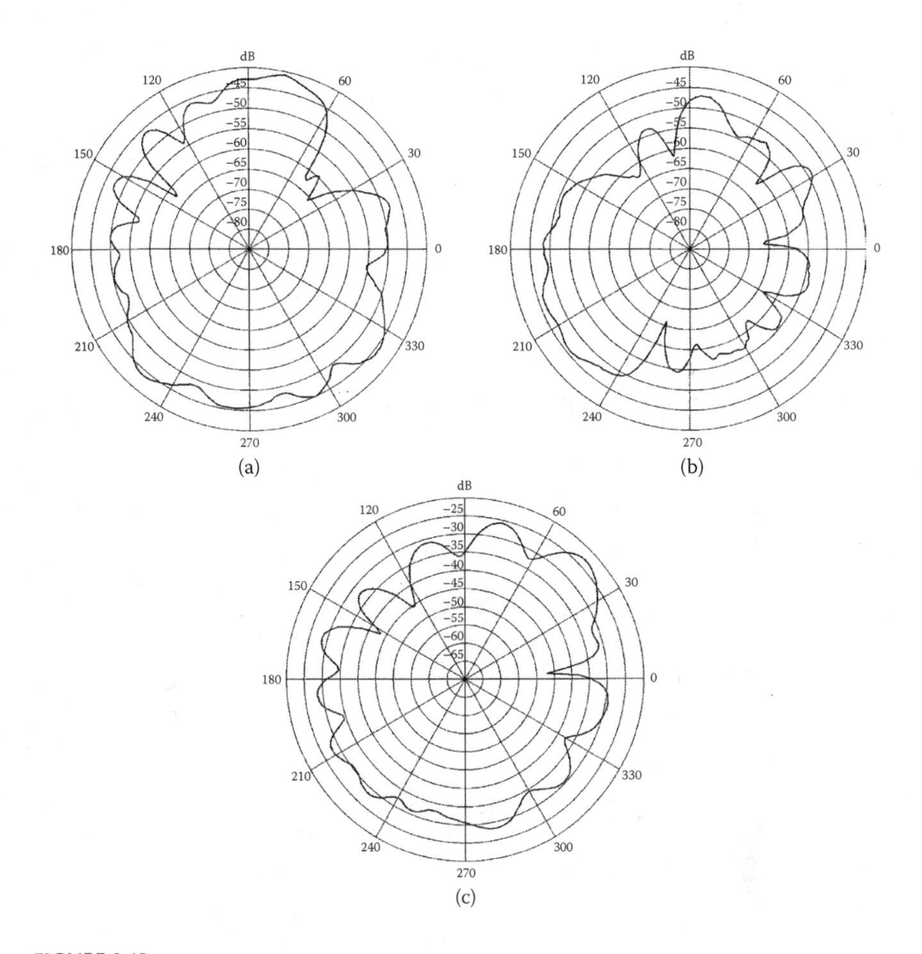

FIGURE 9.19
Radiation pattern of meander line antenna: (a) right front support pillar mounting $G_a = -7.1$ dBi and $\sigma = 4.9$ dB); (b) rear roof support pillar mounting $G_a = -11.6$ dBi and $\sigma = 4.9$ dB); (c) active design $G_a = 9$ dBi and $\sigma = 4.1$ dB). (From V. Rabinovich, B. Al-Khateeb, and B. Oakley, *Microwave and Optical Technology Letters*, 43, 2004. Copyright 2004 *Microwave and Optical Technology Letters*. Reprinted with permission of Wiley-Blackwell, Inc.)

the radiation pattern of the active antenna assembly mounted under the front dash. A comparison of the overall active gain value and passive antenna gain G_a shows that the amplifier gain of the active antenna is about 15.4 dB—approximately the same as the gain of the amplifier alone, measured in the laboratory. The overall active gain G_Σ can be presented as $G_\Sigma \approx G_a \cdot G_{amp}$. We did not specify the orientation of the antenna assembly (antenna plus RF cable) in the car. However, such radiation pattern curve fluctuations are typical for any orientation in a vehicle. A car body significantly changes the radiation pattern in comparison with a smooth curve measured in an anechoic chamber.

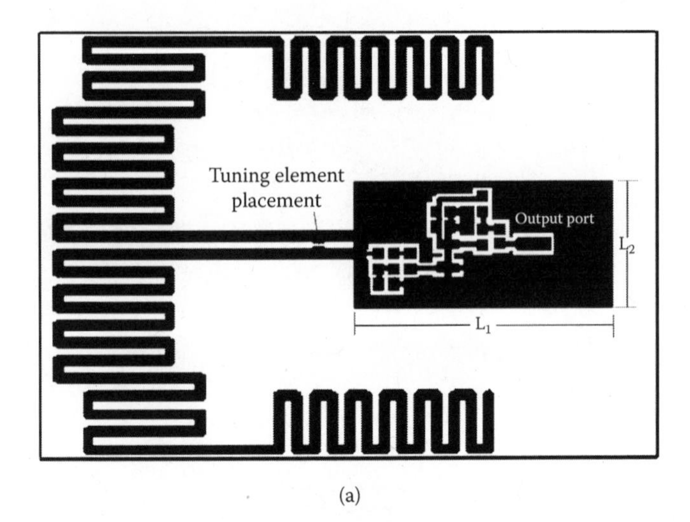

(a)

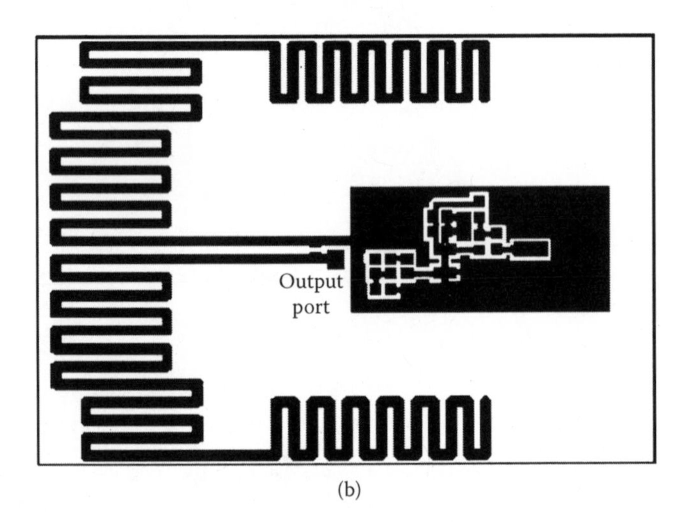

(b)

FIGURE 9.20
Geometry of printed meander line antenna, 433.92 MHz: (a) active topology; (b) passive portion with output port placement.

9.6 Example of 433.9 MHz Antenna

Figure 9.20a presents an example of symmetrical geometry of an active antenna for 433.9 MHz. The total antenna dimensions are 70.4 mm × 47.2 mm—the same as the 315 MHz meander design. The dimensions of the ground plane spot under the electronic components are L_1 = 29.9 mm × L_2 = 14.1 mm. The

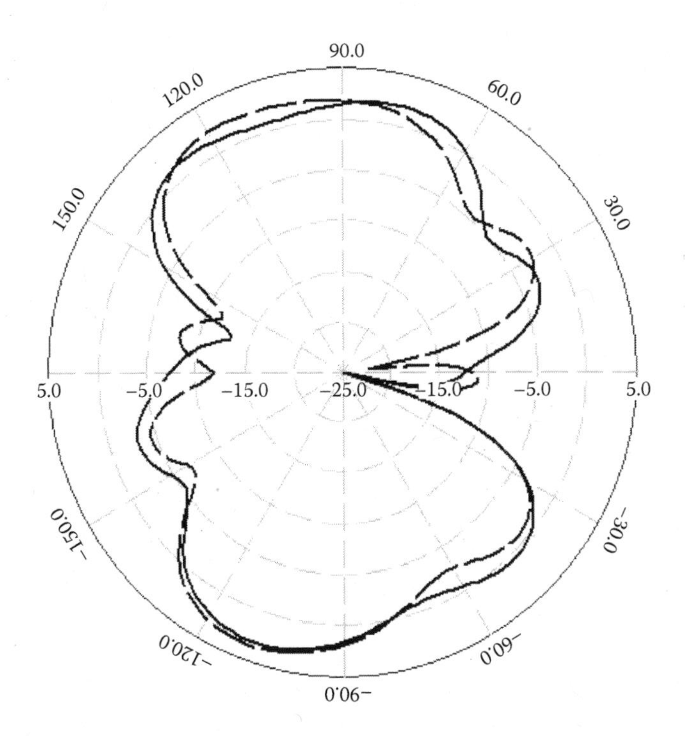

FIGURE 9.21
Measured HP radiation pattern of 433.9 MHz antenna in horizontal plane. Horizontal polarization.

total length including a parallel tuning inductor (15 nH) provides 50 ohm output impedance of the passive antenna portion shown in Figure 9.20b.

Figure 9.21 presents two HP directionality plots of the antenna in horizontal plane. Solid directionality is for the cable direction perpendicular to the antenna board plane. The second radiation pattern is measured with 1 m of cable placed at the plane of the antenna circuit board for comparison with a simple dipole. We see good agreement between these curves; this indicates minimal cable effect on the symmetrical 433.92 antenna topology.

9.7 Alternative Antenna Systems for Short Range Communication

As stated earlier, modern vehicles are equipped with luxury electronic devices that create "electrically noisy" interior environments. Packaging small antennas inside a car in this type of noisy environment makes communication system performance a challenge and antennas mounted outside

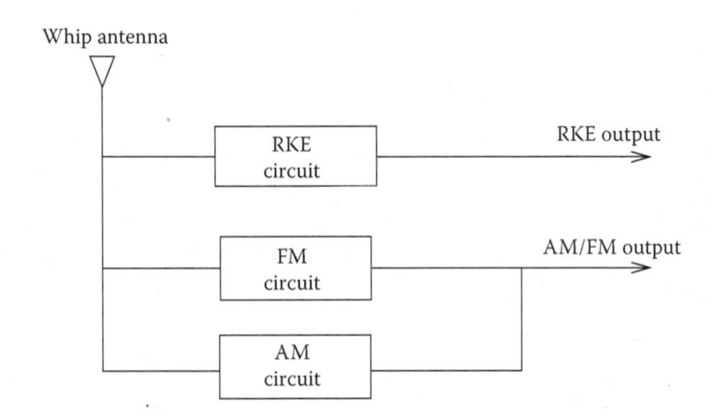

FIGURE 9.22
Block diagram of splitter whip antenna with AM/FM circuit and RKE branch.

a vehicle are highly desirable. The first part of this section shows how to use an AM/FM whip antenna for radio controlled (RK) automotive applications such as RKE, RSE, and TPS. The second part demonstrates a printed-on-glass design particularly suitable for RK systems with extended range.

9.7.1 Splitter Antenna for 315 MHz Frequency Band

A splitter is inserted into a regular coaxial cable that connects the whip antenna to the car radio (Figure 1.10, Chapter 1). Such cable has one input for the connection with an antenna and two outputs: one connected to a car radio and the other connected to the RKE module. This design utilizes a whip antenna for AM/FM reception and RKE, RSE, and TPS functions. The antenna is positioned on the exterior of the vehicle. The splitter consists of three branches: AM, FM and RKE circuits (Figure 9.22). Each branch may be implemented as an active or passive circuit, depending on the system requirements.

Passive FM and RKE circuits must provide minimum insertion losses between an antenna and radio. Active circuits must provide low insertion noise and a gain value that achieves high quality reception for the radio and maximum communication range for the RKE system. Figure 9.23 demonstrates a completely passive splitter design topology. Measurement results show that a properly designed passive splitter has FM frequency losses not more than 0.5 to 0.8 dB and losses at 315 MHz do not exceed 1 dB. Figure 9.24 shows the impedance (VSWR is ~1.2:1 at 315 MHz) at the output of the passive RKE matching circuit branch of the splitter. Practical implementation of the splitter with passive AM/FM branch and active RKE circuit is demonstrated in Figure 9.25. The RKE branch has an amplification of about 15 dB.

Radiation patterns of a whip antenna at 315 MHz mounted on a Silverado pick-up truck were simulated with FEKO software and measured (along

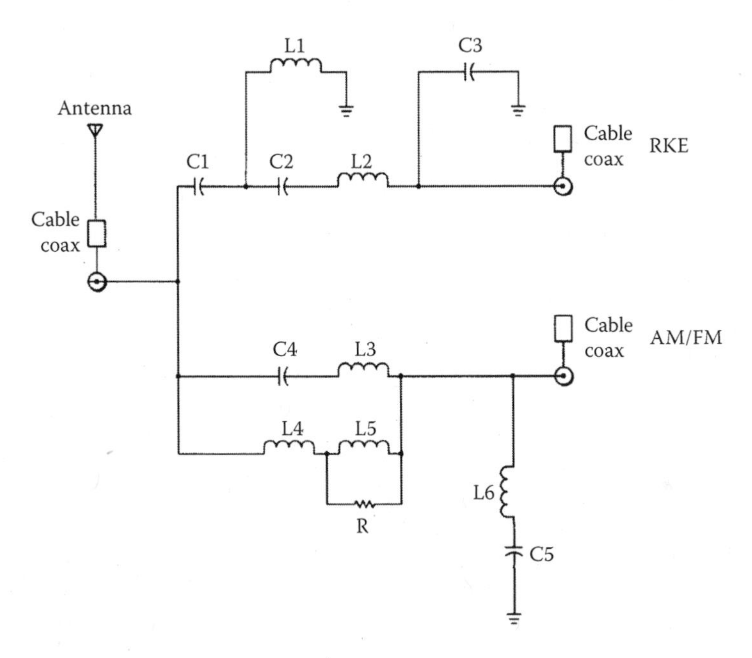

FIGURE 9.23
Passive splitter prototype circuit inserted into RF cable.

with a matching circuit). VP and HP results in horizontal plane are shown in Figure 9.26. Curves presented in Figure 9.26a and c are simulation results; Figure 9.26b and d depict measurement results. We noted satisfactory agreement of simulation and measurement results. The average VP gain over 360 degrees around the car was about 1 dBi and for HP about –2 dBi.

9.7.2 Combined Roof AM/FM/RKE System

To achieve a more omnidirectional radiation pattern, the suggestion was made to use a short mast roof antenna for receiving AM/FM and RKE signals [23]. The antenna design in Figure 9.27 is a helix for AM/FM reception (shown in Figure 1.3) with a second helix that overlaps the first one and is responsible for RKE signal reception. The RKE helix element has a length of about 6 cm, two turns with a diameter D about 1 cm, and is implemented so that it does not significantly reduce the gain in the AM/FM range. The designed antenna is tuned for 50 ohm impedance at 433.9 MHz and has a VP gain value of –3 dBi in horizontal plane.

Results of a comparison of gain in the FM frequency band for the short helix AM/FM antenna with and without an RKE element are shown in Table 9.5. The antenna is mounted on the car roof and the polarization of the received signal is vertical. The average gain of the antenna in the RKE frequency range over 360 degrees in the horizontal plane is ~0 dBi. Row 1

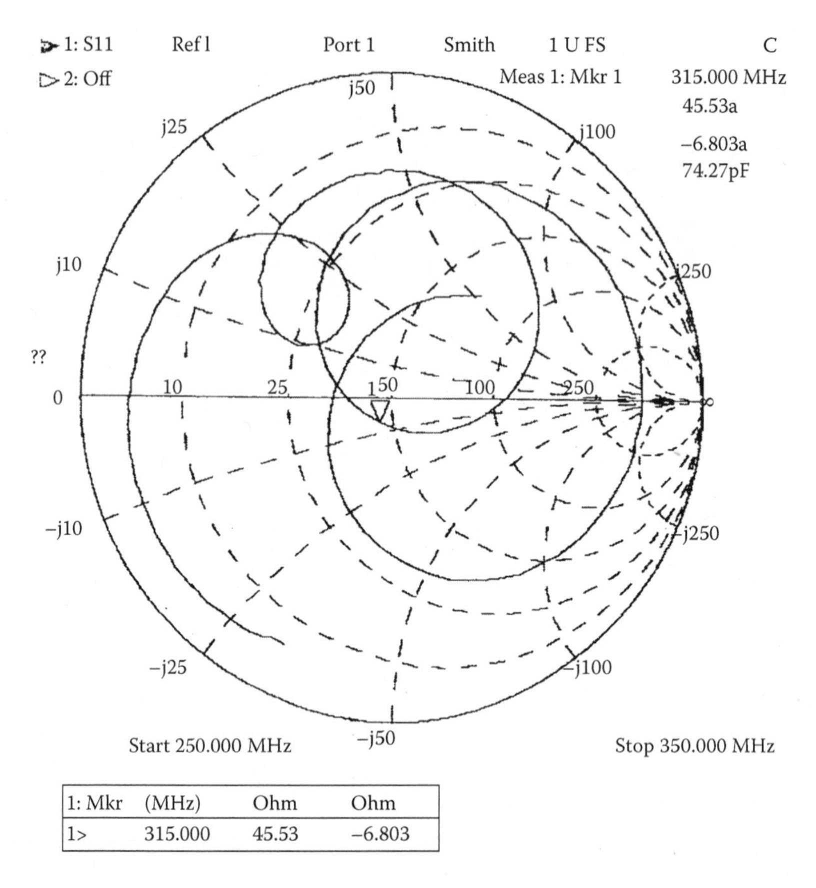

1: Mkr	(MHz)	Ohm	Ohm
1>	315.000	45.53	−6.803

FIGURE 9.24
Impedance of whip antenna matched to 315 MHz.

determines the power (dBm scale) received by the helix without the additional RKE element (dBm scale). Row 2 data shows the output signal for combined AM/FM and RKE antenna (433.9 MHz) elements (dBm scale). Row 3 demonstrates the difference between measurements of rows 1 and 2 (dB scale). The last column lists averages over frequency levels. The average difference between row 1 and row 2 values over the entire FM frequency band did not exceed 1.4 dB. Testing revealed that the communication range for the RKE system with such an antenna with an amplifier is more than 150 m for 90% of the angles over 360 degrees around a car.

9.7.3 Combined RKE/PCS/GPS Design

A combined antenna system [24] based on the helix design for RKE and cellular phone use and a patch for a GPS application is described in Section 8.11

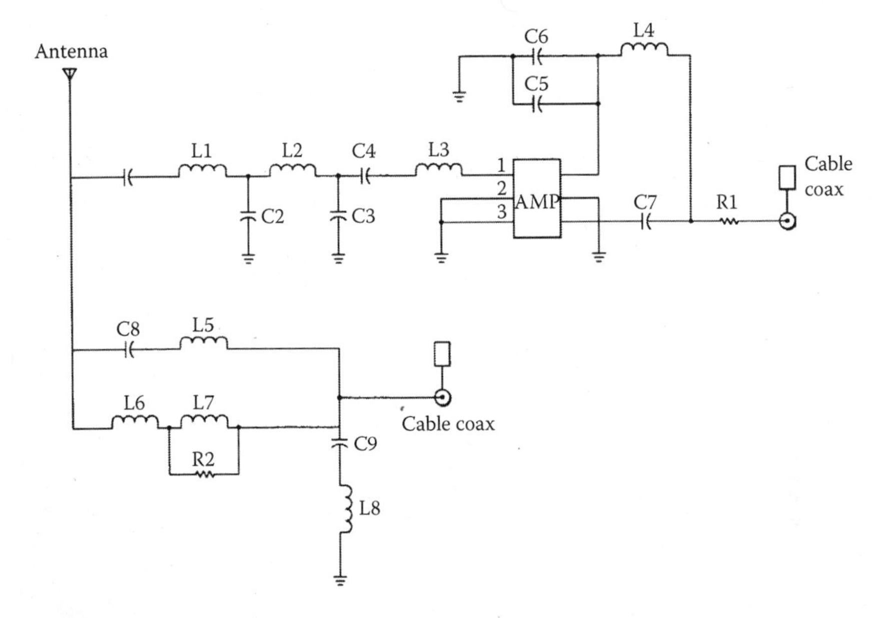

FIGURE 9.25
Electrical circuit of splitter: passive AM/FM; active RKE.

and shown in Figure 8.28. This combined system in one package is mounted on the roof of a car. The RKE antenna supports a 447.7 MHz frequency range; the cellular frequency range covers 1750 to 1870 MHz. Table 9.6 shows the measured performances of this integrated design.

9.7.4 Printed-on-Glass RKE Antenna Design

More than 90% of printed-on-glass vehicle antennas are used to receive broadcast signals. Only a few patents [7,25–27] are devoted to the application of printed-on-glass antennas at RKE frequency bands. This type of antenna is a good candidate for use in an RKE system when packaging and costs are considered. Achieving increased communication range becomes more feasible and less challenging than with interior hidden antennas.

A simple 315 MHz printed-on-glass antenna configuration (Figure 9.28a) has one horizontal strip line printed on the vehicle window. The critical design factor is the length of the strip needed to achieve 50 ohm matching impedance. A similar design for 433.9 MHz is shown in Figure 9.28b. Figure 9.29a and b shows the measured output impedance for the design presented in Figure 9.28a and 9.28b. Radiation patterns for the 315 MHz antenna are shown in Figure 9.30a and b. The average 360 degree around-the-car antenna gain is –4 dBd for HP and –5dBd for VP. However, as the directionality curves reveal, this type of antenna has significant dips (more than 15 dB) in some directions. This highlights the main disadvantage of a

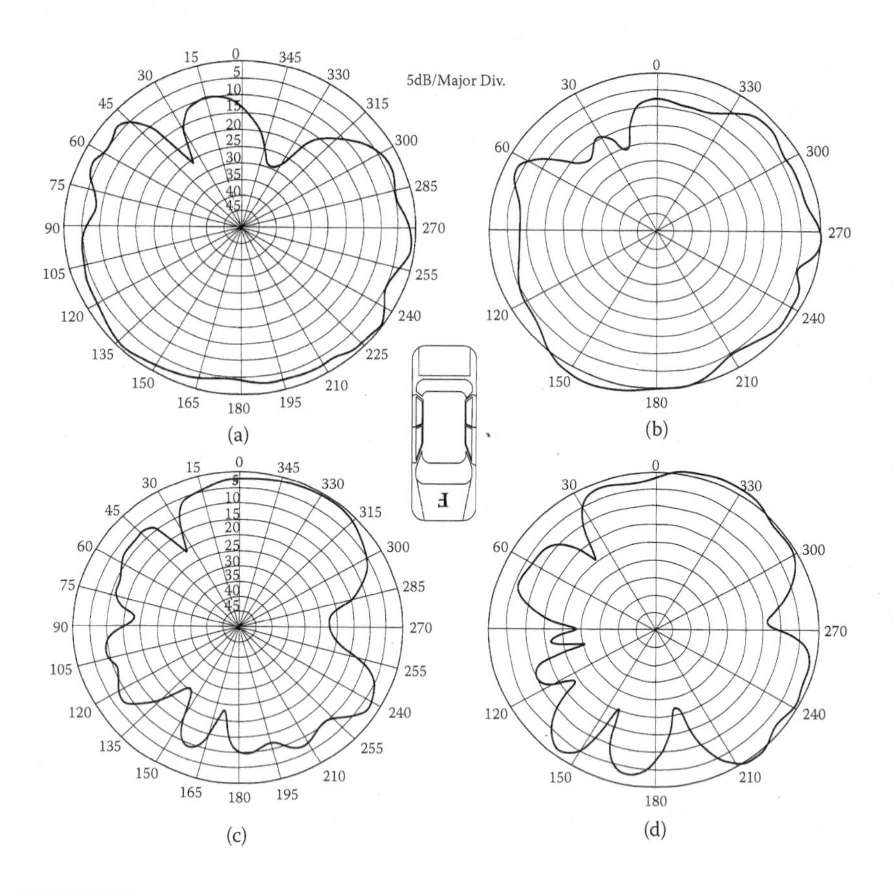

FIGURE 9.26
Radiation pattern of whip antenna, 315 MHz: (a) simulation results, VP; (b) measurement results, VP; (c) simulation results, HP; (d) measurement results, HP.

printed-on-glass antenna and the necessity for a diversity system to cover the resultant dips.

9.7.5 Comparison of 433.9 MHz Antennas

Table 9.7 compares measurement data for averaged VP and HP received signals obtained for the different kinds of antennas used for remote keyless applications. Data are measured for the 433.9 MHz frequency band on a 2006 Chrysler Durango. A printed passive meander dipole antenna shown in Figure 9.20 was placed inside the glove box. The RF cable length was ~1 m. The regular passive whip antenna intended for AM/FM reception was ~76 cm long. The RF cable connecting the antenna and car radio includes a passive splitter with an additional output connected with an RKE receiving module. A short active helix ~about 20 cm for AM/FM reception was installed on the car roof and an additional helical wire overlapped the main helix for RKE

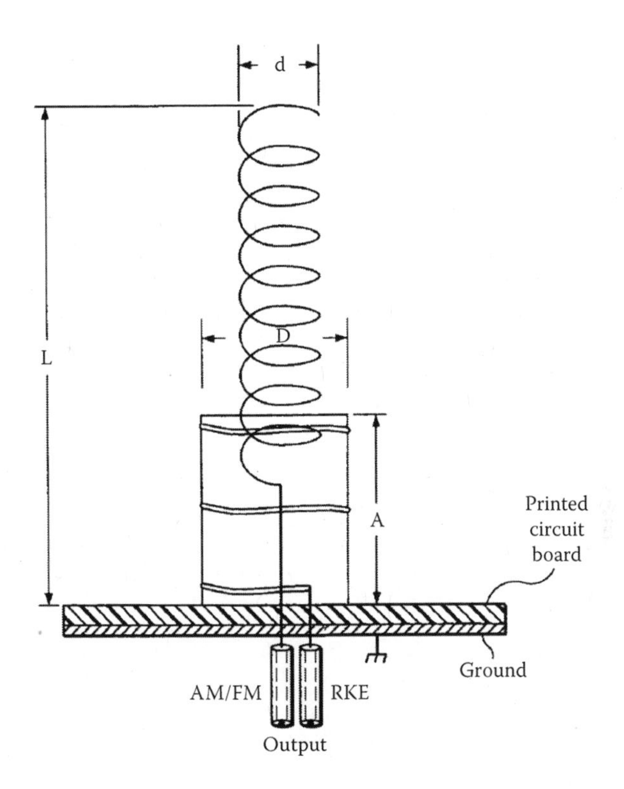

FIGURE 9.27
Two helical antennas combined for AM/FM/RKE application.

reception. Figure 9.28b shows an investigated passive printed-on-glass RKE antenna. This type exhibited the highest average gain for both VP and HP.

Regular whip and helix roof antennas have one disadvantage: reduced HP receiving signals that lessen communication range when the horizontally oriented transmitting antenna (key fob) is held randomly. This problem can be solved using an RKE helix roof antenna designed to receive waves with circular polarization [28]. These antennas can receive incoming VP and HP waves with the same power level (only 3 dB less than a power with the polarization of an incoming wave matched with antenna design polarization). An antenna for

TABLE 9.5
Comparison Gain Results in FM Frequency Band for Short Helix AM/FM Antenna with (row 2) and without (row 1) RKE Element

Frequency	88	92	95	97	100	104	108	Average
1	–38.5	–39.4	–38.9	–39.8	–38.4	–40.6	–41.9	–39.6
2	–39.8	–41.6	–41.7	–40.2	–39.3	–41.1	–43.2	–41
3	1.3	2.2	2.8	0.4	0.9	0.5	1.3	1.4

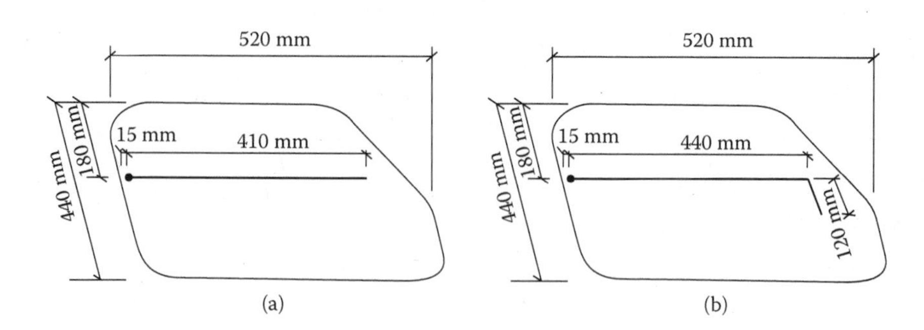

FIGURE 9.28
Printed-on-glass RKE antenna: (a) 315 MHz; (b) 433.9 MHz.

TABLE 9.6

Measured Performance of Integrated GPS/RKE/PCS Antennas

Service	GPS	RKE	Cellular
Frequency	1575 MHz	477.7 MHz	1750 to 1870 MHz
Bandwidth for VSWR (<2:1)	70 MHz	28 MHz	428 MHz
Gain	4.1 dBi	0.87 dBi	2.65 dBi

Source: K. Oh, B. Kim, and J. Choi, *IEEE Microwave and Wireless Component Letters*, 15, 2005. Copyright 2005 IEEE. With permission.

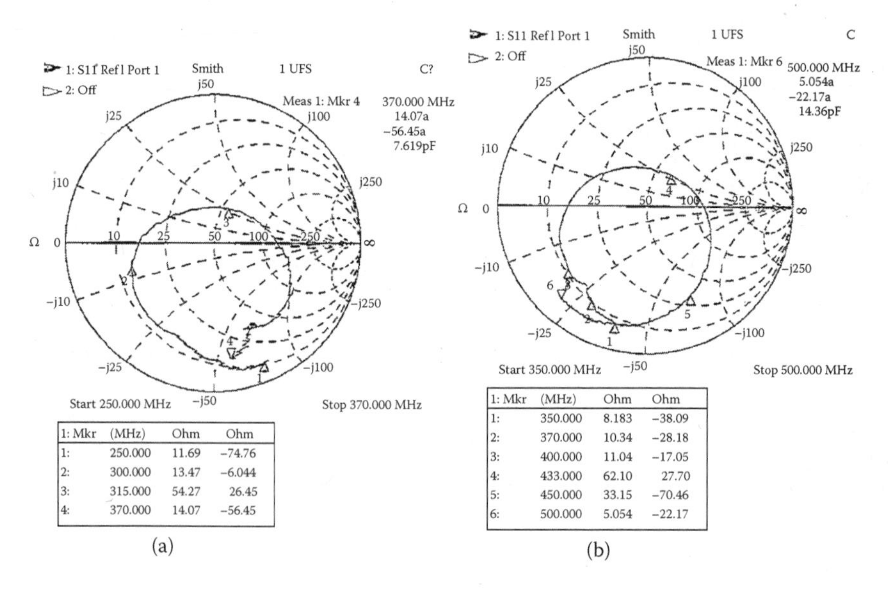

1: Mkr	(MHz)	Ohm	Ohm
1:	250.000	11.69	−74.76
2:	300.000	13.47	−6.044
3:	315.000	54.27	26.45
4:	370.000	14.07	−56.45

1: Mkr	(MHz)	Ohm	Ohm
1:	350.000	8.183	−38.09
2:	370.000	10.34	−28.18
3:	400.000	11.04	−17.05
4:	433.000	62.10	27.70
5:	450.000	33.15	−70.46
6:	500.000	5.054	−22.17

FIGURE 9.29
Smith chart for printed-on-glass antenna: (a) 315 MHz; (b) 433.9 MHz.

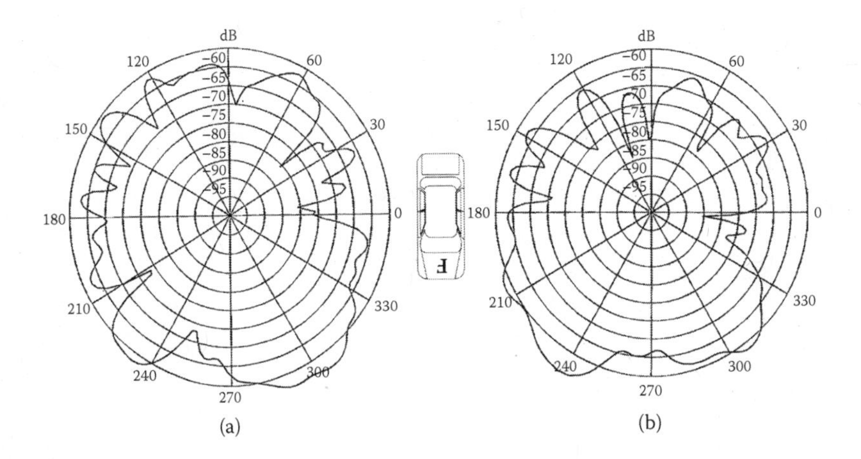

FIGURE 9.30
Radiation pattern for printed-on-glass antenna, 315 MHz: (a) HP, horizontal plane; (b) VP, horizontal plane.

315 MHz has three turns with a helix diameter ~3.25 cm and spacing between helical turns ~0.55 cm. The total wire length is ~28 cm; height is ~1.7 cm.

The printed-on-glass antenna that showed the highest gain averaged over 360 degrees suffered from multiple dips in different directions, thus reducing communication range in the directions of the dips. The solution to this problem is a diversity system.

9.8 Maximum Detection Range of RKE/RSE System

According to the Friis transmission equation and a pass/loss measurement-based propagation model [29], signal power at an antenna receiver output is equal to:

$$P_r = \frac{ERP \cdot G_{act} \cdot G_{rec} \cdot \alpha}{D^n} \tag{9.2}$$

where: $\alpha = (\frac{\lambda}{4\pi})^2$, $ERP = P_t \cdot G_t$ = effective radiated power, P_t = transmitting power, G_t = transmitter antenna gain, $G_{act} = G_a \cdot G_{amp}$ = active antenna

TABLE 9.7
Comparison of Measurement Data for RKE Antennas (433.9 MHz)

Polarization	Printed-on-Glass	Printed Meander	Regular Whip	Helix on Roof
Vertical (dBm)	–44	–49.2	–46.5	–45
Horizontal (dBm)	–44	–50.7	–52	–59

gain (for passive antenna $G_{amp} = 1$), G_a = gain of passive portion of antenna, G_{amp} = amplifier gain (for active antenna), G_{rec} = RKE receiver gain, D = distance between transmitting and receiving antennas; and n = real constant that is a function of the wireless communication environment [29]. For free space propagation, $n = 2$. Residual noise power at the receiver output [30,31] for the UHF band can be estimated as:

$$N_1 = k \cdot T_0 \cdot B \cdot G_{amp} \cdot G_{rec} \cdot F_\Sigma \quad (9.3)$$

where $F_\Sigma = F_{amp} + (F_{rec} - 1)/G_{amp}$ = wireless system noise figure, F_{amp} = amplifier noise figure, F_{rec} = receiver noise figure, B = receiver frequency bandwidth, $T_0 = 290\ K$, and $k = 1.38 \cdot 10^{-23}$ J/K is the Boltzmann constant. Equation (9.3) assumes that the source of the noise in the RK system is only thermal noise caused by the amplifier and receiver active components. External noise generated by electronic car components surrounding the RK system is negligible. Otherwise the total noise value is:

$$N_{total} = N_1 + N_{ext} \quad (9.4)$$

Using Equation (9.2) and noise signal value from Equation (9.3), we can define the SNR at the receiver output as:

$$SNR = \frac{ERP \cdot G_a \cdot \alpha}{D^n \cdot k \cdot T_0 \cdot B \cdot F_\Sigma} \quad (9.5)$$

According to Equation (9.5), the maximum range of the wireless communication system is:

$$D_{\max} = \sqrt[n]{\frac{ERP \cdot G_a \cdot \alpha}{SNR_{\min} \cdot k \cdot T_0 \cdot B \cdot F_\Sigma}} \quad (9.6)$$

Equation (9.6) in dB scale format can be expressed as:

$$10 \log D_{\max} = (10 \log ERP + 10 \log G_a + 10 \log \alpha - 10 \log \mathrm{SNR}_{\min} - 10 \log(K \cdot T_0 \cdot B) - 10 \log F_\Sigma)/n \quad (9.7)$$

where, $SNR_{\min}$ is the minimum SNR required to detect a signal with a certain probability of detection. The ratio of the maximum operating distance of the active versus the passive antenna provides an estimate for the improvement factor q_1:

$$q_1 = \frac{D_{\max}(active)}{D_{\max}(passive)} = \sqrt[n]{\frac{F_{rec}}{F_{amp} + (F_{rec} - 1/G_{amp})}} \quad (9.8)$$

When using a passive dipole antenna instead of a passive printed antenna, the ratio improvement will be designated q_2 and Equation (9.7) becomes:

$$q_1 = \frac{D_{\max}(active)}{D_{\max}(dipole)} = \sqrt[n]{\frac{G_a}{G_{dipole}} \cdot \frac{F_{rec}}{F_{amp} + (F_{rec} - 1/G_{amp})}} \quad (9.9)$$

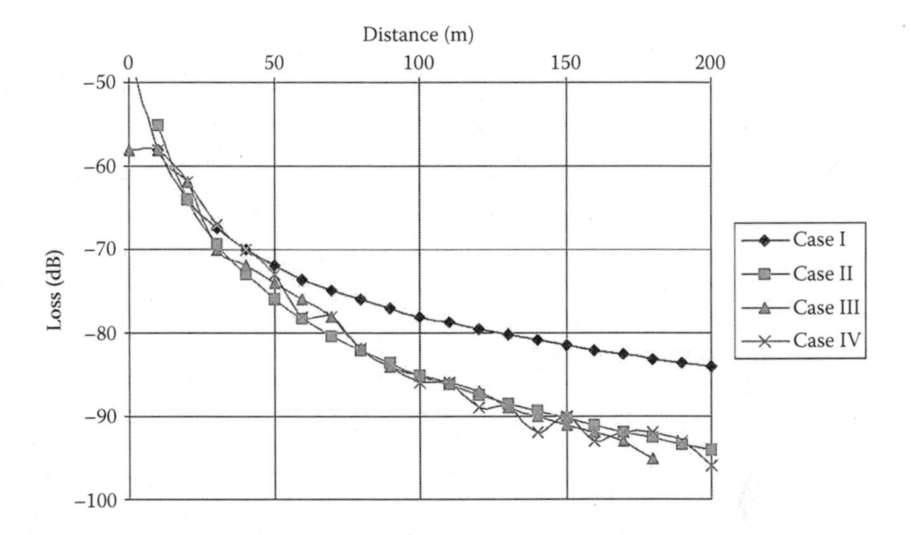

FIGURE 9.31
Dependence of propagation loss on distance. (From V. Rabinovich et al. Compact planar antennas for short-range wireless automotive communication, *IEEE Transactions on Vehicular Technology*, 55, 2006. Copyright 2006 IEEE. With permission.)

As Equation (9.6) indicates, when estimating maximum detection range, key values that must be known are n number, ERP value, SNR_{min}, antenna gain, amplifier gain, and noise parameters of amplifier and RKE receiver. Let us begin with estimation of n. Figure 9.31 reveals measurement results ([11], horizontal polarization) that can be used to help estimate n value in an open area. Four cases are considered:

1. Free space losses vary as C/D^2, where C is a constant that does not depend on the distance D between transmitter and receiver.
2. Losses vary as C/D^3.
3. Measured loss when the key fob transmitter is located 1.5 m above the ground from the front of a car in an open area and the RKE system is in the car.
4. Measured loss when the key fob is located 1.5 m above the ground from the rear of a car in an open area and the RKE system is in the car.

Based on measurement results the n value for an open area can equal 3.

Let us assume that $ERP = -15$ dBm and $\text{SNR}_{min} = 10$ dB. Typical receiver bandwidth B is 300 KHz. Therefore $10 \cdot \log(k \cdot T_0 \cdot B)$ is equal to –119.2 dBm. A typical noise figure for the amplifier is 2 dB and a typical gain value is 15 dB. For a frequency range of 315 MHz, α is equal to –22 dB. If the receiver noise figure is about 6 dB, Equation (9.6) can be expressed as $30 \log D_{max} = -15\text{dBm} + 10 \log G_a - 22\text{dB} - 10\text{dB} + 119.2 \text{ dBm} - 2\text{dB}$ or, after simplifying:

$$30 \log D_{max} = 10 \log G_a + 70.2 \text{ dBm} \tag{9.10}$$

Assume that the antenna gain value in the car is about –7dBi. Based on Equation (9.10), the maximum range should be about 130 m. However, if we use a passive instead of an active antenna, we will obtain a maximum range of 95 m instead of 130 m. This difference in range results indicates how the active antenna exceeds the performance of the passive antenna.

9.9 Effects of Auto Electronic Components on Communication Range

As discussed previously, the range of communication is affected by reflections from the ground, car, and surrounding objects. The receiver "sees" a combined signal with components that have different amplitudes and phases. The reflections cause a phase reversal. The noisy emissions of electronic devices surrounding an antenna are also relevant for estimating signal strength at a receiver. An emission signal level at the operating frequency band can almost cancel the desired signal ("almost" because the emission amplitude will be slightly less than direct signal amplitude). This range reduction occurs because each electronic part in close proximity transmits as a source of secondary wave emission.

The following comparison shows how the noise from devices inside a car dramatically decrease signal strength. Assume that, in addition to the amplifier and wireless control module receiver (WCM) thermal noise, an RKE system also receives varied types of external noise, for example, noise from the external environment or noise from electronic components of the car. For example, an improperly designed automatic HVAC (heating, ventilation, and air conditioning) system has been shown to produce noise $N_{ext} = -74.5$ dBm—some 20 to 25 dB higher than the noise of the amplifier at the input of the RK module. The SNR when external noise is dominant can be expressed as:

$$SNR = \frac{ERP \cdot G_{act} \cdot \alpha}{D^n \cdot N_{ext}} \tag{9.12}$$

or

$$D_{max} = \sqrt[n]{\frac{ERP \cdot G_{act} \cdot \alpha}{SNR_{min} \cdot N_{ext}}} \tag{9.13}$$

For $n = 3$, $ERP = -15$ dBm, $G_{act} = 10$ dB, $\alpha = -22$ dB, $SNR_{min} = 10$, and $N_{ext} = -74.5$ dBm, the range $D_{max} = 18$ m.

As seen from the calculation in Equation (9.13), the range with noise is reduced to 18 m. This drop shows how much high levels of external noise can dramatically reduce the communication range from more than 100 to 18 m. Therefore the first priority for designing an RKE system is to investigate noise interference sources from electronic components of a vehicle and their critical effects on wireless system operation. As proven experimentally, many conventional packaged devices operating in close proximity to an antenna can inadvertently, kill wireless communication in a car.

9.10 Signal and Noise Measurements for Antenna/Receiver Combination

As it was mentioned before, communication range depends on an antenna gain, receiver sensitivity, noise of a car's electronic components etc. Some communication range-related parameters (noise level, path loss) are unpredictable in a real environment. For example, the presence of multiple microprocessor noise sources inside a car can dramatically decrease signal range. To address such concerns, this section estimates signal and noise values received by the antenna and RF receiver of an RKE system in a real environment.

As seen in Figure 9.1, a typical RKE module includes an antenna, an RKE RF receiver, a microprocessor, and a command circuit. The RKE RF receiver captures the RF signal, demodulates it, and sends the data stream to the microprocessor that decodes it and sends commands to the command module. Most RF receivers have output pins that record voltage as a function of the input receiver power.

Figure 9.32 shows the calibration function [14] between the signal generator power injected into the RF receiver input (typical RKE receiver from Infineon) and the output receiver voltage.

The RKE system test setup is shown in Figure 9.33. This system uses an external antenna installed separately from the RF module under the front dash. Tests are performed at a distance $R = 100$ m between transmitter and the vehicle. The CW (continuous wave) transmitter provides an HP signal with an effective radiated power equal to –12 dBm. The RKE receiver, microprocessor, and command circuit are located on the steering column and connected to the active antenna via the RF cable. The car with the installed RKE system is positioned on an outdoor turntable. A receiver output data point is taken every 2 degrees as the turntable rotates through a full 360 degree azimuth. The antenna range instrumentation is controlled by a computer that drives the turntable rotation and transfers measured data to a hard drive and printer.

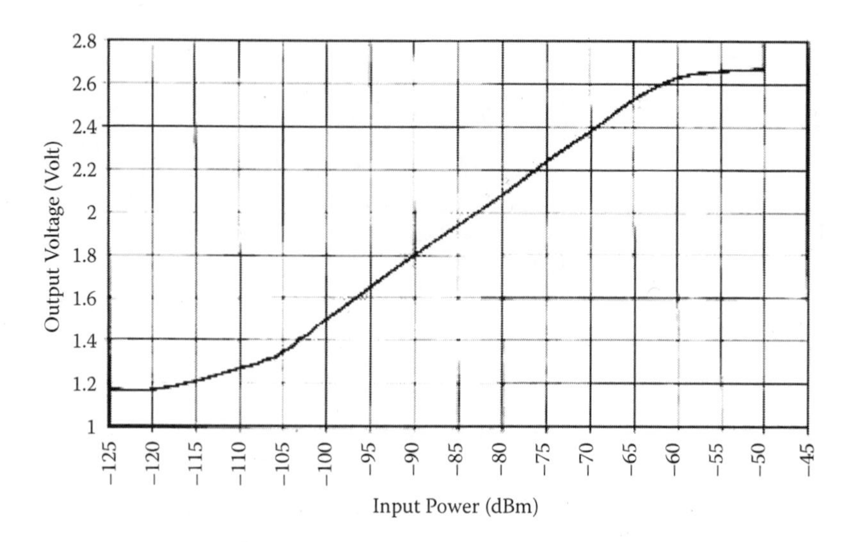

FIGURE 9.32
Calibration curve of RF power injected to RF receiver (Infineon TDA 5211) and output voltage level. (From V. Rabinovich et al., *Microwave and Optical Technology Letters*, 47, 2005. Copyright 2005 *Microwave and Optical Technology Letters*. Reprinted with permission of Wiley-Blackwell, Inc.)

9.10.1 Measurement Results

A 2004 Chrysler Durango was chosen as the base vehicle for the measurements. As a preliminary step, the DC response for the noise received by the RF receiver was measured with an active antenna and found to be 1.35 V. According to the calibration curve, this corresponds to a total noise power level at the receiver input of –102.5 dBm, including the thermal noise of the antenna amplifier and noise received by the antenna from the surrounding environment (car equipment noise and receiver thermal noise). The thermal

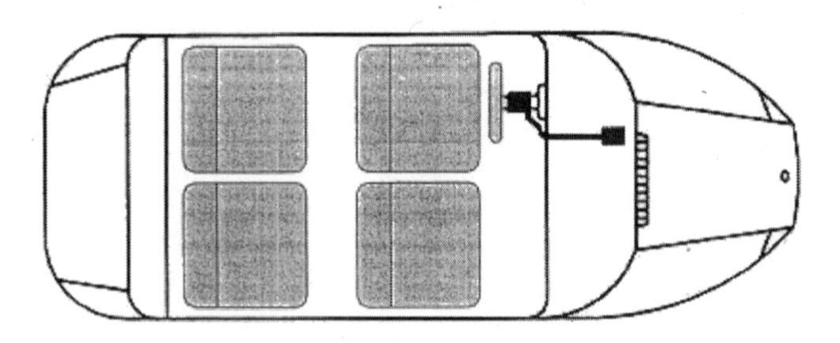

FIGURE 9.33
In situ RKE system in vehicle. (From V. Rabinovich et al., *Microwave and Optical Technology Letters*, 47, 2005. Copyright 2005 *Microwave and Optical Technology Letters*. Reprinted with permission of Wiley-Blackwell, Inc.)

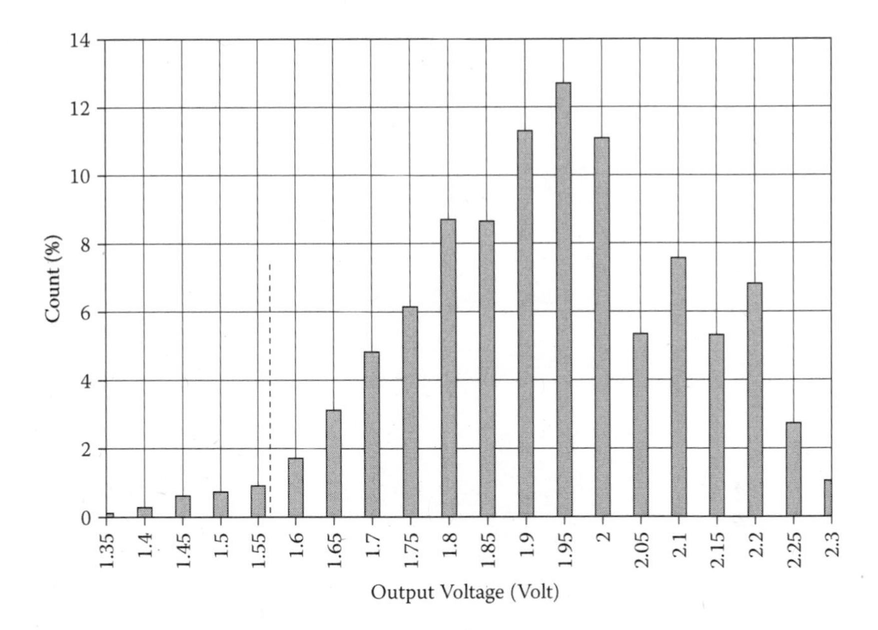

FIGURE 9.34
Measured voltage data as function of specified measured voltage level (statistical distribution of output voltage). (From V. Rabinovich et al., *Microwave and Optical Technology Letters*, 47, 2005. Copyright 2005 *Microwave and Optical Technology Letters*. Reprinted with permission of Wiley-Blackwell, Inc.)

noise for an antenna amplifier can be calculated from Equation (9.3), with the equivalent bandwidth of the measuring device taken as 300 KHz.

Laboratory measurements revealed that the antenna amplifier gain was 15 dB and the amplifier noise figure was 2 dB. Signal levels were measured as the car on the turntable rotated a full 360 degrees. Under "real" use conditions, the gain of an antenna in a vehicle and the resulting signal at the RF output vary as a function of relative vehicle orientation. Therefore, we present the measurement results for different orientations relative to transmitter position as a statistical chart (Figure 9.34). This histogram shows the relationship of voltage level and data count at certain voltage levels (expressed as percentage of total data points). The dashed vertical line shows the voltage level (1.57 V) for which the input RF power (–92.5 dBm) is above the noise level by 10 dB. This signal value provides a 0.99 probability of detection, as our example shows. The chart shows that 97% of the measured values are above the threshold of 1.57 V. This means that the RKE system must detect signals for 97% of the vehicle angle positions if the person activating the system is 100 m from the car.

A field experiment with an RKE system showed 95% successful communication between the key fob and RKE module (successful opening or closing of car door). A second set of measurements explored situations in which the

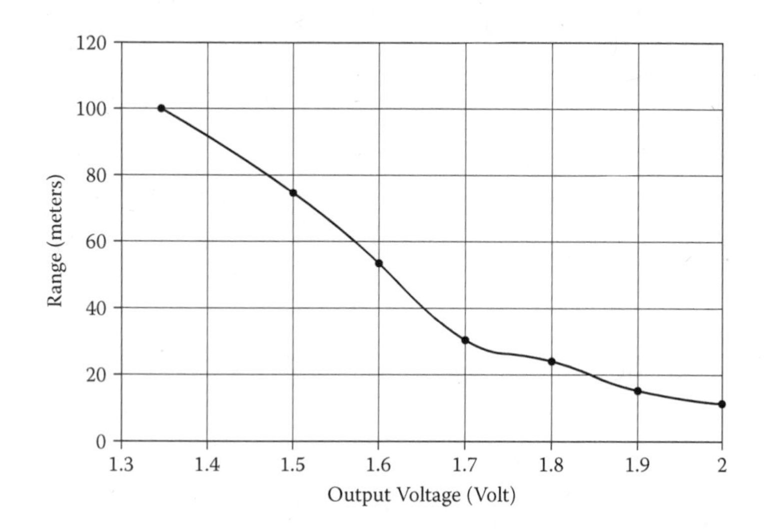

FIGURE 9.35
Relationship of noise voltage and RKE system range. (From V. Rabinovich et al., *Microwave and Optical Technology Letters*, 47, 2005. Copyright 2005 *Microwave and Optical Technology Letters*. Reprinted with permission of Wiley-Blackwell, Inc.)

external noise exceeded the thermal noise. We placed a wide band jammer in the car to increase the voltage level at the receiver output to 1.9 V (when the CW transmitter was off). This value corresponds to a noise power at the receiver input of –86 dBm. The threshold level for stable RKE operation (q_0 = 10 dB) increased to 2.2 V (–76 dB). According to the histogram, only 11% of the angle-measured points exceeded 2.2 V. Laboratory measurements of the RKE system showed 7% successful transmission. Figure 9.35 is a practical graph that allows estimation of an RKE system range as a function of the voltage measurements at the receiver output. The curve shown reveals the relationship between the noise voltage level and the system range.

9.10.2 Electromagnetic Emission Interference

Two 2006 vehicles were chosen to investigate the electromagnetic emission interference (EMI) noise effect on remote start engine (RSE) range. The noise was generated by electronic modules packaged near the extended range antenna and the RKE RF receiver. The vehicles were chosen because one exhibited a better range (more than 100 m on average) and the range of the other was below 80 m. Both cars were measured for a comparison of EMI levels radiated by their packaged electronic components and field communication range measurements were also taken. The electrical and electronic components and subsystems in the 2006 premium (fully equipped) vehicle with lower range were tested as emission disturbances. Testing was performed in a vehicle-shielded test room (VSTR) with a standard vehicle antenna tester installed on the roof. To measure potential electromagnetic interference, magnetically mounted antennas were placed on the body of the vehicle. The results are presented below.

For the first car tested, modifications were made to a wireless control module (WCM). An external connection at the intermediate frequency (IF) output was made for measurements of signal received by WCM. An ambient reference for the WCM was measured with all electronic modules disconnected. The emissions of individual devices were traced and plotted. Individual devices were connected to corresponding harnesses with other devices disconnected. Narrow band emissions (10.685 to 10.715 MHz) were measured with an IF of 10.7 MHz. Broad band emissions (0.15 to 960 MHz) were also measured. All measurements were made in a shielded room. The bottom curves indicate ambient reference noise levels. Based on measurements,

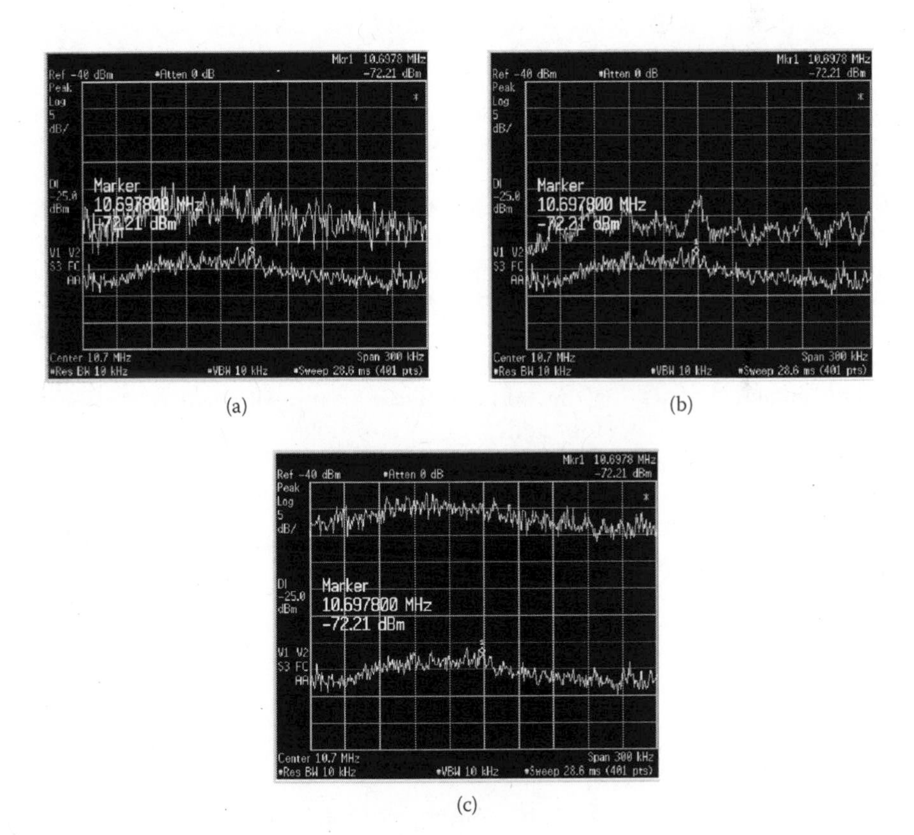

FIGURE 9.36
Noise emissions from electronic components in vehicle: (a) cluster control node (CCN); (b) audio amplifier (AMP); (c) heating, ventilation and air conditioning (HVAC) with automatic temperature control (ATC).

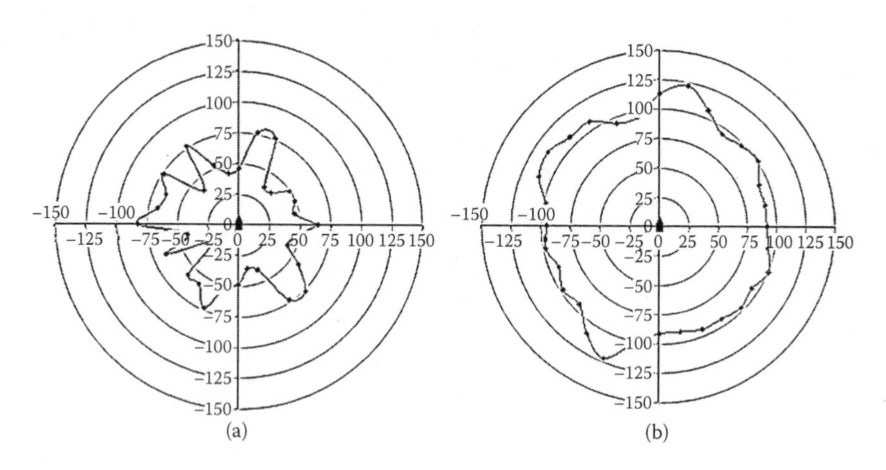

FIGURE 9.37
Communication range test: (a) shorter range vehicle; (b) better range vehicle.

several devices served as sources of noise: the cluster control node (CCN) emitted 12 dB above the WCM noise floor as shown in Figure 9.36a; the audio amplifier (AMP) emitted 10 dB above the floor as shown in Figure 9.36b; and the HVAC system with automatic temperature control (ATC) emitted 25 dB above the noise floor as shown in Figure 9.36c. A field range test was performed to measure average communication range on the vehicle exhibiting an 80 m range. The tester held the key horizontally near his hip about 1 m above the ground and pressed the key button every 10 feet as he walked away from the car. The results are shown Figure 9.37a.

The measurements for the better-range Durango identified emissions from some electronic components. The main noise source was the CCN, emitting 8.39 dB above the WCM noise floor. The AMP emission was 0.87dB above the noise floor, and the HVAC module emission was 1.84 dB above the noise floor. Results of field experiments with this vehicle are shown in Figure 9.37b.

9.11 Compact Diversity Antenna System for Remote Control Applications

As stated in earlier sections, the communication range of a radio system depends strongly on the antenna gain. Hidden antennas invariably suffer from performance problems due to shielding created by the vehicle. The extended range for such hidden antenna systems for some angle directions can be a problem. Maximum-to-minimum antenna directionality values over

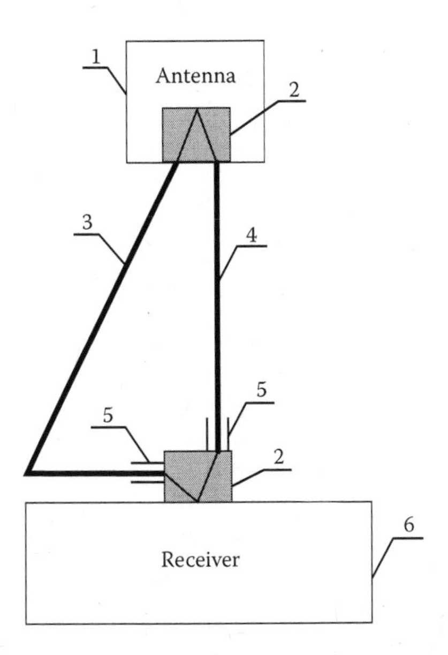

FIGURE 9.38
Configuration of diversity system with single output. (From V. Rabinovich et al., *IEEE Antennas and Propagation Society International Symposium*, 2005. Copyright 2005 IEEE. With permission.)

360 degrees can show differences up to 30 dB. This represents the equivalent of about ten times the range ratio. If we assume that maximum directionality corresponds to 100 m and the minimum directionality is less than 10 m, one of the most effective ways to increase the range is to use the antenna diversity technique. Diversity can help to reduce degradation in the communication link. We investigated a small but highly efficient diversity antenna [32]. The antenna may be integrated with a low noise amplifier to extend short range wireless range.

A meander line antenna (Figure 9.12a) was chosen as the base of the proposed space diversity system. Figure 9.38 shows the configuration of the active antenna element consisting of a meander line antenna (1), electronically controllable switch (2), two cables (3 and 4), and two ferrite beads (5). The antenna (1) has dimensions smaller than 50 mm × 70 mm. The pin diode switch (2) can connect either cable (3) or (4) to the receiver (6). Antenna directionality can be controlled by using different lengths and locations of the cable between the antenna and receiver.

The receiver provides control functions: it electronically opens or closes the vehicle doors or starts the engine. To start the diversity operation the transmitting key fob located some distance from the vehicle activates the receiver (6). The antenna (1) is connected with the receiver through cable (3) or (4), depending on switch position. The antenna is connected to the receiver about half the activated time through cable (3) and about half the activated time

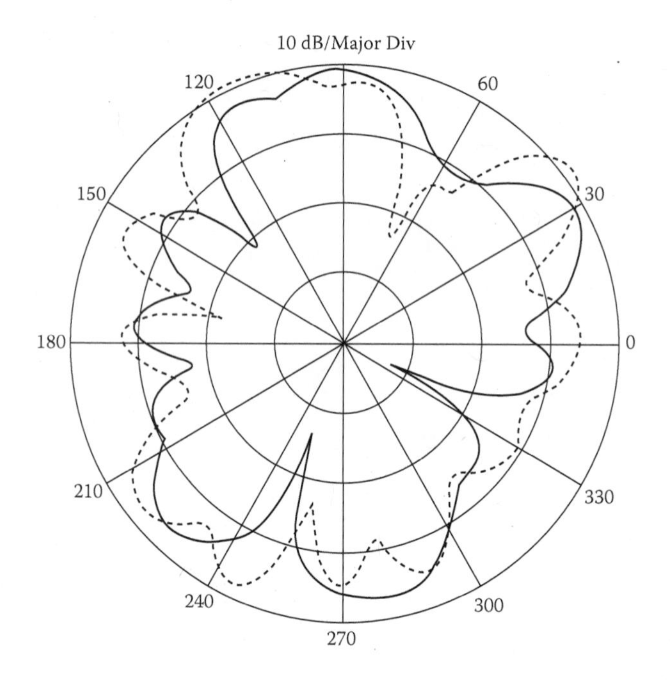

FIGURE 9.39
Measured radiation pattern on vehicle. (From V. Rabinovich et al., *IEEE Antennas and Propagation Society International Symposium*, 2005. Copyright 2005 IEEE. With permission.)

through cable 4. The equivalent antenna directionality can be represented as the overlap of curve 1 and curve 2 in Figure 9.39 (azimuth plane, VP). Figure 9.40 is the graph.

Let us compare the range of wireless communications for two antenna directionality curves shown in Figure 9.39 (curve 1) and Figure 9.40. Assume that the average directionality value corresponds to 100 m; range is described by Equation (9.6); $n = 2.5$. For curve 1 in Figure 9.39 we have 28% of the angle positions with a range below 63 m, 8% with a range below 40 m, 3% below 25 m, 1.5% below 16 m, and 0.5% below 10 m. For the Figure 9.40 curve, we have 12% angle points with a range below 63 m and no angle points with ranges below 40, 25, 16, or 10 m. Thus, we see the significant improvements in range performances with an antenna diversity system. Figure 9.41 shows a design with two RKE antennas installed under the front dash panel.

Generally, the quality of a keyless entry system can be estimated using the probability $P(D > L)$ that the range D is more than a certain value L. This probability is the ratio of the number M of the points for which the range D is more than L to the total number M_0 of the measuring points. Figure 9.42a shows the probability $P(D > L)$ [32] percentage as a function of the range L (in meters) calculated from the measurements of VP of a transmitting wave (horizontal plane). Similar plots for HP are shown in Figure 9.42b. Assume

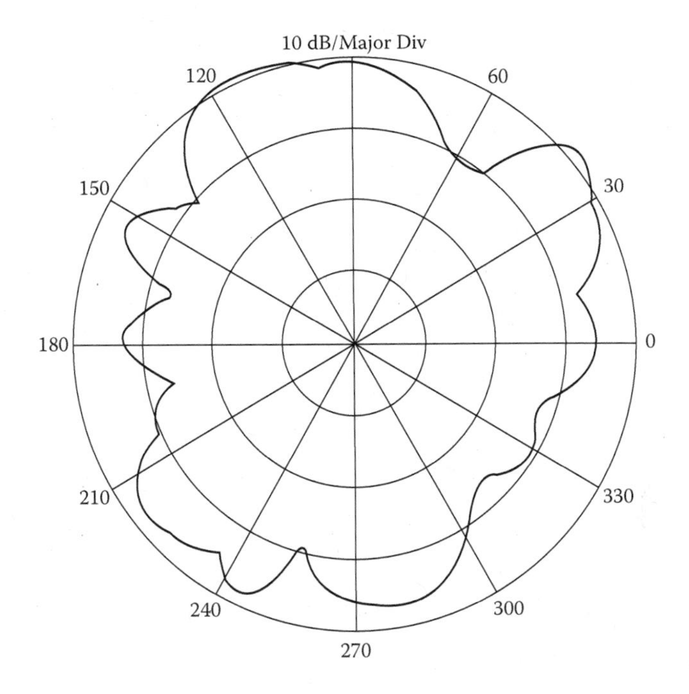

FIGURE 9.40
Diversity antenna pattern. (From V. Rabinovich et al., *IEEE Antennas and Propagation Society International Symposium*, 2005. Copyright 2005 IEEE. With permission.)

that we want to estimate the probability that the maximum range would exceed 100 m for the antenna (1). Based on Figures 9.42a and b, the probability $P(D > 100)$ is 87% for HP and 72% for VP. For diversity system, we can see that P for 100 m is increased to 98% for HP and 90% for VP.

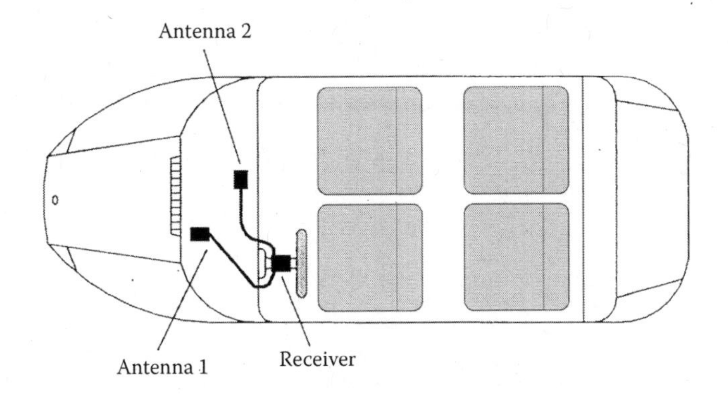

FIGURE 9.41
Diversity design utilizing two antennas. (From V. Rabinovich et al. Compact planar antennas for short-range wireless automotive communication, *IEEE Transactions on Vehicular Technology*, 55, 2006. Copyright 2006 IEEE. With permission.)

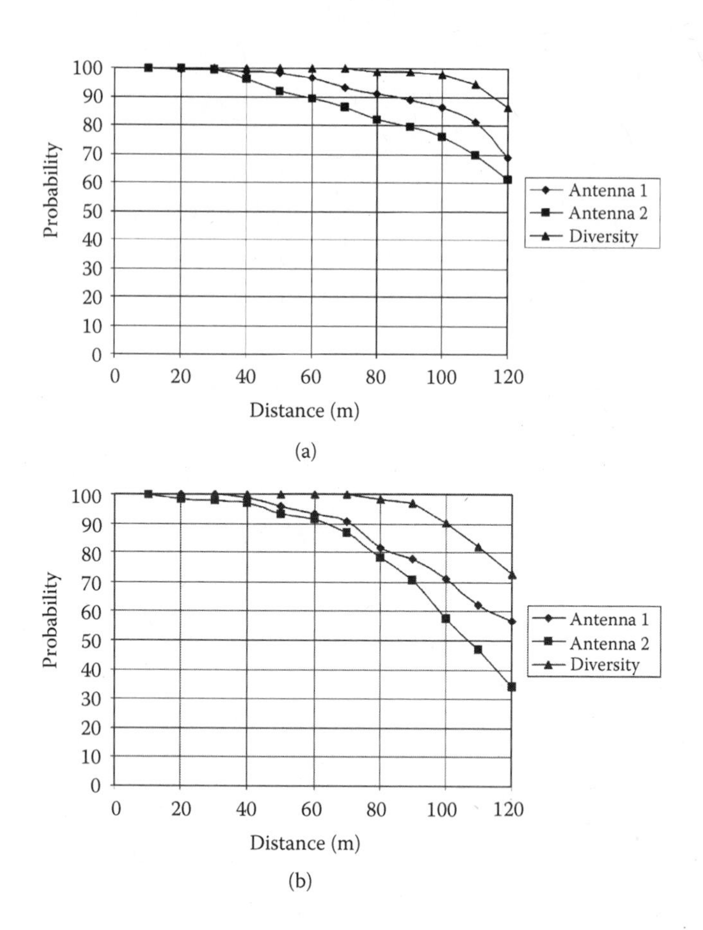

FIGURE 9.42
Probability of detection versus distance between transmitter and RKE system: (a) HP; (b) VP. (From V. Rabinovich et al. Compact planar antennas for short-range wireless automotive communication, *IEEE Transactions on Vehicular Technology*, 55, 2006. Copyright 2006 IEEE. With permission.)

9.12 Compact Antennas: Literature and Patent Review

Figure 9.43 presents a few new antennas that function at 300 to 450 MHz and are used for short range wireless automotive applications. Figure 9.43a depicts a miniaturized compact planar folded dipole antenna printed on a circuit board [33] and operating at 336 MHz. The dimensions are about $0.06 \cdot \lambda_0 \times 0.065 \cdot \lambda_0$ (λ_0 = wavelength in free space). The measured gain of this antenna fabricated on a 0.762 mm thick RT Duroid 5880 substrate with a dielectric constant of $\varepsilon_r = 2.2$ and loss of tan = 0.0009 is about –6.5 dBi (simulation shows –7 dBi). The antenna does not require a matching circuit.

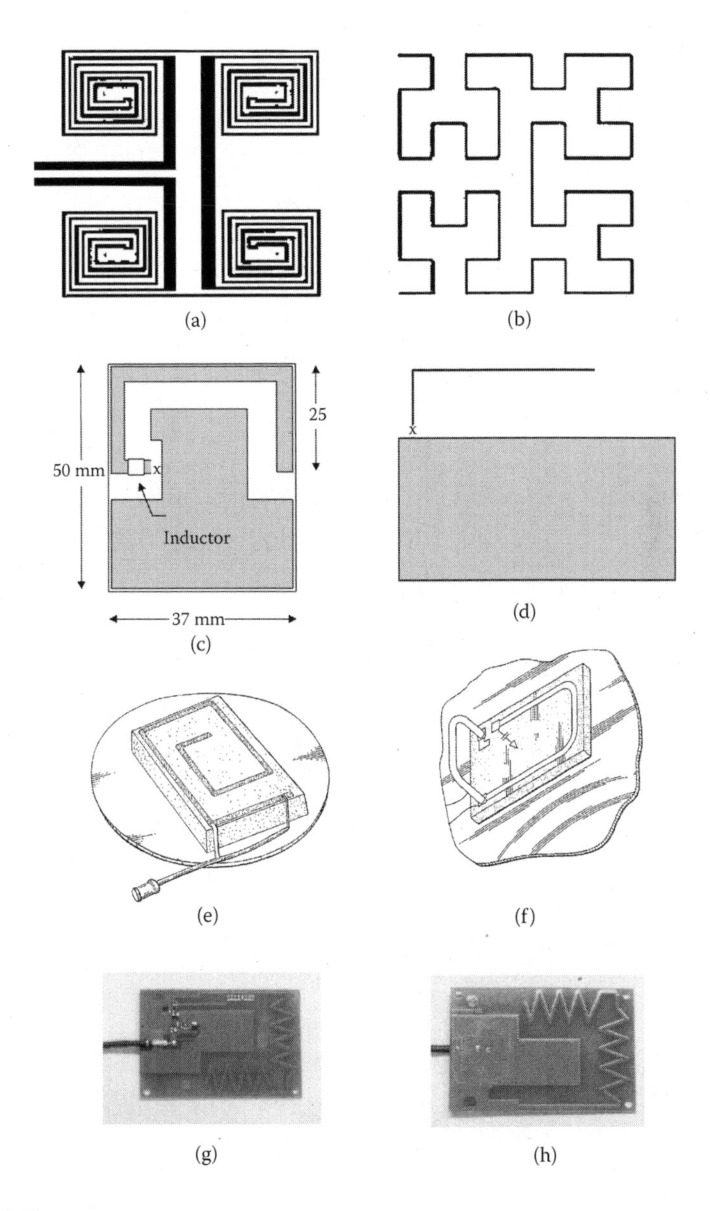

FIGURE 9.43
(a) Miniaturized compact planar folded dipole antenna printed on circuit board. (From R. Azadegan and K. Sarabandi, A compact planar folded-dipole antenna for wireless applications, *IEEE Antennas and Propagation Society International Symposium*, 2003. Copyright 2003 IEEE. With permission.) (b) third order of center-fed matched fractal planar antenna.(From J. Zhu et al., *IEEE Antennas and Wireless Propagation Letters*, 2, 2003. Copyright 2003 IEEE. With permission.) 2003. (c) printed 433.9 MHz band antenna 7 mm × 50 mm. (d) F type antenna. (e) low profile antenna for remote communication. (f) antenna for tire pressure monitoring system. (g) and (h) active 315 Mhz band antenna.

Figure 9.43 b illustrates the interesting Hilbert antenna topology [34]. This third-order center-fed matched fractal planar antenna has 70 × 70 mm dimensions; the real part is only ~2 ohms at the frequency where the imaginary part is zero. If this antenna is fed as a monopole above the ground plane (the feed point is at the end of the antenna), the input impedance is very small at the resonance frequency. However, a properly chosen off-center feed point may provide about 50 ohms of real input impedance at the resonant frequency point.

RFM Corporation distributes the omnidirectional antenna shown in Figure 9.43c. This 433.9 MHz band printed antenna is 37 mm × 50 mm, with a gain around –16 dBi. It requires a 56 nH inductor to match the 70 mm long line.

Figure 9.43d depicts an F antenna from Freescale Semiconductor. The antenna should not be nearer than 1/10 wavelength from the ground or efficiency will diminish. At this close spacing, radiation resistance is low (~1 ohm).

Figure 9.43e shows a low profile antenna for vehicle remote communication from Lear Corporation [35]. A linear antenna trace is dispersed on an upper surface of a dielectric board. Electronic components are mounted on the bottom side of the second layer, separated from the linear antenna trace by the ground plane. This design improves RF isolation between components and antenna, improves the electronic noise protection, and increases antenna gain by 1 or 2 dBi.

An antenna for a tire pressure monitoring system (TPMS) mounted near a metal tire rim [36] is shown in Figure 9.43f. The assembly is installed on a printed circuit board (PCB). An antenna element is printed on the top side; the bottom surface of the board serves as a ground plane. The first segment of the antenna element is positioned in the plane of the top surface and a second portion is oriented perpendicular to the top side of the PCB. The antenna portions are C- or U-shaped, connected in series, and approximately equal in length. The system reduces the coupling of antenna element and metal rim, thus improving efficiency and performance when compared with conventional antenna approaches.

The active antenna design manufactured by Marquardt and shown in Figure 9.43g (top side) and h (bottom side) has the same dimensions as the antenna depicted in Figure 9.12 and Figure 9.20. All these antennas have narrow bandwidths and some of them require matching circuits between antenna elements and receivers to provide maximum reception power.

9.13 Meander Line Antennas for Multifrequency Applications

This section discusses use of compact meander line antennas (Figure 9.12) for different frequency range applications. Figure 9.44 shows simulated (IE3D software) and measured return losses in a wide frequency band from 200 to 2400 MHz. Note the satisfactory agreement of the simulated and measured results. The antenna has many resonant frequencies, for example, the

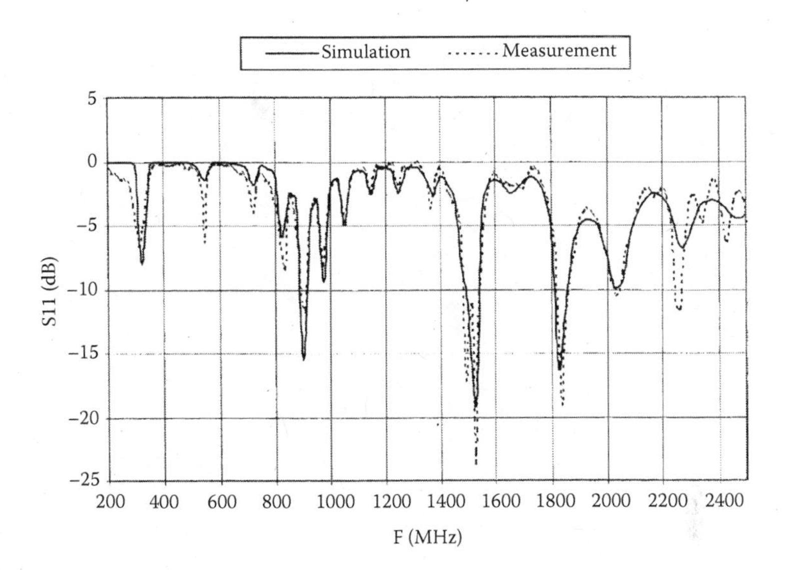

FIGURE 9.44
Simulation and measurement results for meander line antenna in wide frequency range: (a) Solid line indicates simulation; (b) dashed line indicates measurement.

frequency bands for which the measured return S11 losses (Figure 9.44) were below –5dB were 315, 880 to 925, 1450 to 1550, 1800 to 1900, 1970 to 2100, 2220 to 2280, and 2400 to 2450 Mhz.

Note that this antenna may be used for several applications without modifications: remote keyless entry (RKE) for automotive applications in the 315MHz frequency band; RFID applications in North America (the authorized band in North America including Mexico is 902 to 928 MHz); L-band digital radio in Canada (1452 to 1492 MHz); GSM in 1800 to 1900 MHz band; and WLAN application fields at 2400 to 2450 MHz.

Measurement results show also that meander length can be modified to vary the resonant frequency without redesigning the antenna topology. We must tune the antenna for a certain predetermined frequency for wireless communication. Figure 9.45 shows a meander line antenna with zero ohm resistance (breaking point) inserted in positions where the lengths of the lines can be adjusted according to the intended application. If we assume that resistance (1) is removed from the meander line, the antenna resonates at a frequency of 406 MHz—the perfect frequency for the Search and Rescue Satellite-Aided Tracking (SARSAT) system.* VSWR measurement of the meander antenna with resistance (1) removed is shown in Figure 9.46. With

* SARSAT is an international, humanitarian satellite-based distress alerting system credited with saving over 16,000 lives worldwide and over 4,600 lives in the U.S. since its inception in 1982 (totals as of January 1, 2004). The system operates 24 hours a day, 365 days a year, and detects transmissions from emergency beacons carried by ships, aircraft, and individuals.

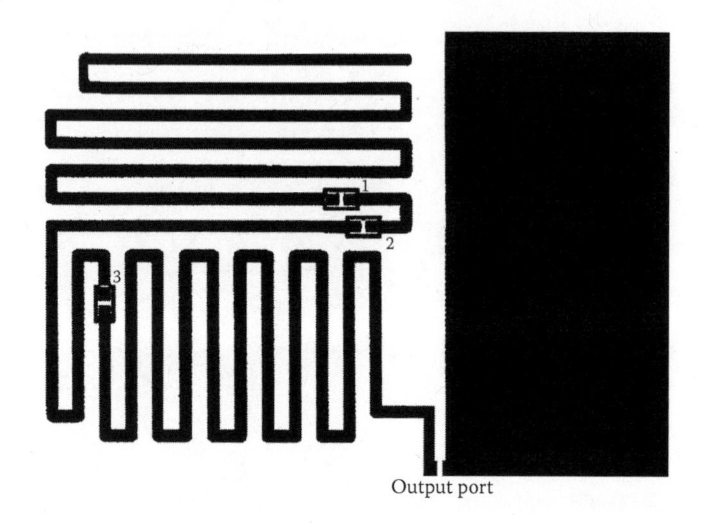

FIGURE 9.45
Geometry of antenna with reconfigurable shape.

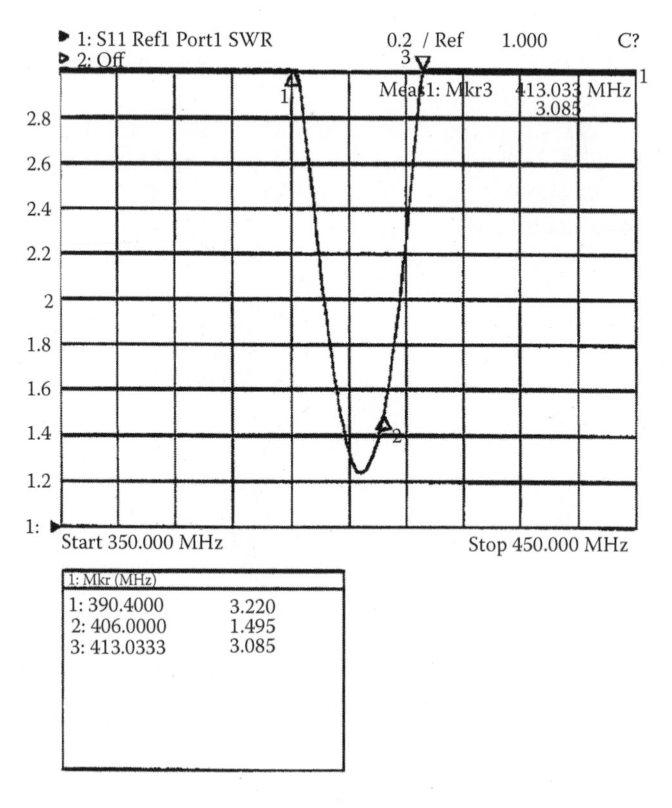

FIGURE 9.46
VSWR (resistor at point 1 in Figure 9.45 is disconnected).

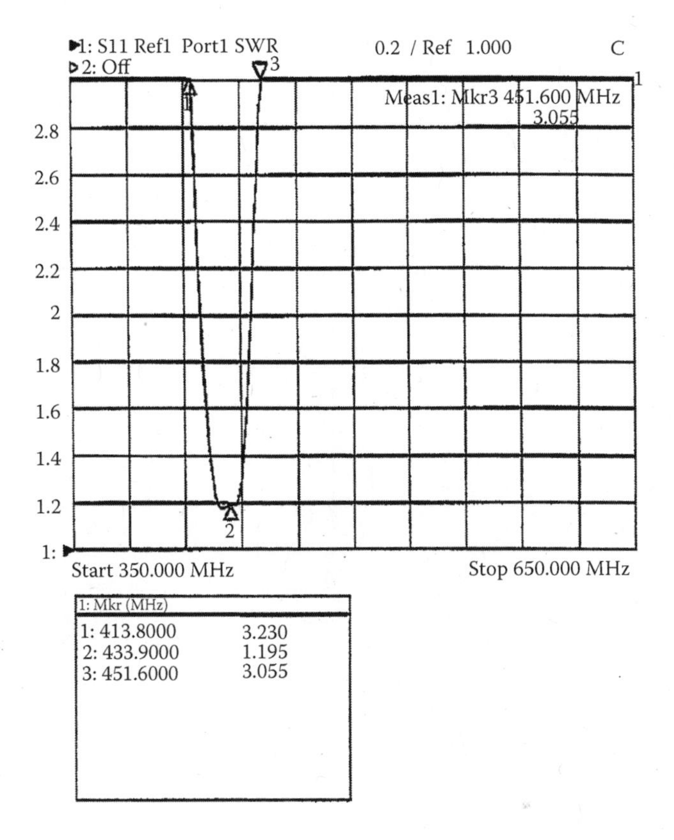

FIGURE 9.47
VSWR (resistor at point 2 in Figure 9.45 is disconnected).

this minor modification, the VSWR value becomes 1.5 at 406.00 MHz, indicating a very good resonance frequency, tuning, and effective performance.

When resistance (2) is removed, the meander line antenna becomes tuned to 433.9 MHz, an RKE frequency for European automotive applications. The measured VSWR for a tuned antenna with this resistance removed appears in Figure 9.47. The VSWR is 1.2, making this antenna a good candidate for such a frequency range application. Finally, by removing resistance (3), we can tune the antenna to a DCS frequency (1710 to 1880 MHz). Figure 9.48 presents the corresponding curve for VSWR. The VSWR is optimum for both frequencies (1710 and 1880 MHz), confirming the validity of this design for this frequency range. These measurements clearly show the ease of use, validity, and flexibility of the overall antenna design for multi-application uses.

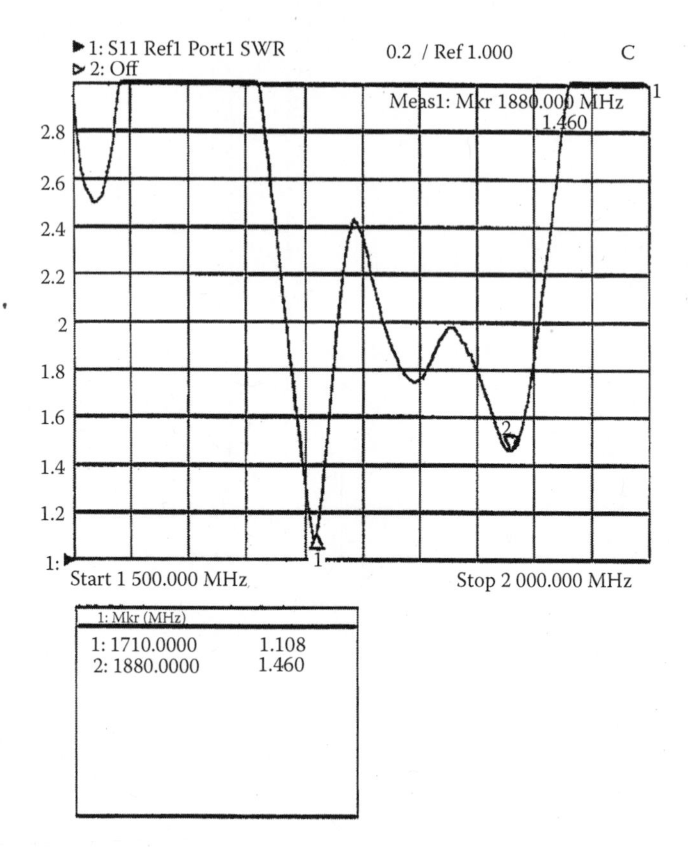

FIGURE 9.48
VSWR (resistor at point 3 in Figure 9.45 is disconnected).

References

1. A. Bensky, *Short-Range Wireless Communications*, Eagle Rock, VA: LLH Technologies, 2000.
2. F.L. Dacus, Design of short-range radio systems, *Microwaves and RF*, pp. 73–80, September 2001.
3. H. Morishita, Y. Kim, and K. Fujimoto, Design concept of antennas for small mobile terminals and the future perspective, *IEEE Antennas and Propagation*, 44, 30–43, 2002.
4. H. Blaese, Inside Window Antenna, U.S. Patent 5,027,128, June 1991.
5. M. Wiedmann et al., Antenna for a Central Locking System of an Automotive Vehicle, U.S. Patent 6,937,197, August 2005.
6. R. Drozd and W. Joines, Comparison of coaxial dipole antennas for applications in the near-field and far-field regions, *Microwave Journal*, May 2004.
7. K. Nishikawa et al., Automotive Window Glass Antenna, U.S. Patent 5,416,491, May 1995.

8. V. Rabinovich, B. Al-Khateeb, and B. Oakley, An active receiving antenna for short-range wireless automotive communication, *Microwave and Optical Technology Letters*, 43, 293–297, 2004.
9. H. Wheeler, Fundamental limitations of small antennas, *IRE Proceedings*, 35, 1479–1488, 1947.
10. S. Best, A discussion on the properties of electrically small self-resonant wire antennas, *IEEE Antennas and Propagation*, 6, 9–22, 2004.
11. V. Rabinovich et al. Compact planar antennas for short-range wireless automotive communication, *IEEE Transactions on Vehicular Technology*, 55, 1425–1435, 2006.
12. Zeland Software, Inc., IE3D electromagnetic simulation and optimization software. www.zeland.com
13. Eagleware Corporation, Genesys: high-speed linear circuit simulation, active and passive matching network synthesis, L-C filter synthesis. www.eagleware.com
14. V. Rabinovich et al., A signal and noise measurement procedure for an antenna/RF receiver combination in a short-range automotive communication system, *Microwave and Optical Technology Letters*, 47, 116–119, 2005.
15. K. Puglia, Application notes: electromagnetic simulation of some common balun structures, *IEEE Microwave Magazine*, 3, 56–61, 2002.
16. V. Rabinovich et al., Small printed meander symmetrical and asymmetrical antenna performances including the RF cable effect in 315 MHz frequency band, *Microwave and Optical Technology Letters*, 48, 1828–1833, 2006.
17. O. Staub, J. Zurcher, and A. Scrivervik, Some considerations on the correct measurement of the gain and bandwidth of electrically small antennas, *Microwave and Optical Technology Letters*, 17, 156–160, 1998.
18. C. Icheln et al., Optimal reduction of the influence of the RF feed cables in small antenna measurements, *Microwave and Optical Technology Letters*, 25, 194–196, 2000.
19. C. Icheln et al., Use of balun chokes in small-antenna radiation measurements, *IEEE Transactions on Instrumentation and Measurements*, 53, 498–506, 2004.
20. L. Saario et al., Analysis of ferrite beads for RF isolation on straight wire conductors, *Electronics Letters*, 33, 1359–1360, 1997.
21. Y. Chow, K. Tsang, and C. Wong, An accurate method to measure the antenna impedance of a portable radio, *Microwave and Optical Technology Letters*, 23, 349–352, 1999.
22. K. Hirasawa, M. Haneishi, and K. Fujimoto, *Analysis, Design, and Measurement of Small and Low-Profile Antennas*, Boston: Artech, 1992, Chap. 9.
23. V. Rabinovich et al., Antenna and Splitter for Receiving Radio and Remote Keyless Entry Signals, U.S. Patent Application 20090002246, January 2009.
24. K. Oh, B. Kim, and J. Choi, Novel integrated GPS/RKES/PCS antenna for vehicle application, *IEEE Microwave and Wireless Components Letters*, 15, 244–246, 2005.
25. M. Ohnishi et al., Automotive Window Glass Antenna, U.S. Patent 5,461,391, October 1995.
26. A. Miller, Slot Antenna with Reduced Ground Plane, U.S. Patent 5,646,637, July 1997.
27. L. Nagy et al., Automotive Radio Frequency Antenna System, U.S. Patent 6,266,023, July 2001.

28. M. Ahrabian et al., Remote Keyless Entry System Having a Helical Antenna, U.S. Patent 5,723,912, March 1998.
29. T.S. Rappaport, *Wireless Communication*, Short Hills, NJ: Prentice Hall, 2002, pp. 138–139.
30. J.D. Kraus and R.J. Marhefka, *Antennas*, New York: McGraw-Hill, 2002, pp. 409–414.
31. J. Salter, Specifying UHF active antennas and calculating system performance, British Broadcasting Corporation R&D White Paper WHP066, 2003.
32. V. Rabinovich et al., Compact diversity antenna system for remote control automotive applications, *IEEE Antennas and Propagation Society International Symposium*, 2005, pp. 379–382.
33. R. Azadegan and K. Sarabandi, A compact planar folded-dipole antenna for wireless applications, *IEEE Antennas and Propagation Society International Symposium*, 2003, pp. 439–442.
34. J. Zhu et al., Bandwidth, cross polarization, and feed-point characteristics of matched Hilbert antennas, *IEEE Antennas and Wireless Propagation Letters*, 2, 2–5, 2003.
35. R. Ghabra et al., Low Profile Antenna for Remote Vehicle Communication System, U.S. Patent 7,050,011, May 2006.
36. J. Nantz et al., Antenna for Tire Pressure Monitoring Wheel Electronic Device, U.S. Patent 6,933,898, August 2005.
37. V. Rabinovich et al., Antenna System for Remote Control Automotive Application, US Patent 7,564,415, July 2009.

Index

Q

R

S

T